Advances in

Heterocyclic
Chemistry

Volume 52

Advances in

HETEROCYCLIC CHEMISTRY

Edited by

ALAN R. KATRITZKY, FRS

Kenan Professor of Chemistry
Department of Chemistry
University of Florida
Gainesville, Florida

Volume 52

ACADEMIC PRESS, INC.
Harcourt Brace Jovanovich, Publishers
San Diego   New York   Boston
London   Sydney   Tokyo   Toronto

Copyright © 1991 by ACADEMIC PRESS, INC.
All Rights Reserved.
No part of this publication may be reproduced or transmitted in any form or by any means, electronic or mechanical, including photocopy, recording, or any information storage and retrieval system, without permission in writing from the publisher.

Academic Press, Inc.
San Diego, California 92101

United Kingdom Edition published by
Academic Press Limited
24–28 Oval Road, London NW1 7DX

Library of Congress Catalog Number: 62-13037

International Standard Book Number: 0-12-020752-4

PRINTED IN THE UNITED STATES OF AMERICA
91 92 93 94   9 8 7 6 5 4 3 2 1

Contents

v

JA 2.7 '92

3(*2H*)-Isoquinolinones and Their Saturated Derivatives
LÁSZLÓ HAZAI

Directed Metalation of Pi-Deficient Azaaromatics: Strategies of
Functionalization of Pyridines, Quinolines, and Diazines
GUY QUEGUINER, FRANCIS MARSAIS, VICTOR SNIECKUS,
AND JAN EPSZTAJN

Preface

Volume 52 of *Advances in Heterocyclic Chemistry* comprises three chapters. M. A. E. Shaban, M. A. M. Taha, and E. M. Sharshira have systematically reviewed those tri-ring systems in which another heterocyclic ring is fused onto the pyrimidine ring of quinazoline. No previous survey is available for these compounds, many of which show significant biological activity.

L. Hazai provides an overview of 3(2*H*)-isoquinolinones and their saturated derivatives, a group of compounds which has seen considerable recent synthetic activity.

Last, but certainly not least, G. Queguiner, F. Marsais, V. Snieckus, and J. Epsztajn have brought a French–Canadian–Polish collaboration to a fruitful end by reviewing the directed metalation of pi-deficient azaaromatics, a field which has recently exploded (half of the references are 1983 or later) and is of importance to all engaged in heterocyclic synthesis.

A. R. Katritzky

ADVANCES IN HETEROCYCLIC CHEMISTRY, VOL. 52

Synthesis and Biological Activities of Condensed Heterocyclo[n,m-*a*,*b*, or *c*]Quinazolines

MOHAMMED A. E. SHABAN, MAMDOUH A. M. TAHA,
AND ESSAM M. SHARSHIRA

Department of Chemistry,
Faculty of Science,
Alexandria University, Alexandria, Egypt

I. Introduction

The title compounds will be classified into the following three types according to the positions of fusion of the heterocycle and quinazoline rings:

(1) Condensed heterocyclo[n,m-a]quinazolines (**1**)
(2) Condensed heterocyclo[n,m-b]quinazolines (**2**)
(3) Condensed heterocyclo[n,m-c]quinazolines (**3**)

(n,m are the numbers indicating positions of linkage of the heterocycle to the quinazoline ring). The presence of one nitrogen bridgehead is a common feature of the three classes. Accordingly, this review will not cover the synthesis of heterocyclo-quinazolines of the types **4–6.**

(1)   (2)   (3)

(4)   (5)   (6)

In this review, the synthesis and biological activities of the title compounds are systematically arranged according to the complexity of the heterocyclic ring directly fused to the pyrimidine ring of the quinazoline nucleus (irrespective of other rings which might be fused to it), starting with those having one nitrogen atom in a three-membered ring and going to more complex ones. The heterocycles have been arranged according to the type of their heteroatom in the following order: nitrogen, oxygen, sulfur, and selenium. Consideration has been given to the alternative nomenclature of some heterocyclo-quinazolines, also known as quinazolino-heterocycles, in order to comply with the nomenclature rules of the IUPAC. The literature has been searched up to the end of June 1990.

It is not unusual to find that some of the various possible structures of a particular heterocyclo-quinazoline ring system have not been synthesized. The absence of some of these possible structures in this review indicates that, to the best of our capability, they have not been synthesized as yet.

The diverse biological activities of the title compounds have certainly contributed to the extensive efforts directed towards their synthesis.

II. Condensed Azirino-quinazolines

A. Azirino[1,2-*a*] quinazolines

(7)   (8)

Cyclization of 3-substituted 2-chloromethyl-4-oxo-1,2,3,4-tetrahydroquinazolines (**7**) with potassium *tert*-butoxide in *tert*-butanol afforded **8** [66TL2609; 74CPB601, 74JAP (K) 7431687, 74JAP (K) 7431697]. Blocking *N*-3 of the quinazoline nucleus is necessary to direct the cyclization towards *N*-1.

B. Azirino[2,1-*b*]quinazolines

Cyclization of 2-bromomethyl-6-chloro-1-methyl-4-phenyl-1,2,3,4-tet-rahydro-quinazolines (**9**) with sodium hydroxide gave the azirino [2,1-*b*] quinazolines (**10**) (72BRP1279842).

(9)   (10)

C. Azirino[1,2-*c*]quinazolines

A reported example (**12**) of this ring system was prepared by Woerner *et al.* (71CB2789) by the cycloaddition of the ketenimine **11** to 3-phenyl-2*H*-

(11)   (12)

azirine. All of the aforementioned azirino-quinazolines were obtained as intermediates by ring expansion during the synthesis of 1,4-benzodiazepines.

III. Condensed Azolo-quinazolines

A. Pyrrolo-quinazolines

1. *Pyrrolo[1,2-a]quinazolines*

In an approach to synthesizing these compounds, both of the diazine and azole rings were formed when anthranilic acid derivatives were cyclized with four-carbon cyclizing compounds such as γ-chlorobutyraldehyde (61AP556), γ-oxocarboxylic acids (68JOC2402, 68USP3375250; 69G715, 69USP3441566; 74USP3843654; 75USP3883524), γ-chlorobutyroyl chloride (68JOC1719; 69G715), γ-butyrolactone [76IJC(B)879; 86MI6], γ-halobutyronitriles (77UKZ711; 79KGS1427), γ-bromocrotononitriles [80EGP (D) 142337; 81JPR647], 2-formylsuccinonitrile (60JCS4970), succinic anhydride (68JHC179; 69USP3475432), 4,4-dimethoxybutyronitrile (68JHC179; 72USP3707468), benzoin (82H249), ethyl 3-ethoxymethylene-2,4-dioxovalerate (78H1729; 80JHC945), or levulinoyl chloride (67ZC456) (13) to give pyrrolo [1,2-a] quinazolines (e.g. 14).

(13)   (14)

Reaction of anthranilonitrile or methyl anthranilate with 3-hydroxy-2-butanone followed by malononitrile gave the pyrrolo [1,2-a] quinazoline 16 (79AP552). Both of the diazine and azole rings of pyrroloquinazolines were also simultaneously formed by cyclization of the anilide 17 derived from 3-chloropropionic acid and 2-aminobenzophenone with potassium cyanide to afford the pyrrolo [1,2-a] quinazoline 18 (68JHC185; 71USP3595861).

(15)   (16)

(17)   (18)

The anilide 19 was also cyclized with ethyl cyanoacetate to give 20 (68JHC185; 69USP3459754). Acid-catalyzed cyclization of suitably 2-substituted anilides such as 21 gave 22 [76IJC(B)705; 88M1405]. The pyrrolo [1,2-a] quinazolone 24 was obtained by double ring closure of the succinic acid anilide (**23**) derived from 1-aminobenzylamine [35JCS1277; 36CB(B)2052].

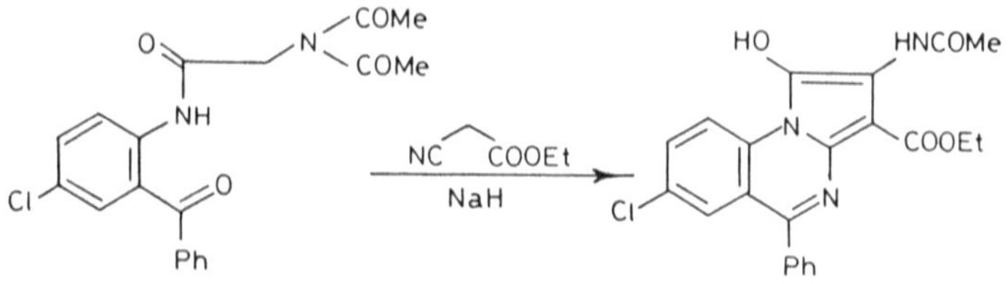

(19)   (20)

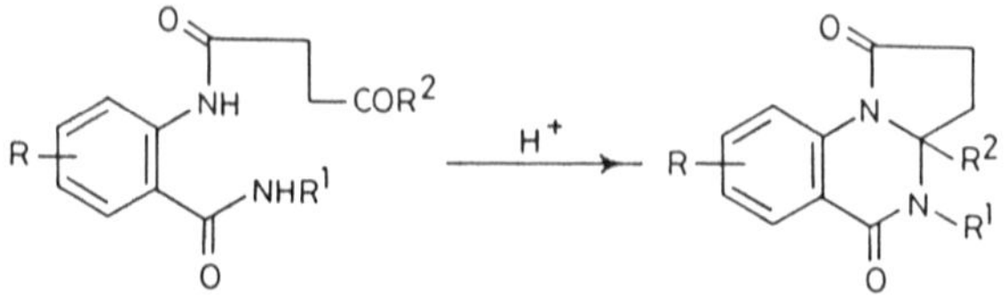

(21)   (22)

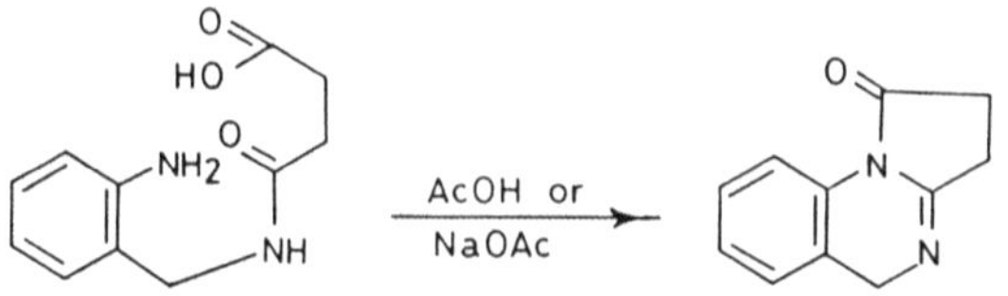

(23)   (24)

In another approach, the pyrrole ring of the title compounds (e.g. **26**) was formed by acid- or base-catalyzed cyclization of suitably functionalized quinazolin-2-yl derivatives (71JIC743; 72JIC1185; 73JMC633) such as 2-(2-cyanoethyl)quinazoline-3-oxide (**25**) (70USP3506663). When

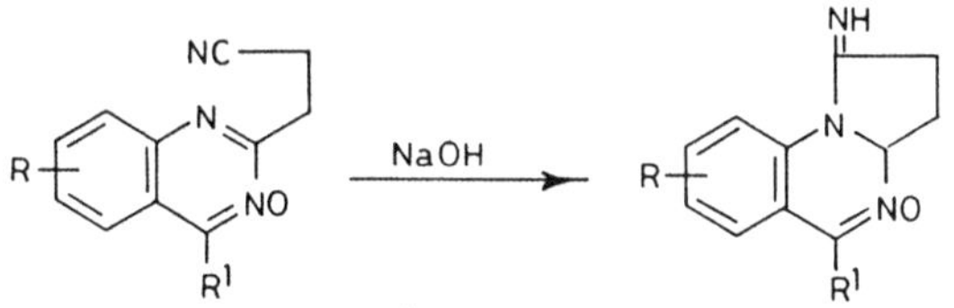

(25)   (26)

N-3 of the quinazoline derivative was not blocked, a mixture of the angular pyrrolo[1,2-*a*]quinazoline and the linear pyrrolo[2,1-*b*]quinazoline was obtained (73JMC633). Yamada *et al.* (83CPB2234) used a Dieckmann-type reaction to prepare the pyrro[1,2-*a*]quinazoline (**28**) from the diester **27**.

In a third approach, the title compounds were prepared through the formation of the pyrimidine ring by starting with an appropriately *N*-(substituted phenyl)pyrrole. Thus Garcia *et al.* (68JOC1359) cyclized the *anti* oxime **29** to the pyrrolo[1,2-*a*]quinazoline-4-oxide (**30**) with bromine. Thermal cyclization of the *N*-(substituted phenyl)pyrrole (**31**) gave **32** (84TL4309). Moehrle *et al.* (78AP586; 79AP838) obtained the pyrrolo-quinazoline **34** by the oxidative cyclization of **33** with Hg(II)-ethylenediaminetetracetate (EDTA).

2. Pyrrolo[2,1-b]quinazolines

Synthesis of these compounds (e.g. **35**) through the formation of both of their diazine and azole rings was achieved by cyclocondensation of anthranilic acid derivatives with γ-aminoaldehydes [36LA1; 85IJC(B)789] or γ-aminoacids (57CLY2122; 68P307) as well as by cyclocondensation of anthranilamides with γ-ketoacids (67ZC456; 82CPB1036).

(35)

The pyrrole rings of this system were formed by cyclization of 2-(3-substituted propyl)quinazolines. Whereas cyclization of ethyl 3-(quinazolin-2-yl)propionate (78H1375) or 2-(3-bromopropyl)-3,4-dihydroquinazoline **36** (35JA921) was reported to afford only the corresponding pyrrolo[2,1-*b*]quinazoline (e.g. **37**), dehydrative cyclization of 2-(3-hydroxypropyl)-3,4-dihydroquinazoline (**38**) gave (73JMC633) the logically expected mixture of pyrrolo[1,2-*a*]quinazoline and pyrrolo[2,1-*b*]quinazoline (**37** and **39**) in the ratio of 2 : 1, respectively.

(36)   (37)

(38)   (39)

Base catalyzed cyclization of 1-substituted-2-(3-haloacetonyl)quinazoline-4-ones (**40**) gave only pyrrolo[2,1-*b*]quinazolinones (**41**) [35JA951; 77JAP(K)77144697]. A Japanese group (80CPB702) found that allowing a chloroform solution of 2-(3-chloropropyl)-1-phenyl-4(1*H*)-quinazolin-4-one (**42**) to stand at ambient temperature afforded a quantitative yield of the mesoionic pyrrolo[2,1-*b*] quinazolinone (**43**).

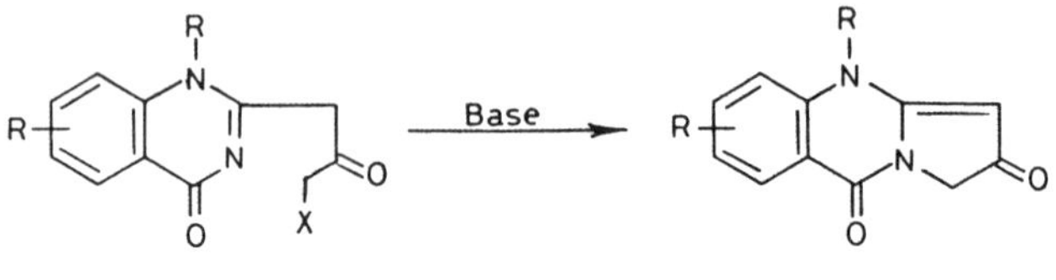

(40)   (41)

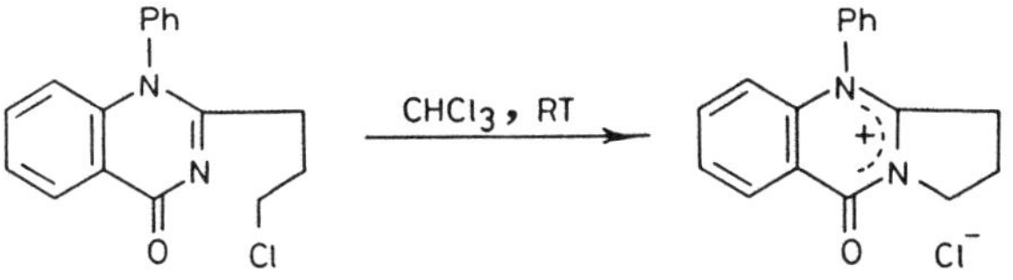

(42)   (43)

Oxidation of the 1-aminoquinazolinones **(44)** with lead tetraacetate culminated in the intramolecular addition of the resulting *N*-nitrene intermediate to the triple bond to give the pyrrolo[2,1-*b*]quinazoline **(45)** [85CC544; 86JCS(P1)1215]. The pyrrole ring of **47** was also formed by cyclization of 2-methyl-3-phenacylquinazolin-4-one **(46)** with dilute aqueous alkali followed by fusion (86T4481).

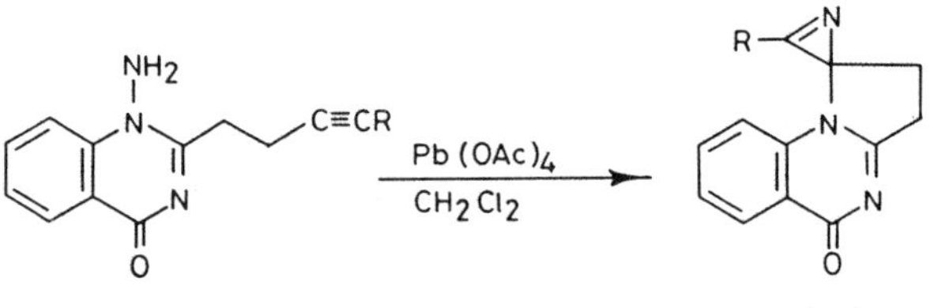

(44)   (45)

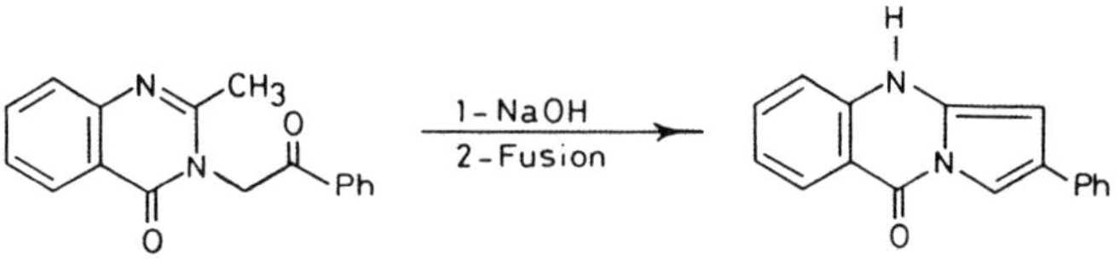

(46)   (47)

Pyrrolo[2,1-*b*]quinazolines **(49)** were obtained through the formation of their diazine ring by condensation of an anthranilic acid derivative with a pyrrole derivative such as *O*-alkylbutyrolactimes **(48)** [60GEP1088968;

68ZOB2030; 71TL4387; 76H1487, 76IJC(B)354, 76JA6186; 77JA2306, 77JPR919; 78H1375; 79JCS(P1)1765; 80MI3; 86JHC53] or pyrrolidones [35CB(B)2221; 67TL2701; 68JCS(C)1722; 74KPS681; 75KPS435; 76KGS1564; 77JAP(K)7777093; 78JAP(K)7877075; 82KPS498].

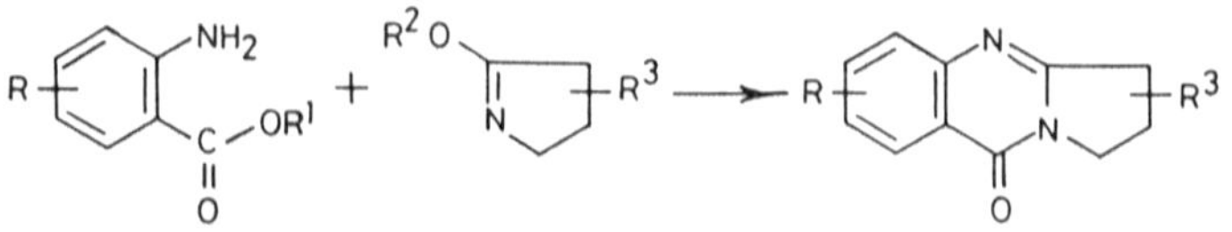

(48)   (49)

2-Alkoxy-1-(2-nitrobenzoyl)pyrrole (50) (74FES579; 78FES271) or 1-(2-nitrobenzyl)pyrrolidones [35CB(B)497, 35CB(B)699, 35MI1; 58JA1168; 72YZ1184; 75KPS809; 88JOC1873] underwent reductive cyclization to pyrrolo[2,1-b]quinazolines (e. g. 51).

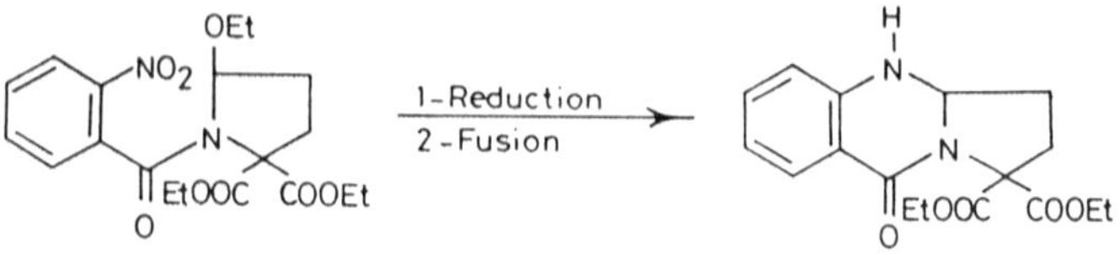

(50)   (51)

In a series of publications, by Moehrle and his group (70TL997, 70TL3249; 73AP541; 79CZ266; 80M627, 80PHA389), a number of pyrrolo-[2,1-b]quinazolines (53) were synthesized via formation of their 1,3-diazine ring by oxidative cyclization of 1-(2-aminobenzyl)pyrroles (52) with Hg(II)-EDTA. The two nitrogens of the pyrimidine ring of 56 originated from 2-amino-3-cyano-4-phenylpyrrole (55), which, upon cyclo-condensation with 2-acylcyclohexanones (54), gave 56 (70KGS428).

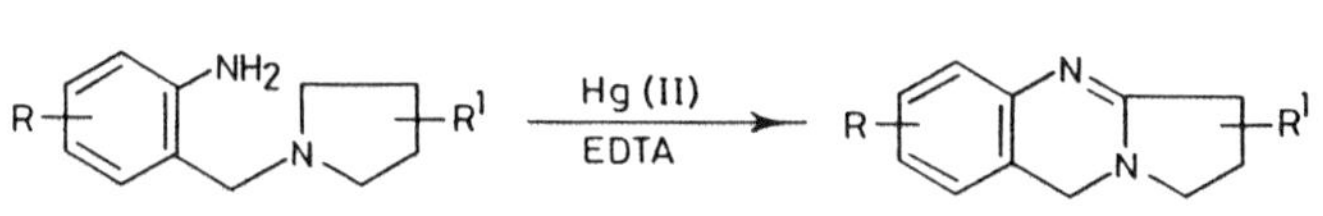

(52)   (53)

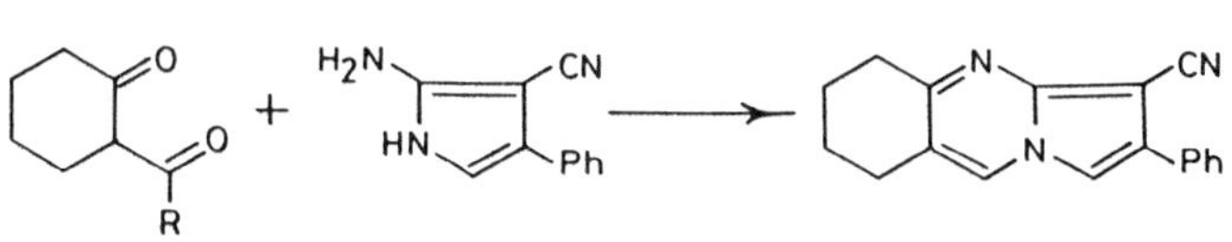

(54)   (55)   (56)

3. *Pyrrolo[1,2-c]quinazolines*

Compounds of this type were synthesized by constructing the pyrrole ring onto a quinazoline nucleus, as exemplified by the reaction of 4-methylquinazolines with α-haloketones to give **57** (78USP4129653; 79JHC623, 79JHC1497). The parent quinazoline reacted with 4-nitrophenacyl bromide to give the ylide **58**, which gave **59** upon cycloaddition to dimethyl acetylenedicarboxylate (DMAD) (81RRC109; 86RRC365). DMAD was also added to 3-benzoyl-4-cyanoquinazoline to give the corresponding pyrrolo[1,2-c]quinazoline (85CPB950, 85RRC611). Cyclization of the dianion **60**, derived from 2,4-diphenylquinazoline with 1,3-dihalopropanes, gave the pyrrolo[1,2-c]quinazoline **61** (76JOC497).

Alternatively, pyrrolo[1,2-c]quinazoline (**63**) was prepared through formation of the pyrimidine ring by cyclocondensation of 2-(2-amino-phenyl) pyrrole precursors (e.g. **62**) with one-carbon cyclizing agents (69MI1).

(62)   (63)

Spiro{1-methylsulfonyl-3H-indoline-3,2'-[3',5'-di(ethoxycarbonyl)-4'-methyl-2'H-pyrrole]} (**64**) underwent molecular rearrangement with 100% H_3PO_3, 60% H_2SO_4, or by heating at 200° to 1,3-di(ethoxycarbonyl)-2-methyl-6-methylsulfonyl-4,5-dihydropyrrolo[1,2-c]quinazoline   (**65**) (70AJC781).

(64)   (65)

Various biological applications have been reported for pyrroloquinazolines such as being antihypertensive [73JMC633; 77JAP(K)7777093; 78USP4129653; 81JMC1455], antiarrhythmic (78USP4129653), antibiotic (69USP3459754), central nervous system depressant [66USP3271400; 69USP3459754, 69USP3475432; 70USP3506663; 77JAP(K)77144697, 77JAP(K)7777093], antihistaminic [77JAP(K)77144697; 78USP4129653; 85IJC(B)789], antiinflammatory [74USP3853858; 77JAP(K)77144697, 77JAP(K)7777093], analgesic [75USP3883524; 77JAP(K)77144697], antipyretic [75USP3883524; 77JAP(K)77144697], antitussive [77JAP-(K)77144697], diuretic [77JAP(K)77144697], hypnotic [80EGP(D)142337], and anticonvulsant agents [80EGP(D)142337].

B. INDOLO-QUINAZOLINES

1. Indolo[1,2-a]quniazolines

Indolo[1,2-a]quinazolines (**67**) were prepared through the formation of their diazine rings, by cyclizing 2-acylamino-1-phenyl-indoles (**66**) with phosphoryl chloride (81KGS844).

(66)   (67)

2. Indolo[2,1-b]quinazolines

6*H*-Indolo[2,1-*b*]quinazoline-12-thione (**69**) was obtained by cyclodehydrohalogenation of 2-bromomethyl-3-phenylquinazoline-4(3*H*)-thione (**68**) with liquid ammonia (73IJC500); the expected amination product was not isolated.

(68)   (69)

Laser irradiation of isatin (**70**) gave **71** as a result of bimolecular condensation (78TL3007). The diazine ring of **73** was formed during the cyclocondensation of anthranilic acids with the imidate esters derived from indolinone (**72**) (81AP271).

(70)   (71)

(72)   (73)

The diazine and pyrrole rings of **75** were formed when methyl *N*-(anthraniloyl) anthranilate (**74**) was cyclized with formamide (60JCS4970).

(74)   (75)

3. Indolo[1,2-c]quinazolines

Acylation of 2-(2-aminophenyl)indoles (76) gave the 6-substituted-indolo[1,2-c]quinazolines (77) [56JCS1319; 71GEP(O)2051961; 75URP-481613]. The 12-acylaminoindolo[1,2-c]quinazolines (79) were synthesized by heating the 1,3-diacylindoles (78) with arylhydrazines in acetic acid followed by acetic anhydride (79KGS832, 79MIP1). Reductive rearrangement and cyclization of 1-acyl-3-(arylazo)indoles by treatment with zinc and a mixture of AcOH-Ac$_2$O-NaOAc also afforded 79 (74URP451698; 75KGS1096; 76URP515836, 76URP539885; 77KGS377).

(76)   (77)

(78)   (79)

Some indoloquinazolines have been patented for their sedative [71GE-P(O)2051961] and antifugal activities [80JAP(K)8047684].

C. ISOINDOLO-QUINAZOLINES

1. Isoindolo[2,1-a]quinazolines

Dihydroisoindolo[2,1-a]quinazoline (e.g. 82) was claimed (54JCS2354; 80AP729) to be the product of the reaction between 2-aminobenzylamine and phthalaldehyde (80AP729) or phthalaldehydic acid (54JCS2354). The angular structure **82** rather than the linear isoindolo[1,2-*b*]quinazoline **80** was assigned on the basis of its identity with the product resulting from the cyclodehydration of 2-(2-carboxyphenyl)quinazoline **81**. Obviously, this rationale is unsatisfactory since **81** may dehydrate to **80** or **82** or both.

Synthesis of the title compounds from isoindole precursors was also reported (85KGS1368). According to this route, 1-aminoisoindole was cyclocondensed with 2-ethoxycarbonylcyclohexanone to give a 2 : 1 mixture of the isoindolo[1,2-*b*]- and -[2,1-*a*]quinazolinones (**83** and **84**), respectively.

The two heterocyclic rings of **85** were formed during its synthesis from anthranilamide and 2-benzoylbenzoic acid or its chloride (7OUSP-3509147; 71USP3609139) or from anthranilamide and phthalic anhydride (69JOC2123). Heating anthranilic acid and 2-cyanobenzyl chloride also led to the formation of both heterocyclic rings of **86** (76URP527423).

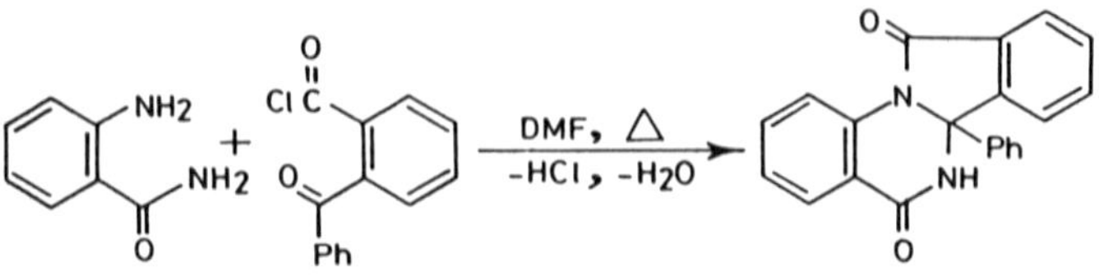

(85)

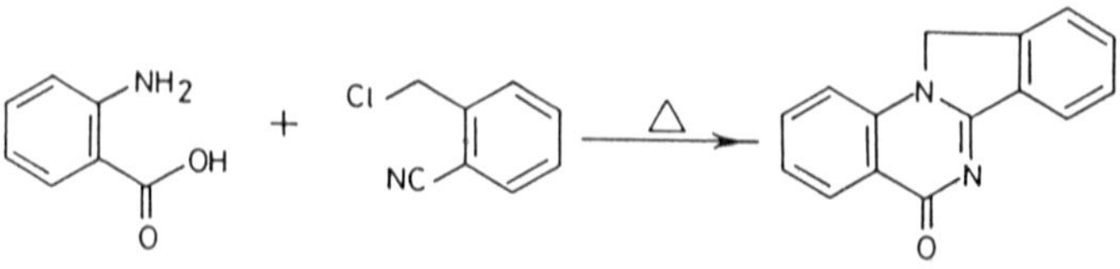

(86)

2. Isoindolo[1,2-b]quinazolines

Methyl anthranilate and phthalimide reacted in the presence of phosphorus trichloride to give isoindolo[1,2-*b*]quinazoline-10,12-dione **87** (27JCS1708). The latter was also obtained by thermal molecular rearrangement of isoindolo[2,1-*a*]-quinazoline-5,11-dione (**88**) (69JOC2123). Reduction and cyclization of **89**, obtained from 2-nitrobenzylamines and phthalic anhydride, with triethyl phosphite gave **90** together with other products [71JHC1071; 72JAP(K)7242750].

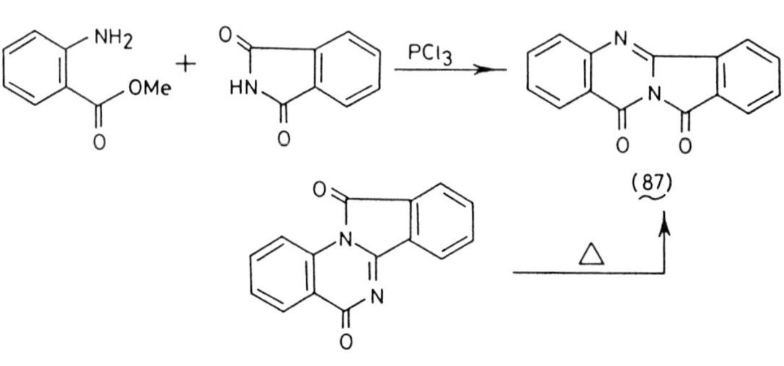

(87)

(88)

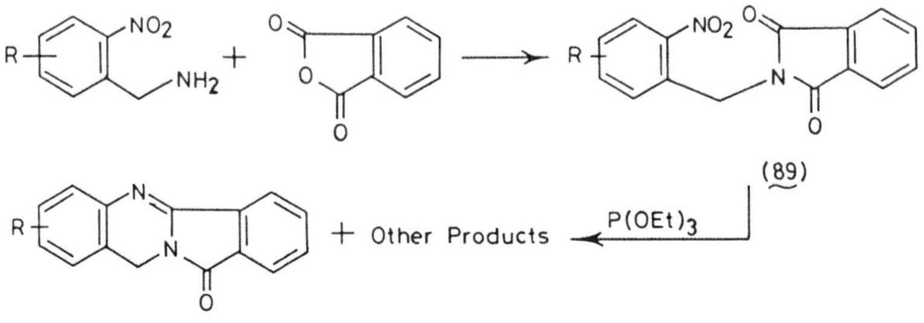

(89)

(90)

IV. Condensed Diazolo-quinazolines

A. PYRAZOLO-QUINAZOLINES

1. *Pyrazolo[1,5-a]quinazolines*

Both heterocyclic rings of pyrazolo[1,5-a]quinazolin-5(4H)ones (**91**) were formed during the condensation of 2-carboxyphenylhydrazines with α,β-unsaturated nitriles (69JHC947; 78USP4105766; 89JHC713). In an almost similar synthesis, both heterocyclic rings of **93** were also formed by base-catalyzed cyclocondensation of the aroylhydrazidoyl chlorides **92** with ethyl cyanoacetate or malononitrile (57G1191; 59MI1; 84JHC1049).

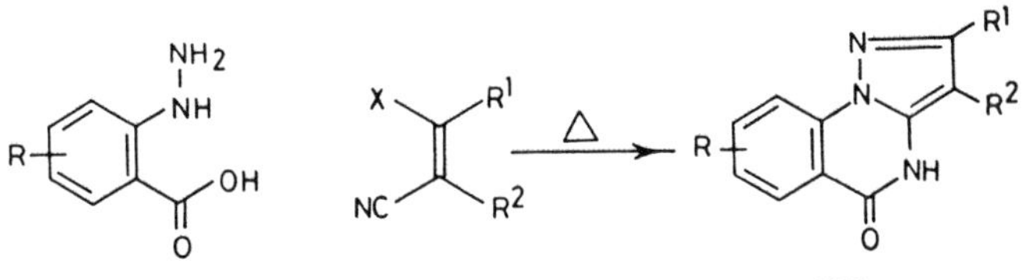

(91)

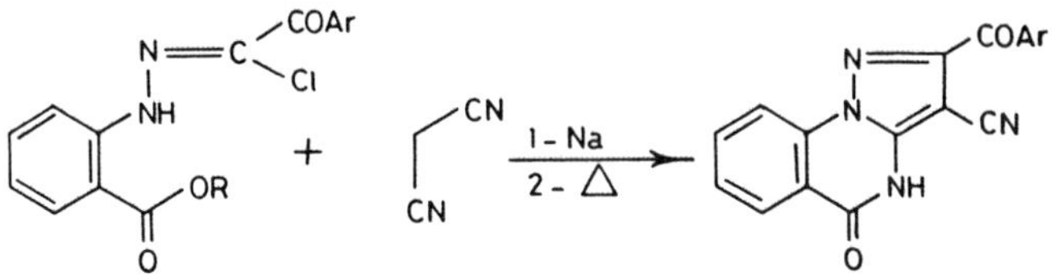

(92)   (93)

Cycloaddition of 4-phenyl-1,2,3-benzotriazine (**94**) to diphenylcyclo-propenone gave, among other products, the pyrazolo[1,5-*a*]quinazoline (**95**) (80CC808).

2. *Pyrazolo[5,1-b]quinazolines*

Synthesis from quinazoline precursors was achieved by carrying out a Vilsmeier–Haack reaction on 3-amino-2-methyl-4-quinazolone (**96**) to give the intermediate diformyl derivative **97** that cyclized to 3-formylpyra-zolo[5,1-*b*]-quinazolin-9-one (**98**) [73IJC532; 84IJC(B)161].

Synthesis through the formation of the pyrimidine ring was made by condensation of 2-hydroxymethylenecyclohexanone with 3-amino-4-cyanopyrazole (**99**) in the presence of 4-toluene sulfonic acid (TSA) to give 3-cyano-5,6,7,8-tetrahydropyrazolo[5,1-*b*]quinazoline (**100**) [79JCS(P1)3085]. Reaction of isatoic anhydrides with 2-ethoxycarbonyl-1,4-dihydropyrazol-5-one (81JHC117, 81JMC735, 81USP4261997) as well as reaction of anthranilic acids with 3-amino-1,4-dihydropyrazol-5-one (**101**) [73GEP(O)2305172] gave the title compounds (e.g. **102**).

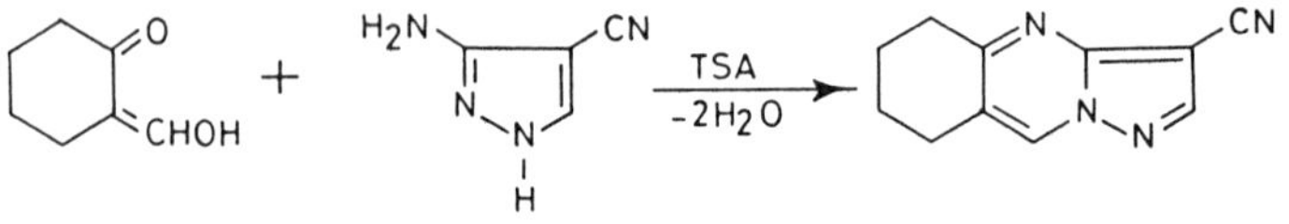

(99)   (100)

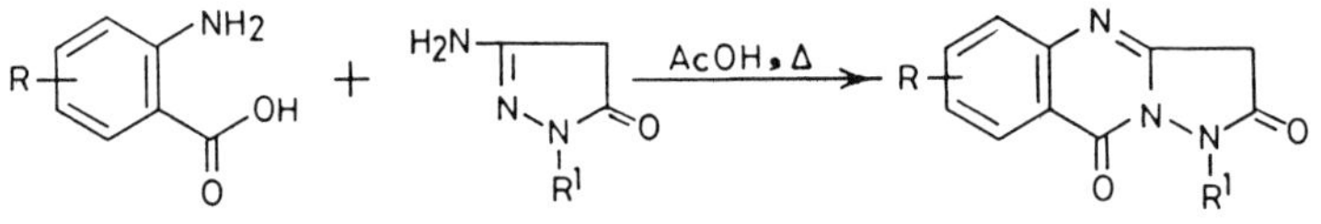

(101)   (102)

Formation of both heterocyclic rings of **104** was accomplished upon reaction of acetonylbenzoxainones (**103**) with hydrazine (62GEP1139123).

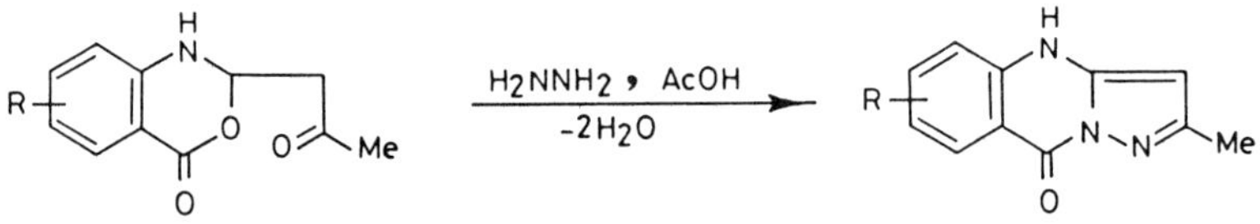

(103)   (104)

A general method for synthesizing pyrazolo[5,1-*b*]quinazolin-9-ones (e.g. **105**), which involves the formation of both heterocyclic rings, is the cyclocondensation of anthraniloylhydrazines with β-ketoesters [60GEP-1111505; 61GEP1120455; 80EUP15065; 81JMC735, 81USP4261996; 84M12; 88IJC(B)342], diethyl malonate [83IJC(B)496], or diethyl acetylenedicarboxylate (79JHC957; 82FES719). The 4,9-dihydropyra-zolo[5,1-*b*]quinazolines (**107**) were prepared by the reaction of phenyl isocyanate with the α-substituted phosphoranes **106** (84JOC1964; 87JOC1810).

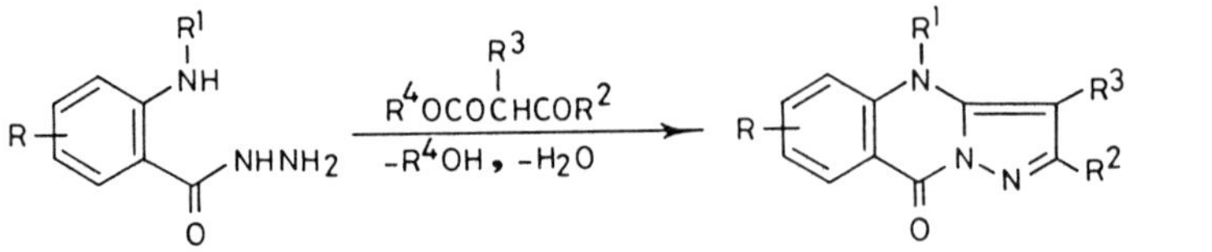

(105)

(106)   (107)

3. Pyrazolo[1,5,c]quinazolines

Cyclization of 3-(2-aminophenyl)pyrazoles (108) with triethyl orthoformate [62AG249; 67USP3313815; 80GEP(O)3019019), phosgene
(67USP3313815; 81EUP23773), or carbon disulfide [67USP331815;
70USP3531482; 79GEP(O)2916992; 81USP4282226] gave the corresponding pyrazolo[1,5-c]quinazolines (109).

(108)   (109), Y= H, OH, or SH

Reaction of 2-isothiocyanato-*trans*-cinnamaldehyde (110) with hydrazine gave the pyrazolo[1,5-c]quinazolines (111), presumably via the corresponding hydrazone and conjugate addition to the pyrazoline followed by
cyclization [76JCS(P1)653].

(110)   (111)

Reaction of acetylenic aldehydes [78GEP(O)2726389], acetylenic esters
(73TL1417; 75USP3897434; 78USP4110452, 78USP41122096, 78USP-
4112098), or acetylenic nitriles (78USP4128644) with 3-diazooxindoles
(112) gave the corresponding polysubstituted pyrazolo[1,5-c]quina-
zolinones (113).

(112)   (113)

Pyrazoloquinazolines have been reported to possess antiinflammatory [78USP4105766, 78USP4110452; 79GEP(O)2916992], antiallergic [78USP-4105766, 78USP4110452; 79GEP(O)2916992], parasiticidal (78USP-4105766), analgesic [80GEP(O)3019019], diuretic (67USP3313815; 84MI2), hypotensive (67USP3313815), psychomotor depressant (67USP3313815), bacteriostatic (67USP3313815), anxiolytic (78USP4110452), platelet aggregation inhibitory (78USP4110452), phosphodiesterase inhibitory (78USP4110452), and anaphylaxis inhibitory activities (80EUP15065; 81JMC735, 81USP4261996, 81USP4261997, 81USP4282226).

B. INDAZOLO-QUINAZOLINES

1. *Indazolo[2,3-a]quinazolines*

A reported example (**115**) of this ring system was prepared by the reaction of two molar equivalents of 2-cyanocyclohexanone with hydrazine to give the azine **114,** which underwent base-catalyzed cyclization to **115** [63CI(M)709].

(114)   (115)

Thermolysis of the 3-substituted 2-(2-azidophenyl)quinazolin-4(3H)-ones (**116**) or deoxygenation of the 3-substituted 2-(2-nitrophenyl)quinazolin-4(3H)-ones by triethyl phosphite afforded the indazolo[2,3-a]-quinazolin-5(6H)-ones (**117**) [79IJC(B)125].

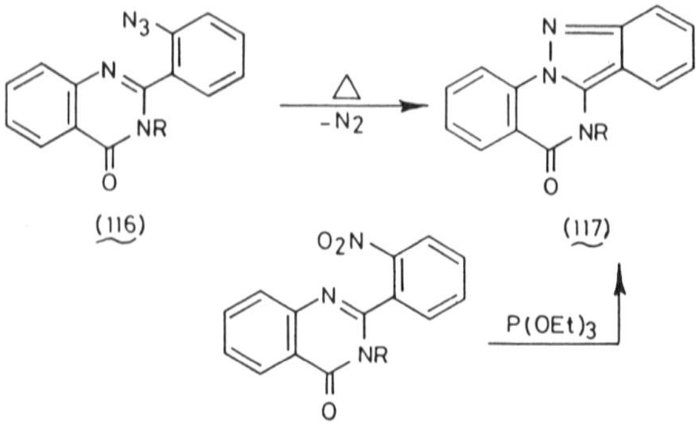

(116)   (117)   (118)

2. Indazolo[3, 2-b]quinazolines

The acid-catalyzed cyclization of the azine **114** gave 1,2,3,4-tetrahydro-indazolo [3,2-*b*]8,9,10,11-tetrahydroquinazol-7-imine (**119**), which was also obtained by cyclization of 2-cyanocyclohexanone with 3-amino-4,5,6,7-tetrahydroindazole (**120**) [63CI(M)709].

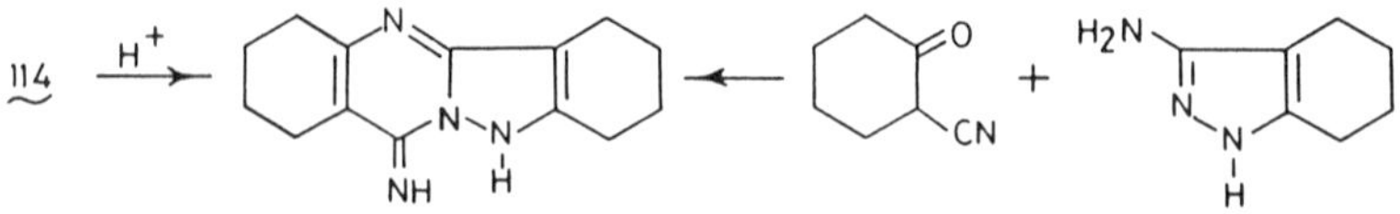

(119)   (120)

3. Indazolo[2,3-c]quinazolines

Cyclization of 3-(2-acylaminophenyl)-4-sulfamoylindazoles (**121**) with a mineral acid gave indazolo[2,3,-*c*]quinazoline-12-sulfonamides (**122**) (70USP3505315, 70USP3505333). Some indazoloquinazolines were found to exhibit antiinflammatory activity (70USP3505333).

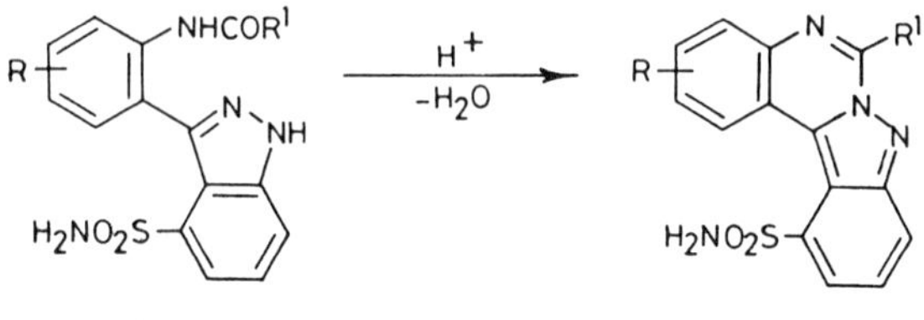

(121)   (122)

C. Imidazo-quinazolines

1. Imidazo[1,2-a]quinazolines

This class of compounds was prepared from quinazoline precursors by cyclization of (a) 2-(2-substituted-ethylamino) quinazolines (e.g. **123** and **125**), [73NKK1944; 78JAP(K)7823997; 86JHC833]. (b) *N*-(quinazolin-1-yl)acetamides carrying a good leaving group at the 2-position of the quinazoline ring (e.g. **127**) (63CB1271; 65ACH357), or (c) 2-(aminoquinazolin-1-yl)acetic acids or their esters (**129**) (76JOC825; 77USP4020062; 82EUP46267; 83EUP73060). Whenever *N*-3 of the quinazoline system was unblocked, a proof was offered, physical and chemical, to establish that cyclization did take place with *N*-1 rather than *N*-3 (63CB1271; 73NKK1944; 86JHC833).

2. Imidazo[2,1-b]quinazolines

Appropriately, 2,3-bifunctionalized quinazolines with one nitrogen less than the skeleton of this condensed ring system, such as 2-chloro-3-(2-

chloroethyl)quinazolines (**130**) [60JCS3551; 80CPB2024; 83GEP(O)323-3766; 84USP4451448, 84USP4452787], esters of quinazolin-3-yl)acetic acid (**132,** LG = Cl or SMe) [63ACH457; 65ACH357; 66CB1532; 76USP3983119, 76USP3983120; 77FZ88; 81ZN(B)366; 83HCA148; 84-EUP129258; 85JAP(K)60152416], or 3-acetonyl-2-chloroquinazolin-4-ones (73URP405895; 76KFZ60), were cyclized with ammonia, primary amines, and hydroxylamine to give the title compounds (e.g. **131** and **133**).

(130), LG= Cl   (131)

(132), LG=Cl or SMe   (133)

Alternatively, 2,3-bifunctionalized quinazolines having the total carbon and nitrogen skeleton of this system, e.g. 1-substituted-2-(2-hydroxy-ethyl)aminoquinazolin-4-ones (**134**) [73GEP(O)2257376, 73SZP532068; 76USP3969506, 76USP3982000, 76USP3984556], 2-(2,2-dimethoxyethy-lamino)quinazolin-4-one (86JHC833), esters of (2-aminoquinazolin-3-yl)acetic acid (**136**) (82CP1137474), or 3-(2-aminoethyl)quinazolin-2-ones (**138**) [75GEP(O)2508543; 79CP1057752], cyclized thermally or chemically to members of the title compounds (**135, 137,** and **139**). Related to the latter methods is the cyclocondensation of 2-aminoquinazolin-4-ones (**140**) with α-haloketones (74URP445665; 77FZ37) to give **141**.

(134)   (135)

(136)   (137)

(138)   Z = O or S   (139)

(140)   (141)

The second general pathway for synthesizing imidazo[2,1-*b*]-quinazolines (**142**) is one through which the pyrimidine ring is formed by condensing anthranilic acids or isatoic anhydrides with imidazolines having a good leaving group at the 2-position, such as 2-methylthio-imidazoline [71USP3621025; 73GEP(O)2234174, 73GEP(O)2257376, 73GEP(O)2319851, 73USP3745216; 74GEP(O)2402454; 75JMC447, 75USP3919210; 76USP3978059; 77USP4025511; 83GEP(O)3220438; 84USP4451464; 86MI2, 86MI3], or 2-methylthioimidazolin-4-one (2-methyl-thiohydantoin) [63ACH457; 70GEP(O)2025248, 70MI3; 72-JMC727; 74SAP7302111; 75USP3875160; 80CPB2024; 81ZN(B)366; 82H1375]. Condensation of 2-halobenzoyl chlorides with 2-methylthio-imidazoles followed by cyclization of **143** with hydrazine derivatives gave **144** (89USP4871732).

LG = Cl or SR^1 (142)

(144)

The third approach to synthesizing this ring system consists of the one-step formation of both the pyrimidine and imidazole rings. Thus, cyclization of N-(2-aminobenzyl)glycine esters (145) [73GEP(O)2305575; 75JMC224; 76USP3988340; 79GEP(O)2832138; 80USP4208521; 82EUP-54180; 84EUP129258; 85EUP133234; 86USP4593029; 87USP4670434] or 2-amino-N-(2,2-dimethoxyethyl)benzylamines (81EUP29559) with cyanogen bromide, guanidine, or alkyl isothiourea (79MI3) as well as condensation of ethyl N-(ethoxy-carbonylphenyl)thiocarbamates (146) with 1,2-diaminoethane (82JHC1117) gave the imidazo[2,1-b]quinazolines 137 and 142.

3. Imidazo[5,1-b]quinazolines

Sherlock prepared the imidazo[5,1-b]quinazoline (148) by the condensation of 2-(methylaminomethyl)-4-phenyl-6-chloro-3,4-dihydroquinazoline (147) with formaldehyde [70GEP(O)1932885]. Condensation of anthranilic acids and 4-thiohydantoin afforded 149 (82H1375).

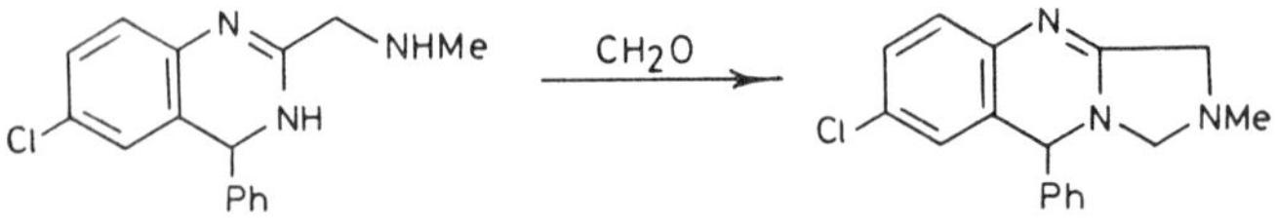

(147)   (148)

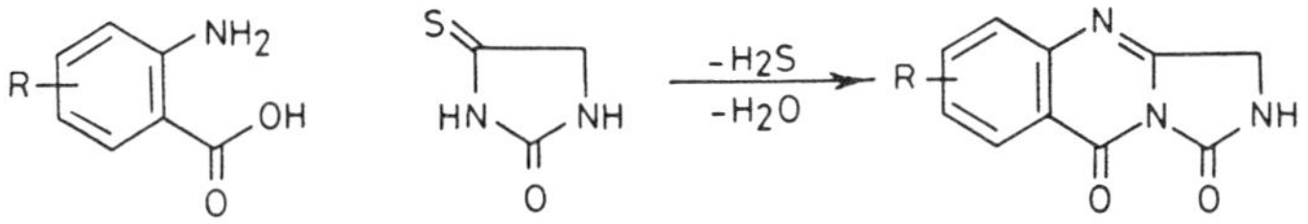

(149)

4. *Imidazo[1,2-c]quinazolines*

Compounds belonging to this system **(151)** wre synthesized from 4-aminoquinazolines **(150)** by cyclocondensation with two carbon cyclizing agents, such as α-haloketones (84MI3; 85MI1) or 2-chloroethanol (86MI1). This cyclocondensation was also performed, in an indirect manner, when quinazolines with good leaving groups at position 4, such as 4-chloroquina-zolines **(152,** LG = Cl) (54JOC699; 64MI1; 74JOC3508, 74URP433-149, 74YZ417; 75MI1; 76KGS834) or 4-mercaptoquinazolines **(152,** LG = SMe) (82PHA605; 84MI5), were cyclized with 2-aminoethanol or 1,2-diaminoethane.

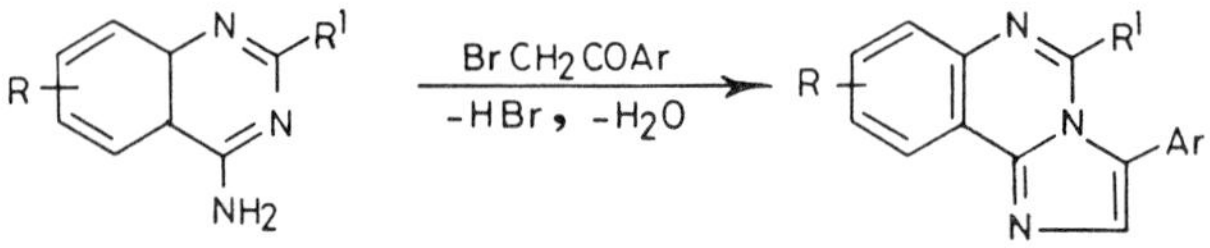

(150)   (151)

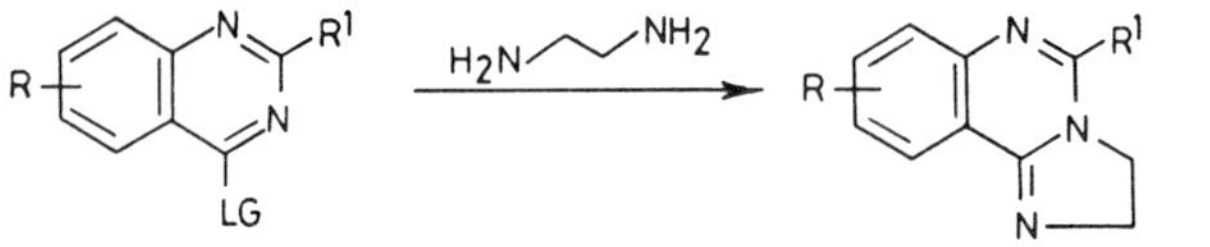

(152), LG = Cl or SMe   (153)

Cyclization of either 3-allyl-4-imino-2-mercaptoquinazoline (**154**) [77IJC(B)751] by heating with polyphosphoric acid or N-(4-quinazolinyl)-α amino acids (**156**) (76KGS1268) by heating with a mixture of acetic anhydride and acetic acid was reported to afford the imidazo[1,2-c]-quinazolines **155** and **157**, respectively.

Reaction scheme: (154) → PPA, Δ → (155)

(156) → Ac₂O, AcOH, Δ → (157)

An elegant method for synthesizing the title ring system is the iodide ion catalyzed isomerization of 4-(1-aziridinyl)quinazolines (**158**). The obtained 2,3-dihydro derivatives (**159**) were easily dehydrogenated wtih chloranil to the imidazo[1,2-c]quinazolines (**160**) [70GEP(O)1946188; 74JOC3508, 74JOC3599; 75JMC447].

Reaction scheme: (158) → NaI → (159) → Chloranil → (160)

Imidazo[1,2-c]quinazolines were also prepared from properly substituted imidazole precursors; thus, cyclocondensation of 2-(2-alkyl-aminophenyl)-4,5-dihydroimidazoles (**161**) with one-carbon cyclizing agents, such as aldehydes (65NEP6409191; 68SZP452537; 80JHC155), ketones (75FES536; 82EUP46446), acid chlorides (65NEP6409191), or carbon disulfide (75BSF2118), gave the corresponding compound **162**.

(161)   (162)

2-(2-Substituted-ureido)benzonitriles and 2-(2-substituted-ureido)thio-
benzamides are very useful starting materials for the synthetic route comprising the formation of the two heterocyclic rings of the title compounds.
Cyclization of 2-ureidobenzonitriles (163) with α-aminoacids (83JPR88;
89JPR537), base-catalyzed cyclization of 2-(2-chloroethylureido)benzoni-
trile (165) (80JHC1553; 81JHC515), enzymatic cyclization of 2-(allylurei-
do)benzonitrile (89CC835), and acid- or base-catalyzed cyclization of 2-(2-
chloroethylureido)thiobenzamide (89JHC595) led to the formation of examples of imidazo[1,2-c]quinazolines. Also, belonging to this method is
the reaction of 2-isocyanatobenzonitrile with aminoacetonitrile or glycine
methyl ester to give imidazo[1,2-c]quinazolines (87CZ373).

(163)   (164)

(165)   (166)

Reaction of 2-acylaminobenzophenones (e.g. 167) [72GEP(O)2141616;
73GEP(O)2166380; 74URP433149; 80PHA256; 81CPB2135], N-alkoxy-
carbonylanthranilic acids (67USP3329679), or N-aryloxycarbonylan-
thranilic acids (70YZ629; 82EUP53767) with 1,2-diaminoethane resulted
in double ring closure to give 168.
3,1-Benzoxazin-4-ones (169, Z = O) [84MI1; 88EGP(D)258232], 3,1-
benzothiazin-4-thiones (169, Z = S) (75BSF1411), or 3,1-benzothiazin-
2,4-dithiones (72ZC289; 79PHA390; 83ZC215) were also cyclized with
1,2-diaminoethane to the corresponding imidazo[1,2-c]quinazolines 170.

$$\text{(167)} \quad \xrightarrow[-CHCl_3,\ -H_2O]{H_2N\diagdown\diagup NH_2} \quad \text{(168)}$$

(167)   (168)

$$\text{(169)}, Z = O \text{ or } S \quad \xrightarrow{H_2N\diagdown\diagup NH_2} \quad \text{(170)}$$

(169), Z = O or S   (170)

Numerous biological applications have been attributed to imidazo-quinazolines, such as being narcotic antagonists (80JHC155), antihypertensives [71USP3621025; 73GEP(O)2257376, 73GEP(O)2305575, 73-USP3745216; 74GEP(O)2402454; 75GEP(O)2508543, 75USP3919210; 76USP3984556; 79CP1057752], blool platelet aggregation inhibitors [73-GEP(O)2305575; 75JMC224; 79GEP(O)2832138; 80CPB2024; 81EUP29-559; 82EUP46267; 83EUP73060; 84EUP116948, 84EUP129258; 85CPB33-36, 85EUP133234, 85JMC1387; 86JAP(K)61115083, 86USP4593029], blood sugar depressants (76USP3984556; 82EUP46446; 86MI2), bronchodilators [70GEP(O)1932885; 73GEP(O)2234174, 73GEP(O)2257376; 74GEP(O)2402454; 75JMC447, 75USP3919210; 76USP3969506, 76USP3978059, 76USP3984556], antidepressants [75GEP(O)2508543, 75USP3919210; 79CP1057752]; central nervous system stimulants (67USP3329679; 71USP3621025; 75USP3919210; 86MI2); tranquilizers [75USP3919210; 83GEP(O)3233766; 84USP4451464, 84USP4452787; 88JMC1220], analeptics (76USP3969506), circulatory disease treatments (82EUP53767), gastric secretion inhibitors (83HCA148; 84EUP129258); antiallergics (83HCA148), cancer metastasis inhibitors [84EUP116948, 84EUP129258; 85JAP(K)601512416; 86USP4593029], antiphlogistics [83-GEP(O)3220438], analgesics [83GEP(O)3220483], cardiotonics [81EUP-29559; 84EUP129258; 86JAP(K)61115083], vasodilators [81EUP29559; 86JAP(K)61115083], anticonvulsants (86MI2; 88JMC1220), antithrombotics (86USP4593029; 87JMC295), cyclic adenosine monophosphate phosphodiesterase inhibitors (87JMC295, 87USP4670434), cardiovascular agents (79M13), and antibacterials (77FZ84).

D. BENZIMIDAZO-QUINAZOLINES

1. Benzimidazo[1,2-a]quinazolines

Examples (**172**) of the title compounds were obtained from the reaction of 2-chlorobenzoyl chloride and 2-aminobenzimidazoles (**171**) (89K-GS272).

2. Benzimidazo[2,1-b]quinazolines

Benzimidazo[2,1-*b*]quinazolines **174** were prepared by condensing benzimidazoles carrying a leaving group at position 2 (**173**, LG = Cl, NH$_2$, SMe, SO$_3$H) with anthranilic acids [71GEP(O)2058185, 71JHC141; 75JMC447; 76USP3963720, 76USP4000275; 79JCS(P1)3085, 79PHA138; 83JPR88; 85URP1182043; 87KGS1673; 89KGS272]. *N*-Anthraniloyl-2-phenylenediamine (**175**) gave benzimidazo[2,1-*b*]quinazoline (**176**), through double ring formation, when treated with thiophosgene [81IJC-(B)579].

3. Benzimidazo[1,2-c]quinazolines

Cyclization of 2-(2-aminophenyl)benzimidazoles (177) with carboxylic acids [62JCS945; 63BSB365; 74KGS1405; 77MI1; 81IJC(B)579; 82AP866; 83JIC1071; 86FES852, 86KFZ690], aldehydes (62JCS945), or ketones [77IJC(B)1100] gave examples (178) of the title compounds. Iminophosphorane (179) derived from 2-(2-azidophenyl)-benzimidazole and triphenylphosphine underwent aza-Wittig reaction with alkyl isocyanates to give (180) (89T4263).

(177)   (178)

(179)   (180)

A general method for synthesizing benzimidazo[1,2-c]quinazolines (e.g. 181) is the cyclocondensation of 2-phenylenediamine with 3,1-benzoxazinones [37JIC411; 62JCS945; 67AG(E)878; 84MI1], their thio analogues (80PHA293), trichloroacetamidobenzophenone [72GEP-(O)2141616; 73GEP(O)2166380], or isocyanatobenzoyl chloride (76-JHC421).

(181)

(182)   (183)

Photolytic molecular rearrangement of the 4-(1,2,3-benzotriazol-1-yl)-2-phenylquinazoline (182) gave (183) (74JHC737).

Some benzimidazoquinazolines have been reported to possess bronchodilating (76USP3963720), tranquilizing (76USP3963720), anticonvulsant (86KFZ690), and immunodepressant (76USP4000275) properties.

V. Condensed Oxazolo-quinazolines

A. ISOXAZOLO-QUINAZOLINES

1. *Isoxazolo[3,2-b]quinazolines*

Performing a Vilsmeier–Haack reaction on 3-benzoyloxy-2-methyl-quinazolin-4-one (184) afforded the isoxazolo[3,2-*b*]quinazolinone (185) [86IJC(B)709]. This ring system (187) was also synthesized by cyclocondensation of anthranilic acids or isatoic anhydrides with the isoxazolin-3-ones (186) (77AF766; 83MIP1), or by condensation of methyl anthranilate with 3-chloropropanoyl chloride followed by cyclization with hydroxylamine hydrochloride (77AF766).

(184)   (185)

2. *Isoxazolo[2,3-c]quinazolines*

Examples (191) of the title compounds were prepared by the base-catalyzed cyclocondensation of 4-methylthioquinazolin-3-oxides (189) with active methylene compounds (190) (87SC1449).

(186)   (187)   (188)

(189)   (190), X = CN, Ac, COOEt   (191), Z = NH or O
         Y = CN, CONH, COOEt

B. Oxazolo-quinazolines

1. Oxazolo[3,2-a]quinazolines

Members of this ring system (194) were synthesized by reduction of the 2-acyloxyquinazolines 192 with alkali metal borohydrides followed by dehydrative cyclization of the resultant hydroxyalkyl derivatives (193) [74JIC453; 85IJC(B)1035].

2. Oxazolo[3,4-a]quinazolines

Acetylation of 2-(1-hydroxyphenylmethyl)quinazolin-4-one (195) with refluxing acetic anhydride gave the expected acetate 196. Refluxing the starting compound (195) or its acetate (196) with acetic anhydride and sodium acetate gave the oxazolo[3,4-a]quinazolinone (197). Spectral as well as chemical evidence favored the assigned structure(197) rather than the alternative oxazolo-[4,3-b]quinazolinone structure (198) that could arise as a result of cyclization with N-3 of the quinazoline nucleus (74JOC3828).

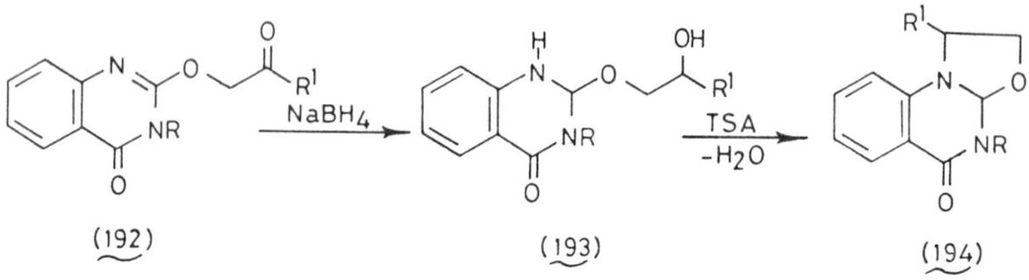

(192)   (193)   (194)

3. *Oxazolo[2,3-b]quinazolines*

The oxazolo[3,2-*a*]quinazoline (**200**) was obtained by cyclization of 3-(2-chloroethyl)quinazoline-2,4-dioine (**199**) with potassium carbonate (60JCS3551). The same compound was also obtained when ethylene oxide reacted with 2-chloroquinazolin-4-one (**201**) in the presence of sodium hydroxide (60JCS3551). 3-Acylalkylquinazoline-2,4-diones (**201**) underwent intramolecular dehydrative cyclization upon heating with polyphosphoric acid to **203** [89IJC(B)274].

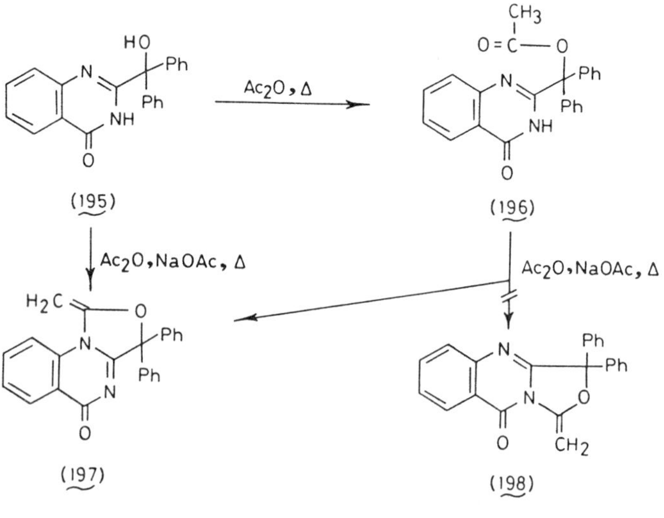

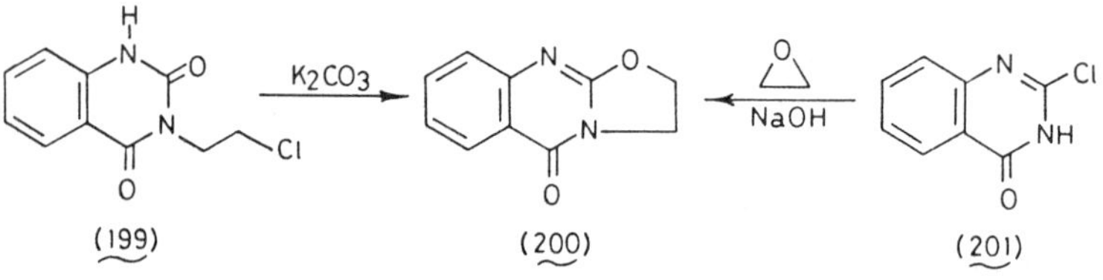

(195)   (196)

(197)   (198)

(199)   (200)   (201)

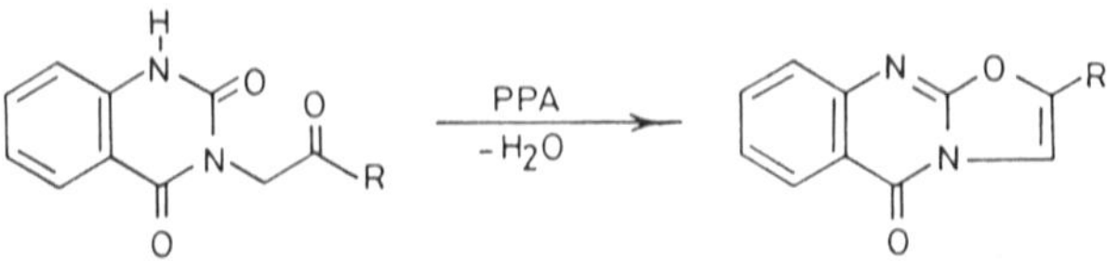

(202)   (203)

Bromination of 3-alkylquinazoline-2,4-dione **204** followed by cyclization of the dibromo derivative (**205**) with potassium hydroxide led to the formation of 2-bromomethyl-2,3-dihydrooxazolo[2,3-*b*]quinazolin-5-one (**206**) (74JIC453).

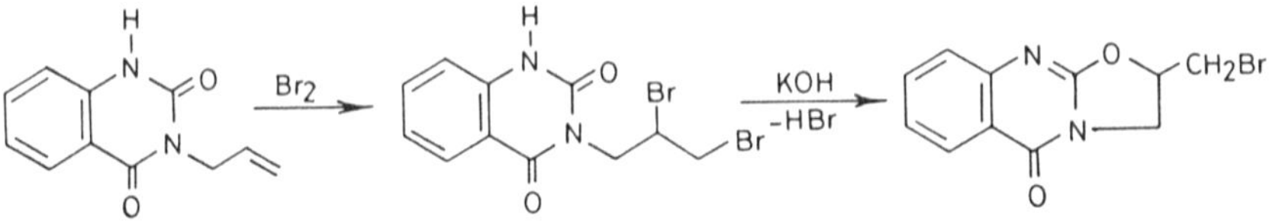

(204)   (205)   (206)

A remarkably simple and efficient method for preparing oxazolo[2,3-*b*]quinazoline (**208**) is the reaction of anthranilic acids or their esters with β-haloisocyanates, followed by base-catalyzed cyclization of the intermediate urea derivatives (**207**) [74GEP(O)2252122; 76S469; 80JHC1553].

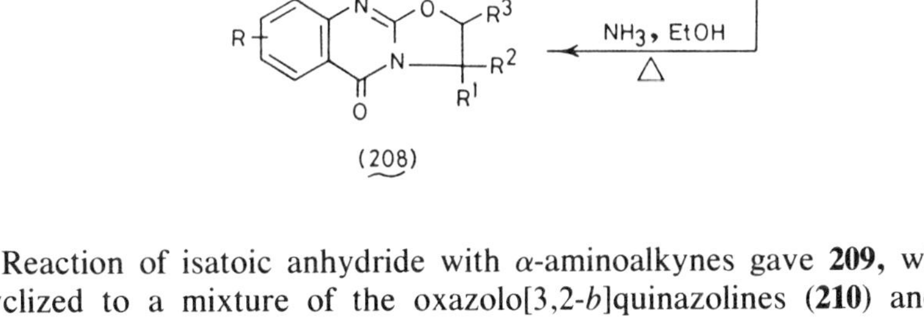

(207)

(208)

Reaction of isatoic anhydride with α-aminoalkynes gave **209**, which cyclized to a mixture of the oxazolo[3,2-*b*]quinazolines (**210**) and 2-aminophenyloxazoles (**211**) upon treatment with phosgene (89JHC1495).

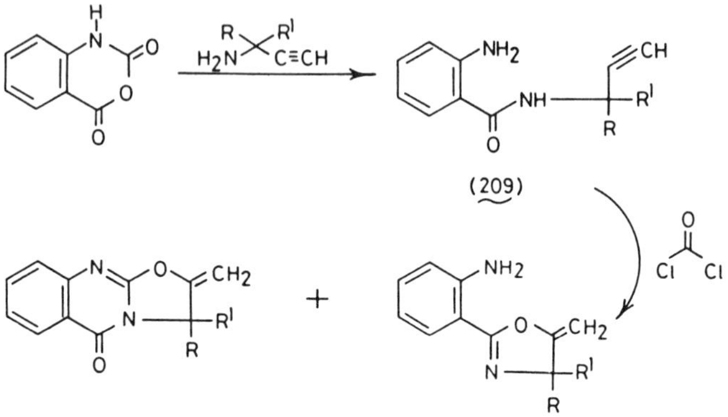

(209)

(210) + (211)

4. Oxazolo[3,2-c]quinazolines

An example of this ring system (212) was prepared by Sinha and Thakur (74JIC453) by cyclization of 3-allylquinazoline-2,4-dione (204) with phosphoric acid. Heating 2-acylaminobenzophenones (e.g. 167) with ethanolamine in dimethyl sulfoxide (DMSO) gave the 10b-phenyl-2,3,6,10b-tetrahydro-5H-oxazolo[3,2-c]quinazolin-5-ones (213) [70YZ629; 72GE-P(O)2141616; 80PHA256; 81CPB2135].

Some oxazolo-quinazolines have been reported to possess antihypertensive activity [74GEP(O)2252122].

(204)   PPA →   (212)

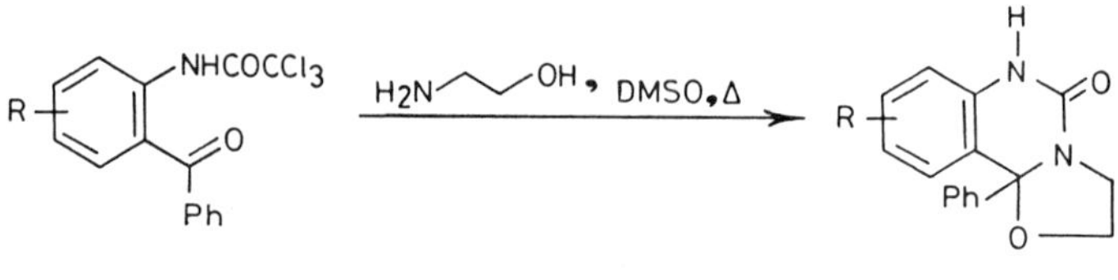

(167)   H₂N⌒OH , DMSO, Δ →   (213)

C. BENZOXAZOLO-QUIANAZOLINES

1. Benzoxazolo[3,2-a]quinazolines

The benzoxazolo[3,2-*a*]quinazolin-5-ones (216) were prepared in excellent yields by two routes. In the first route, the two heterocyclic rings of 216 were consecutively formed when the disodium salts of *N*-(2-hydroxyphenyl)anthranilic acids (214) wre cyclized with two molar equivalents of cyanogen bromide. In the second versatile route, the pyrimidine ring of 216 was formed when *N*-(2-benzoxazolyl)-2-fluorobenzamides (217) were thermally cyclized (81JHC287).

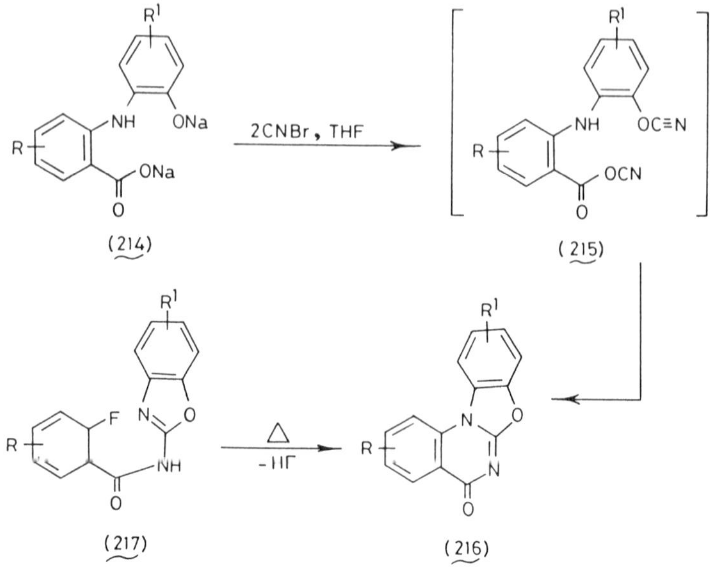

(214)   (215)

(217)   (216)

2. Benzoxazolo[2,3-b]quinazolines

Compounds (219) of this class were prepared from anthranilic acids and 2-chlorobenzoxazines (218) (54JPS583; 68JPS1445; 79PHA138).

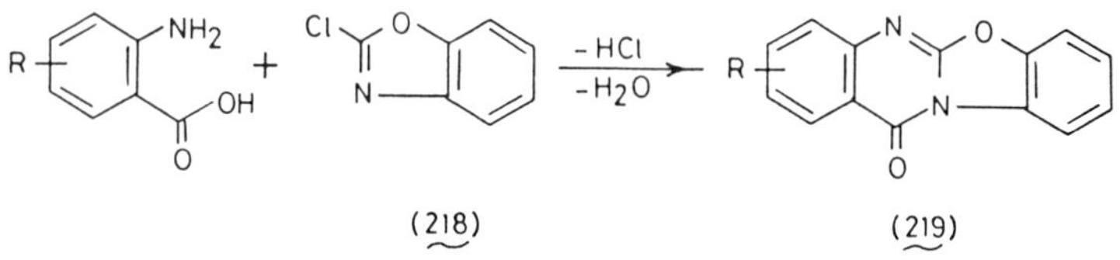

(218)   (219)

VI. Condensed Thiazolo-quinazolines

A. BENZISOTHIAZOLO-QUINAZOLINES

Benzisothiazolo[2,3-a]quinazolines

Reaction of benzodithiol-3-thione (**220**) with anthranilamide gave a complex mixture of products from which benzisothiazolo[2,3-*a*]quinazolin-5-one (**221**) was isolated (75JHC1077).

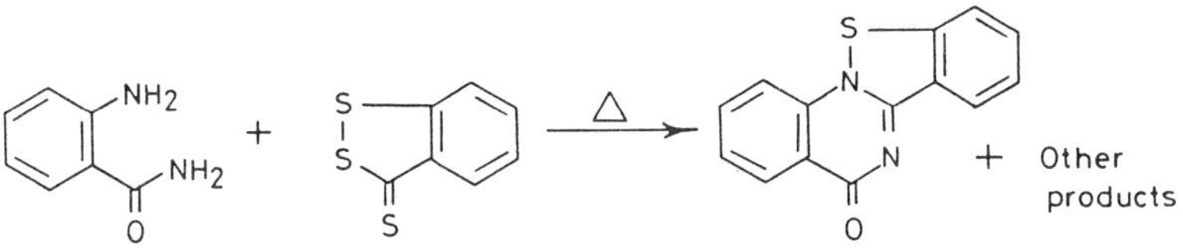

(220)   (221)

B. THIAZOLO-QUINAZOLINES

1. *Thiazolo[3,2-a]quinazolines*

Compounds of this ring system were prepared from quinazoline or thiazole precursors as well as from nonheterocyclic starting materials. 2-Mercaptoquinazolines having suitable substituents at position 1, such as 1-(2-hydroxyethyl (**222**) [63JIC545; 78JAP(K)7844593] or 1-allyl (64JIC715), readily cyclize with acids to the thiazolo[3,2-*a*]quinazolines (**223**.) 3-Substituted(2-quinazolinyl)thioacetic acid (**224,** R′ = OH) [74-JIC457; 77IJC(B)41; 78JIC928] or (2-quinazolinylthio)ketones (**224**) (71JIC443; 74JIC457) also cyclized to 3-substituted thiazolo[3,2-*a*]quinazolines (**226**).

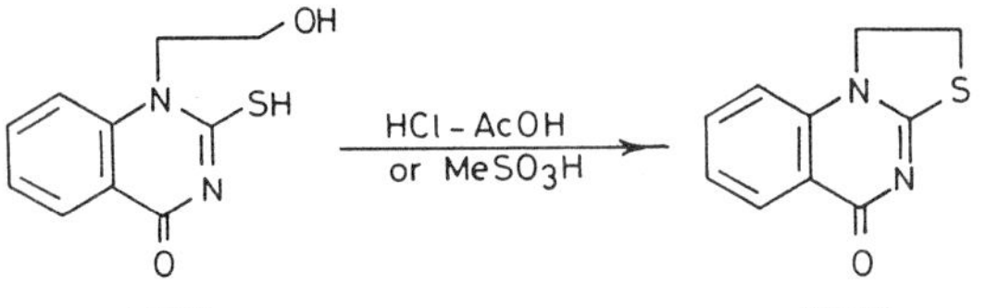

(222)   (223)

Cyclization of quinazolin-2-ylthioketones (227) that are not blocked at N-3 afforded a mixture of thiazolo[3,2-a]quinazolines (228) and thiazolo[2,3-b]quinazolines (229) [72CI(L)255]. Thiazolo[3,2-a]quinazolinones (231) were also prepared by cyclization of 2-(3,4-dihydro-4-oxoquinazolin-2-ylthio)acylacetic esters (230) (69USP3471497; 70MI2; 75JHC1207).

The possible linear isomers wre not obtained during these syntheses, probably because of the greater nucleophilicity of the quinazolinone N-1 compared to N-3. The mesoionic thiazolo[3,2-a]quinazolines (233) were obtained by the acid-catalyzed cyclization of the (3-substituted quinazolin-2-ylthio)ethanoic acids (232) [68JHC185; 70JIC758; 71JIC395; 77IJC(B)41; 79IJC(B)39; 80MI2; 82JIC666] or 2-phenacylthioquinazolines (71JIC395).

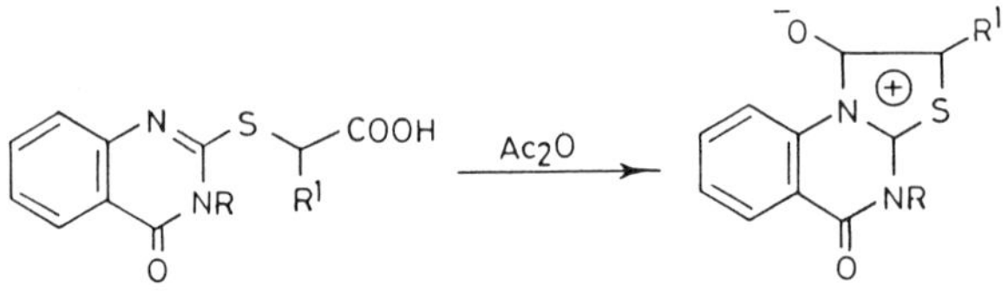

(232)   (233)

A synthesis from thiazole precursors occurred when 2-imino-3,4-diarylthiazolines **(234)** were cyclized with formaldehyde to give **236** [79IJC(B)632].

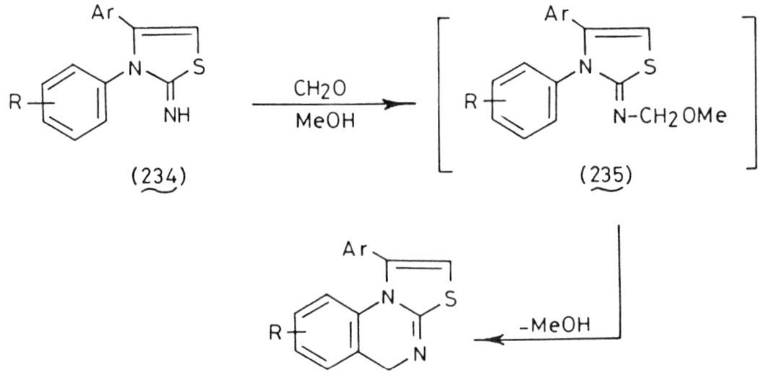

(234)   (235)

(236)

Synthesis from nonheterocyclic precursors involved double ring formation by the reaction of anthranilic acids (59MIP1; 60MI1; 60-MI2; 63JMC450; 66IJC527; 70MI2; 73IJC750; 80MI2), ethyl anthranilate (55JIC644), anthranilonitriles (60MI3), or 2-aminobenzyl alcohol (66IJC527) with α-thiocyanoketones to give the title compounds (e.g.**237**).

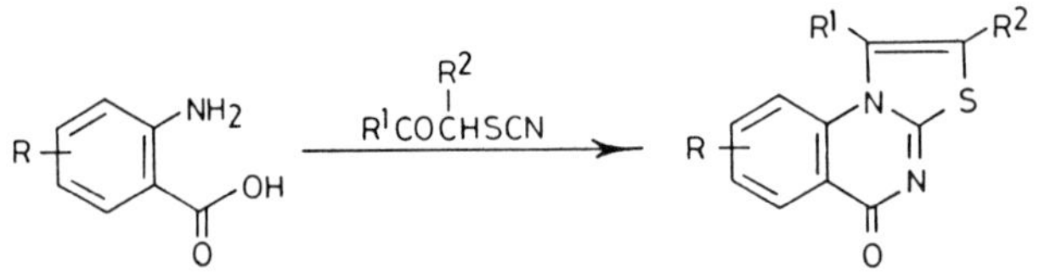

(237)

Double ring closure was also affected by the reaction of *N*-phenacylanthranilic acids **(238)** with potassium thiocyanate or the reaction of *N*-chloroacetylanthranilic acid (64JIC855; 69IJC881) with thiourea. Cy-

clization of 2-ethoxycarbonylanilinium thiocyanate (241) with α-haloketones (71JIC953) or 1,2-dibromopropionic esters (82JIC666) as well as
cyclization of ethyl thioureidobenzoate with α-haloketones (53MI2;
54JIC848; 55JIC589, 55JIC644; 60MI2) gave thiazolo[3,2-a]quinazolines
(e.g. 242).

(238)   (239)   (240)

(241)   (242)

2. Thiazolo[2,3-b]quinazolines

The first general method described for synthesizing this linearly annulated ring system from quinazoline precursors is the cyclization of 2-
mercaptoquinazolines possessing a suitable two-carbon cyclizable substituent at position 3. Thus, acid catalyzed cyclization of 3-allyl- [57MI1;
61T(14)304; 62T1019; 64IJC285; 77IJC(B)751; 77ZC444], 3-(2-hydroxy-
ethyl)- (243, Y = OH) [63JIC545; 67HCA1440; 77ZC444; 78JAP-
(K)7844592, 78PHA185], or 3-(2-aminoethyl)quinazolines (243, Y = NH₂)
(65IJC284) afforded examples of the title compounds (244).

(243) Z=O or S   (244)
Y=OH or NH₂

2-Mercaptoquinazolines (245) were also reported to react with two-
carbon cyclizing agents such as α-haloketones [69IJC765; 72CI(L)255;
81USP4302585; 83AP569; 85AP502; 89IJC(B)274], α-haloacids [77ZN-

(B)94; 78IJC(B)537; 78MI1; 83AP569, 83MI2; 84JIC1050; 85AP502, 87MI1], 1, 2-dihaloalkanes (62JOC3701; 78JIC928; 83AP569; 84JIC1050], 1, 2-dihaloalkenes (87MI1), dimethyl acetylenedicarboxylate [83AP569; 86JHC1359, 86USP4588812; 87EGP(D)251983], or maleic anhydride (83AP569) to give, predominantly, thiazolo[2,3-*b*]quinazolines (e.g. **246**). It is worth mentioning that though the same synthetic pathway was used, some of these results [69IJC765; 72CI(L)255; 83AP569; 85AP502, 89IJC(B)274] are in contradiction with those (69USP3471497; 70MI2; 75JHC1207) described for the synthesis of the angular thiazolo[3,2-*a*]-quinazolines and structural confirmation should be considered. 2-Mercaptoquinazolin-4-one (**247**) was also cyclocondensed with the 2-dichloro compound **248** to give the pentacyclic thiazolo[2,3-*b*]quinazoline **249** (88RRC291).

(245)   (246)

(247)   (248)   (249)

The second general synthetic route to the title compounds involved the cyclocondensation of anthranilic acids with thiazole derivatives having a good leaving group at position 2, such as 2-methylthiothiazoles (**250**) [71BRP1242863; 75JAP(K)7514699; 82MI1; 83AP394] or 2-chloro-thiazoles (**250, LG = Cl**) [55JIC589, 55JIC631, 55JIC647; 56MI1; 61MI1; 63JMC450; 64JIC591; 70JIC793, 70MI2; 77GEP(O)2557425; 79-USP4168380; 80JCS(P1)633; 81EUP27268; 83JMC107; 85EUP142057; 88GEP(O)3634532]. 2-Aminobenzyl alcohol (66IJC527) and 2-amino-acetophenones (69IJC1191) have also been used in place of anthranilic acids in the latter cyclization.

(250), LG = Cl or SMe   (251)

The third route for synthesizing thiazolo[2,3-*b*]quinazolines comprised the consecutive or simultaneous formation of their two heterocyclic rings in one step. Reaction of anthranilic acids, their esters, or amides with 1-halo-2-isothiocyanoethanes (71BRP1242863; 72MI2; 79JHC-391; 87AP1276), α-thiocyanato-ketones [83GEP(O)3142727], or with 1-isothiocyano-4-(diethylamino)-2-butyne (78FRP2393001) gave thiazolo[2,3-*b*]quinazolines (**244**). This reaction was performed in an inverse manner by transposing the functional groups on the reacting entities.

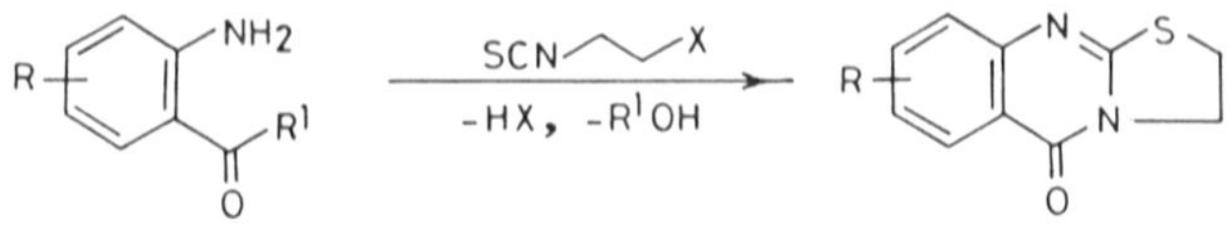

(244)

Thus, reaction of ethyl 2-isothiocyanatobenzoate with α-aminoalkynes led to the formation of **253** [73GEP(O)2212371, 73GEP(O)2228259, 73S426]. Condensation of 2-carboxyphenylthioureas or their esters (**254**) with two-carbon cyclizing agents, such as α-haloketones [52MI1; 53MI1; 55JOC302; 61T(15)53; 83EGP(D)204095], α-haloacids (83HI549), or 1, 2-dihaloalkanes (52MI1; 55JOC302), afforded **251.**

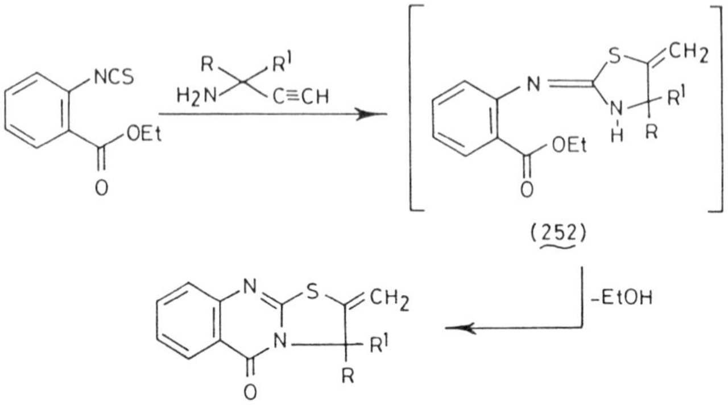

(252)

(253)

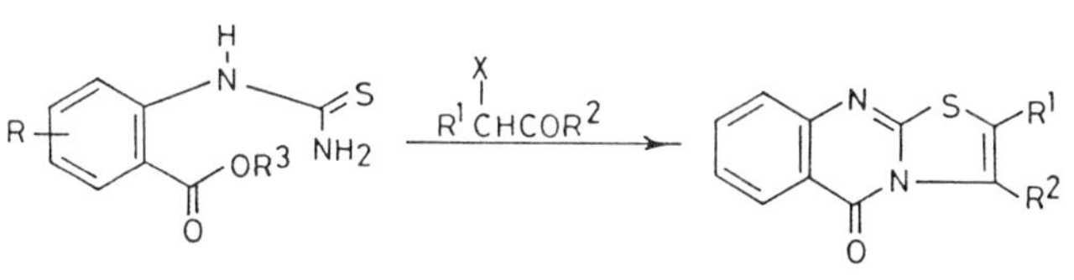

(254)   (251)

   Trithioisatoic anhydrides were also cyclized to thiazolo-[2,3-*b*]quinazolines (**255**) upon reaction with 2-aminoethanol or 1-amino-2-dimethoxyethane (80PHA124; 83ZC215). Double ring closures occured when N-(2-methoxycarbonylphenyl)thiocarbamate reacted with 2-aminothioethanol (82JIC1117) to give **256.** Thermal molecular rearrange-

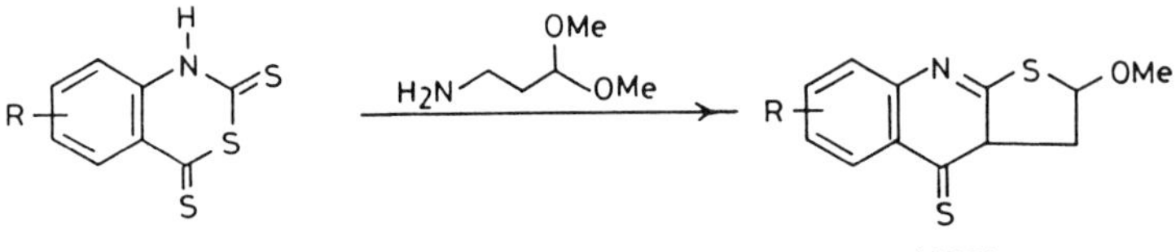

(255)

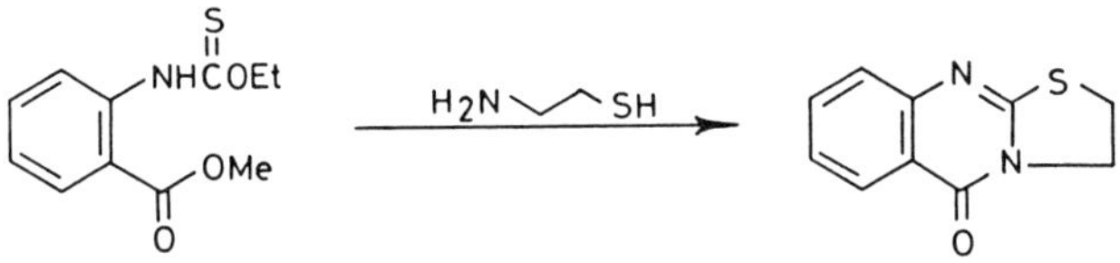

(256)

ment of di[2-(N-2'-aminobenzylformamidopropenyl)]disulfide (**257**) (55-JCS2390) or 3-(thiazol-2-yl)-1,2,3-benzotriazin-4-one (**259**) [80JCS(P1)633] gave the thiazolo[2,3-*b*]quinazolines **258** and **261**, respectively.

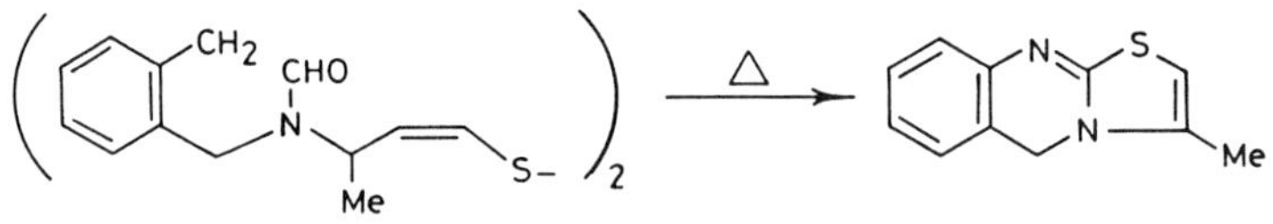

(257)   (258)

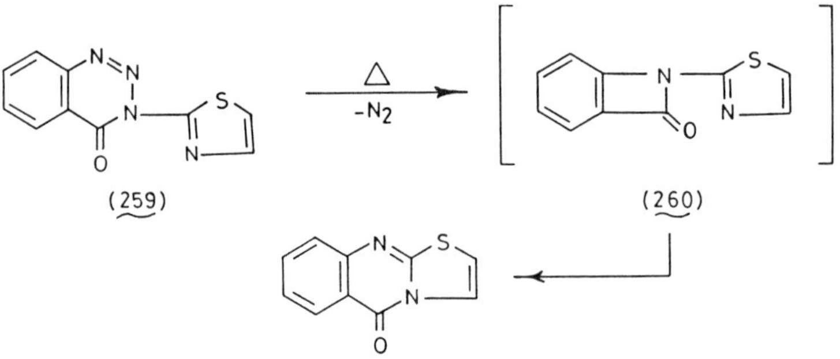

(259)   (260)

(261)

3. Thiazolo[4,3-b]quinazolines

Examples of this ring system (**263**) were prepared by cyclocondensation of anthranilic acid with the 1,3-thiazole derivatives **262** (82MI1; 83AP394). Reductive cyclization of the 4-ethylamino-3-(2-nitrobenzyl)thiazolidines (**264**) was affected by heating with iron filings and acetic acid to give 4H-3,3a-dihydrothiazolo[4,3-b]quinazoline (**265**) (87JHC107).

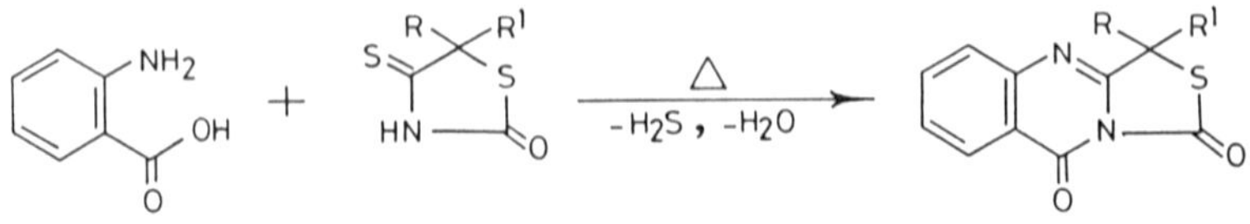

(262)   (263)

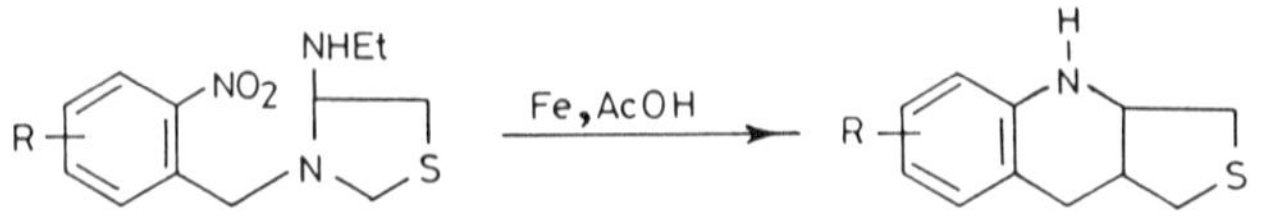

(264)   (265)

4. Thiazolo[3,2-c]quinazolines

Acid-catalyzed cyclization of 3-(2-hydroxyethyl)-2,4-dithioxo-octahydroquinazoline (**266**) was reported to give the thiazolo[3,2-c]quinazoline **267** together with the linear thiazolo[2,3-b]quinazolines (**268**) (78PHA185).

The mesoionic thiazolo[3,2-c]quinazolines (**271**), a resonance stabilized 14π-electron system, were prepared by condensation of the sodium salt of the 4-mercaptoquinazolines (**269**) with α-haloketones (70IJC1065; 80H3), α-haloacids (85JOC1666; 89KGS408), α-haloesters [78IJC(B)331], allyl chloride (70MI1; 87LA103), or tosylhydrazones of α-ketoacid halides (84CC1348) to give **270,** which was cyclized with acids to **271.**

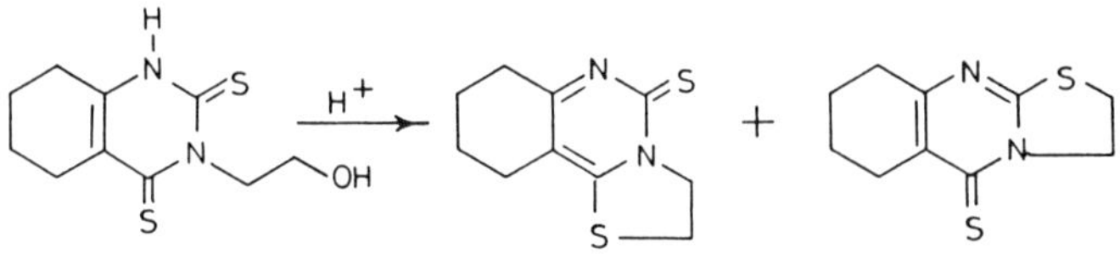

(266)   (267)   (268)

(269)   (270)   (271)

The two heterocyclic rings of thiazolo[3,2-c]quinazolines (**273**) were formed in one step when 2-ethoxycarbonylaminobenzaldehydes (**272**) were cyclized with 2-aminothioethanol or esters of L-cysteine (63IJC318; 65JIC155, 65JIC220).

(272)   (273)

Various biological applications were reported for thiazolo-quinazolines, such as being useful as fungicidals (83H1549; 84JIC1050), antiinflamatory agents [71BRP1242863; 75JAP(K)7514699; 78FRP2393001; 86JHC1359, 86USP4588812], sedatives (71BRP1242863; 78FRP2393001), muscle relaxants [77GEP(O)2557425; 78FRP2393001], central nervous system depressants (78FRP2393001), antibacterials (84JIC1050), cholinesterase inhibitors (80MI2), potential metabolite antagonists (66IJC527), hypotensives [78JAP(K)7844593], blood platelet aggregation inhibitors [78JAP(K)-7844593], analgesics [71BRP1242863; 75JAP(K)7514699], passive cutaneous anaphylaxis inhibitors (79USP4168380; 81EUP27268), antiallergics (83JMC107), antihypertensives (83AP569; 85AP502), hypothermic agents (71BRP1242863), antiasthmatics [77GEP(O)2557425], herbicidals [83GEP-(O)3142727], immunomodulators (86JHC1359, 86USP4588812), and immunosupressants (81USP4302585).

C. BENZOTHIAZOLO-QUINAZOLINES

1. *Benzothiazolo[3,2-a]quinazolines*

Acylation of 2-aminobenzothiazoles (**274**) with 2-fluorobenzoyl chloride followed by thermal cyclization of the intermediate acyl derivatives gave the benzothiazolo[3,2-a]quinazolines (**275**) (81JHC801). Tetrahydrobenzothiazolo[3,2-*a*]quinazolines (**277**) were obtained by condensation of anthranilic acids with 2-thiocyanocyclohexanone (**276**) (72IJC605).

2. Benzothiazolo[2,3-b]quinazolines

Oxidative cyclization of 3-aryl-4-oxo-2-thioxo-tetrahydroquinazolines (**278**) with *N*-bromosuccinimide and sulfuric acid gave examples of the title compounds (e.g. **279**) (87H2371, 87BSB797). Cyclocondensation of 2-thioxoquinazolines (**280**) with 2-chlorocyclohexanone (72IJC605) or 4-chloro-3,5-dinitrobenzotrifluoride (87AP569) also afforded benzothiazolo[2,3-*b*]quinazolines (e.g. **281**).

Synthesis of the title compounds from benzothiazole precursors has also been accomplished by cyclocondensation of 2-chlorobenzothiazole (**282**) with anthranilic acids or their esters (34JIC463; 53JA712, 53JOC1380; 75JIC886; 79CB3424, 79PHA138).

(282)   (283)

Double ring formation occured when 2-aminothiophenol cyclocondensed with 2-isocyanatobenzoyl chloride to yield (**284**) (76JHC421). Double ring formation also took place upon cyclization of 2-(2-aminobenzoylamino)thiophenols (**285**) with ethyl chloroformate to give **279** (69IJC444; 72IJC476).

(284)

(285)   (279)

3. Benzothiazolo[3,2-c]quinazolines

Singh and Lal [73IJC959; 76IJC(B)685] prepared 8,9,10,11-tetrahydro-benzothiazolo[3,2-*c*]quinazolin-7-ium perchlorate (**287**) by dehydrative cyclization of 4-[(2-oxocyclohexyl)thio]quinazoline (**286**) with sulfuric acid and then perchloric acid. 2-(2-Azidophenyl)benzothiazole (**288**) reacted with triphenylphosphine to give the iminophosphorane, which cyclized to **289** on treatment with ethoxycarbonyl isocyanate (89T4263).

(286)   (287)

(288)   (289)

VII. Condensed Triazolo-quinazolines

A. 1,2,3-TRIAZOLO-QUINAZOLINES

1,2,3-Triazolo[1,5-a]quinazolines

Base-catalyzed condensation of 2-azidobenzoic acid or its nitrile with active methylene compounds furnished the 4-substituted 1,2,3-triazolo [1,5-a]quinazolines (290) in good yield [66JCS(C)2290; 69JCS(D)423; 74JCS(P1)534].

(290), R = Ph, CONH$_2$, or CN

B. 1,2,4-TRIAZOLO-QUINAZOLINES

1. 1,2,4-Triazolo[1,5-a]quinazolines

This ring system (e.g. 293) was prepared from 1,2,4-triazole precursors, such as 3-amino-1,2,4-triazole (292) by condensation with the 2-(*N*-phenylamino)methylene-1,3-cyclohexanedione (291) (70MI4). Pertinent to this synthesis is the reaction of 3-amino-1,2,4-triazoles (**294**) with 2-hydroxymethylenecyclohexanone [79JCS(P1)3085] or ethyl 2-oxocyclohexanecarboxylate (89OPP163) to yield a mixture of the angular 1,2,4-triazolo[5,1-*a*]quinazolines (**295**) and the linear 1,2,4-triazolo[5,1-*b*]-quinazolines (**296**); yields of the latter predominated the former.

(291)   (292)   (293)

(294)   (295)   (296)

A synthesis comprising one-step formation of the two heterocyclic rings of this system was reported (80HCA1) in which 2-carboxyphenylhydrazine was cyclocondensed with **297** to give **298**.

(297)   (298)

2. *1,2,4-Triazolo[4,3-a]quinazolines*

A general method for synthesizing 1,2,4-triazolo[4,3-*a*]quinazolines (e.g. **299**) is the cyclocondensation of 2-hydrazino-4-substituted quinazolines [76JAP(K)76100098; 84IJC(B)1293] or 2-hydrazino-3-substituted quinazolin-4-ones [64ZOB1745; 75GEP(O)2508333; 78PHA124, 78PHA-125, 78PHA462, 78PHA507; 80EGP(D)139715, 80PHA800; 83EGP(D)-158549, 83EUP76199, 83PHA25, 83PHA367; 86JCR(S)232; 86MI5] with carboxylic acids or their derivatives. Using 3-unsubstituted-2-hydrazinoquinazolines in place of the 3-substituted isomers led to the formation of a mixture of the angular 1,2,4-triazolo[4,3-*a*]quinazolines (**300**) and the linear 1,2,4-triazolo[4,3-*b*]quinazolines (**301**) (86JHC833). Oxidative cycli-

(299)

(300)   (301)

zation of hydrazones (**302**), derived from 3-substituted-2-hydrazinoquinazolines and aromatic aldehydes, with ethanolic ferric chloride also led to the formation of (**299**) (83BCJ1227; 86JHC833).

(302)   (299)

In one synthesis [76JAP(K)76100098], 1,2,4-triazolo[4,3-*a*]quinazolines (**304**) were prepared from 1,2,4-triazole precursors when 2-(1,2,4-triazol-3-yl)benzophenones (**303**, LG = OR,SH, or SR) were cyclized with ammonia.

(303), LG=OR,SH, or SR   (304)

3. *1,2,4-Triazolo[5,1-b]quinazolines*

Cyclization of the two amino functions of 2,3-diaminoquinazolin-4-ones (**305**) with acid derivatives gave the 1,2,4-triazolo[5,1-*b*]quinazolin-9-ones (**306**) [79JCS(P2)1708; 81EUP34529; 82BRP2086903; 84EGP(D)206996; 85IJC(B)873, 85PHA55; 86AP188, 86JCR(S)232, 86JHC833, 86MI4; 87-MI2; 88EGP(D)253623].

(305)   (306)

1,2,4-Triazolo[5,1-*b*]quinazolines (e.g. **308**) may also be prepared by a Dimroth-type rearrangement of 1,2,4-triazolo[3,4-*b*]quinazolines (**307**) (86JHC833) upon heating with aqueous alkalies.

(307)   (308)

Finally, the reduced 1,2,4-triazolo[5,1-*b*]quinazolin-9-ones (**310**) were prepared from 1,2,4-triazole precursors when 3-amino-1,2,4-triazole (**309**) was cyclocondensed with 2-ethoxycarbonylcyclohexanones (72MI1; 89OPP163).

(309)   (310)

4. *1,2,4-Triazolo[3,4-b]quinazolines*

Reaction of acid derivatives with 2-hydrazinoquinazolines blocked at position 1 (**311**) gave members of this ring system (**312**) [84EGP(D)206-

555; 86JHC833]. Cyclization of the *N*-1 unblocked 2-hydrazino-3,4-dihydroquinazolines (**313**) with acids was claimed [88EGP(D)258815] to yield only 1,2,4-triazolo[3,4-*b*]quinazolines (**314**); the angular 1,2,4-triazolo[4,3-*a*]quinazolines were not isolated.

Synthesis of the title compounds (e.g. **316**) from 1,2,4-triazole precursors was reported as early as 1930, when anthranilic acid was condensed with 4,5-dihydro-3-methyl-1-(4-nitrophenyl)-5-oxo-1,2,4-triazole (**315**, R = 4-NO$_2$-C$_6$H$_4$) (30JIC899).

5. 1,2,4-Triazolo[1,5-c]quinazolines

Thermal (83EUP80176), acid- (83EUP80176), or base-catalyzed [72GEP-(O)2146076] cyclodehdyration of 4-hydrazidoquinazolines (**317**) as well as thermal cyclodehydrogenation (84BCJ1138) of 4-hydrazonoquinazolines (**319**) was accompanied by Dimroth rearrangement to give the 1,2,4-triazolo[1,5-*c*]quinazolines **318**.

Thermal cyclization of 3-guanidino-2,4-thioxoquinazoline (**320**) gave 2-amino-5-thioxo-1,2,4-triazolo[1,5-*c*]quinazoline (**321**) (80PHA582); the corresponding linear isomer, which might have been obtained through the alternative cyclization between the 3-guanidino and 2-thioxo function, was not isolated.

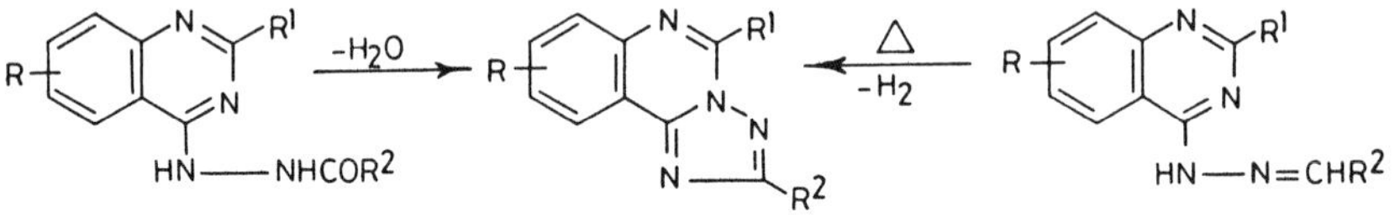

(317)   (318)   (319)

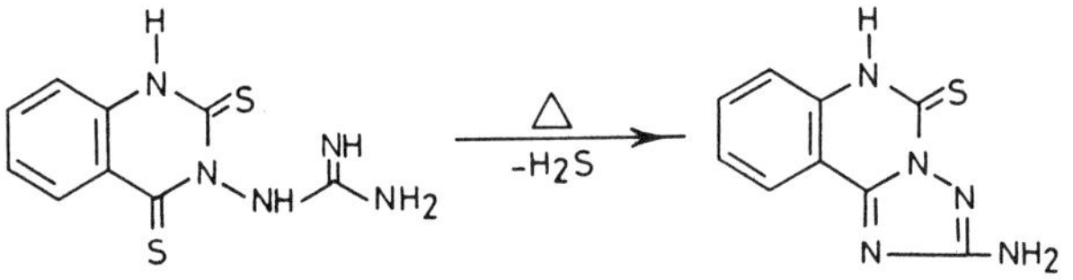

(320)   (321)

Benzoxazinones (**169**, Z = O) or benzothiazinethiones (**169**, Z = S) underwent cyclization with thiosemicarbazide (83MI1), thiocarbohydrazide (86JHC43), or amidrazones (68CB2106; 76MI1; 85H2357) to give 1,2,4-triazolo[1,5-*c*]quinazolines (**322**). The mesoionic 1,2,4-triazolo[1,5-

(169), Z = O or S   (322)

c]quinazolines (**324** and **326**) were prepared from the 3-amino-4-thioxoquinazolines (**323**) by cyclization with isothiocyanates (84S881) or from the 3-(2-aminophenyl)-1,2,4-triazoles (**325**) by cyclization with cyanogen bromide [73TL1643; 79JCS(P2)1708]. Photolysis of the quinazolin-3-yl thioimidate (**327**) in the presence of butylamine gave a mixture of the mesoionic olate (**328**, Z = O) and thiolate (**328**, Z = S) [84JCS(P1)1143].

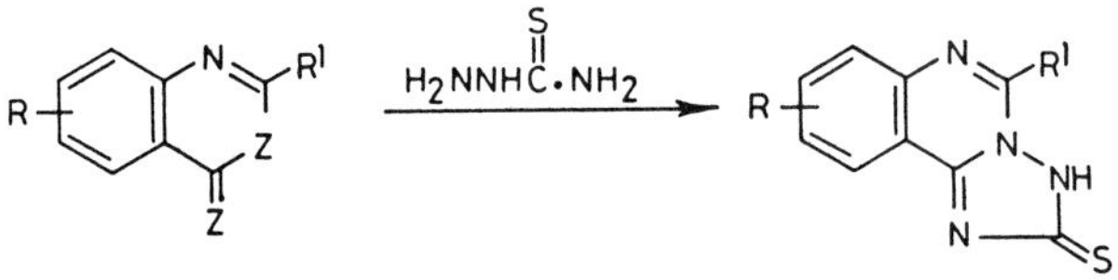

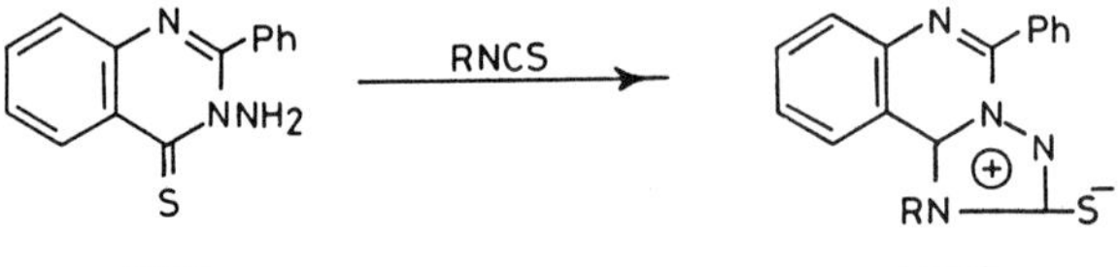

(323)   (324)

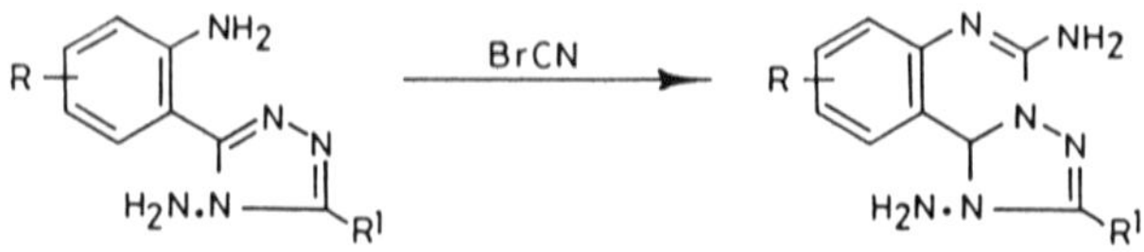

(325) → (326)

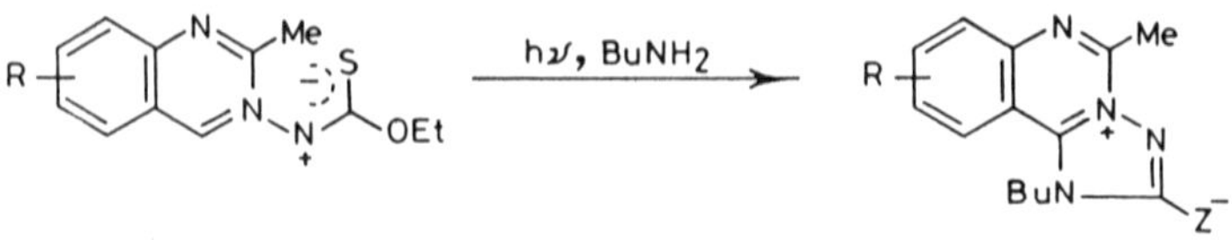

(327) → (328), Z = 0 or S

Synthesis of 1,2,4-triazolo[1,5-c]quinazolines from nonquinazoline intermediates is also known. Thus, the two heterocyclic rings were formed by the reaction of acylhydrazines with 2-ureidobenzonitriles (89JPR-537), 2-ethoxycarbonylaminobenzonitriles (329) (87USP4713383), or 2-isocyanatobenzonitrile (87CZ373) to give the title compounds (e.g. 330).

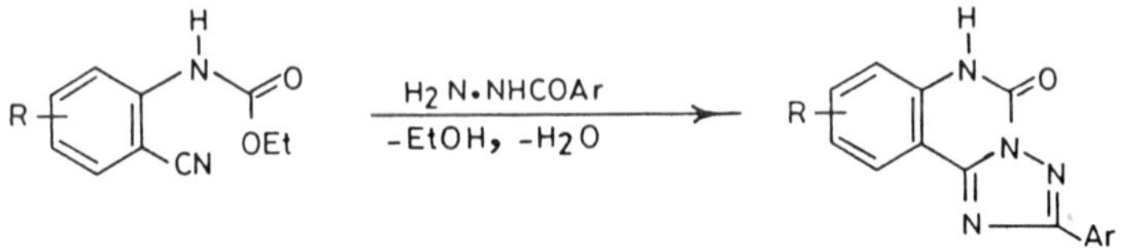

(329) → (330)

Ring transformation of the 3-aryl-1,2,4-triazolo[3,4-a]phthalazines (331) to 1,2,4-triazolo[1,5-c]quinazolines (334) was achieved according to the indicated reaction sequence (86EUP181282). Replacement of the sulfur heteroatom of 1,3,4-thiadiazolo[3,2-c]quinazolium iodide (335) by reaction with butylamine gave the zwitterionic 1,2,4-triazolo[1,5-c]quinazoline 336 [77IJC(B)1110]. The 1,2,4-triazolo[1,5-c]quinazolines (338) were obtained by the base, acid, or thermal isomerization of 1,2,4-triazolo[4,3-c]quinazolines (337) [70JOC3448; 73GEP(O)2261095; 74USP3850932; 76TL1935].

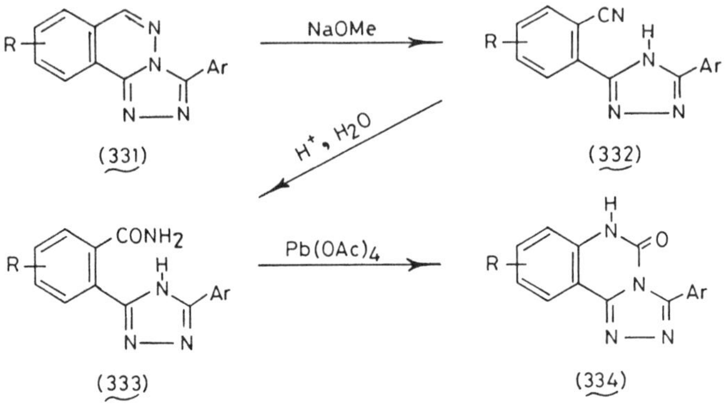
NaOMe
H^+, H_2O
$Pb(OAc)_4$
(331)
(332)
(333)
(334)

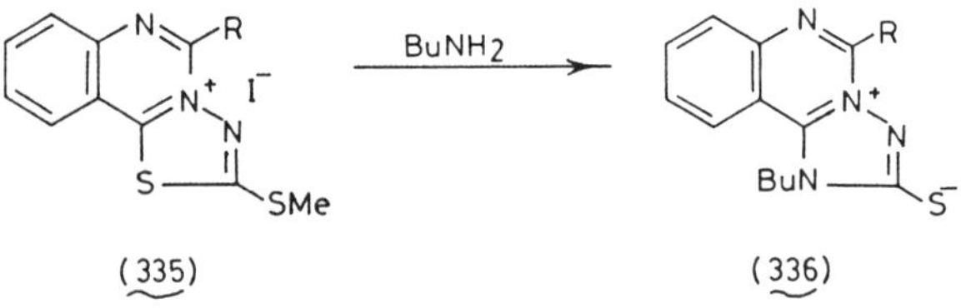
$BuNH_2$
(335)
(336)

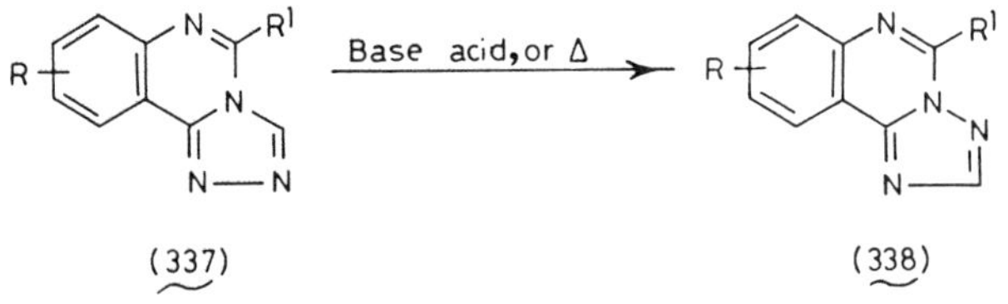
Base acid, or Δ
(337)
(338)

6. *1,2,4-Triazolo[4,3-*c*]quinazolines*

Cyclocondensation of the 4-hydrazinoquinazolines (**339**) with carboxylic acid derivatives gave the title compounds (**340**) [65ZOR1154; 66KGS130; 72GEP(O)2146076; 73GEP(O)2261095; 74USP3850932; 76TL1935; 81PHA62; 84BCJ1138, 84CCC1795, 84CJC2570; 90UP1, 90UP2]. The latter compounds (**340**) were also obtained from hydrazones (**341**) derived from **339** and aldehydes upon oxidative cyclization with bromine, sodium hypochlorite, or ethanolic ferric chloride as well as upon catalytic dehydrogenative cyclization with palladium-on-charcoal (63N732,

63T1587; 89MI1; 90UP1, 90UP2). 2-Substituted-3-carbamoylquinazolines (**342**) cyclize with hydrazine hydrate to give the corresponding 3,5-disubstituted-1,2,4-triazolo[4,3-*c*]quinazolines (**343**) (88CCC329, 88RRC981).

Cyclization of the 3-(2--aminophenyl)-1,2,4-triazoles (**344**) with acid derivatives also afforded the title compounds (**345**) (72URP334219).

Triazolo-quinazolines were reported to possess the following activities: fungicidal (83MI1), bactericidal (83MI1), potential bronchospasmolytic (86JHC833), antiinflammatory [76JAP(K)76100098; 80JAP(K)-8055188], analgesic [76JAP(K)76100098; 80JAP(K)8055188], antiviral [76JAP(K)76100098], antihistaminic (81EUP34529; 83EUP76199; 86EUP-181282), antiallergic (83EUP80176), anxiety modulator (86EUP181-282; 87USP4713383), nervous system stimulant (86EUP181282; 87-USP4713383), hypotensive [72GEP(O)2146076; 86AP188, 86MI4] and anaphylaxis inhibitor [80EGP(D)139715; 83EGP(D)158549].

VIII. Condensed Oxadiazolo-quinazolines

A. 1,2,4-OXADIAZOLO-QUINAZOLINES

1. *1,2,4-Oxadiazolo[2,3-a]quinazolines*

Base-catalyzed cyclization of ethyl 2-hydroxyaminobenzoate with three molar equivalents of methyl isocyanate gave the 1,2,4-oxadiazolo[2,3-a]quinazoline (**347**) (69CB1480; 70CB82).

(346)   (347)

2. *1,2,4-Oxadiazolo[2,3-b]quianzolines*

Treatment of 2-methylquinazoline-4-carboxyhydroxamic acid 3-oxide (**348**) with dicyclohexylcarbodiimide in dioxane caused its rearrangement and cyclization to 5-methyl-1,2,4-oxadiazolo[2,3-*b*]quinazoline-2-one (**349**) (85H623).

(348)   (349)

3. *1,2,4-Oxadiazolo[4,3-c]quinazolines*

Reaction of 2-aminobenzamidoxime (**350**) with two molar equivalents of aldehydes caused the formation of the two heterocyclic rings of the 1,2,4-oxadiazolo[4,3-*c*]quinazoline (**352**) [67CR(C)265; 70BSF2615].

(350)   (351)   (352)

Reaction of 4-hydroximino-1H,3H-quinazolin-2-one (353) or 2-ethoxy-carbonylaminobenzamidoxime (356) with alkyl or arylisocyanates followed by thermal cyclization gave 355 (81KGS1264).

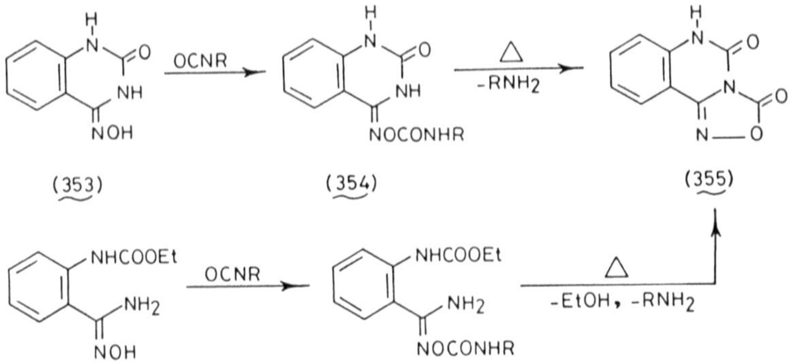

(353)   (354)   (355)

(356)   (357)

In a detailed thermolytic study, 5-methyl-1,2,4-oxadiazolo[2,3-c]-quinazolin-2-one (349), has been found to give the isomeric 1,3,4-oxadiazolo[4,3-c]quinazolinone (358) on heating or on treatment with triethyl phosphite (85H623).

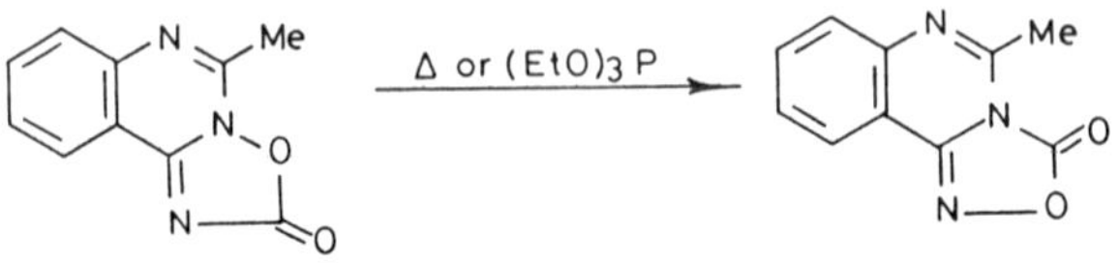

(349)   (358)

B. 1,3,4-OXADIAZOLO-QUINAZOLINES

1,3,4-Oxadiazolo[2,3-b]quinazolines

The 1,3-diazine ring of this system (360) was formed when anthranilic acid derivatives were cyclocondensed with 1,3,4-oxadiazoles having a good leaving group at position 2 (359, LG = Cl, SH, or SMe) (73JPR185; 81FES292; 84JIC436). Some oxadiazolo-quinazolines were found to possess herbicidal and fungicidal properties (84JIC436).

$$(359), LG = Cl, SH,$$
or SMe

$$(360)$$

IX. Condensed Thiadiazolo-quinazolines

Synthesis of 1,2,3- or 1,2,4-thiadiazolo-quinazolines was not reported in the literature.

1,3,4-THIADIAZOLO-QUINAZOLINES

1. *1,3,4-Thiadiazolo[3,2-a]quinazolines*

Reaction of diazotized anthranilic acid or its esters with α-thiocyanatoketones gave the intermediate **361**, which intramolecularly cyclized to the 1,3,4-thiadiazolo[3,2-*a*]quinazolines (**362**) (82JHC73, 82T1527). Reaction of the hydrazidoyl chlorides (**92**) with potassium thiocyanate afforded an alternative synthesis of (**362**) through the same intermediate (**361**) [82JHC73, 82T1527; 89IJC(B)120].

The 1,3,4-thiadiazolo[3,2-*a*]quinazolinium perchlorates (**365**) were prepared by cyclocondensation of 2-chlorocyclohex-1-enecarboxaldehyde (**363**) with 2-amino-1,3,4-thiadiazolium perchlorates (**364**) (73KGS1285).

(363)   (364)   (365)

2. 1,3,4-Thiazolo[2,3-b]quinazolines

Compounds belonging to this ring system (367 and 369) were prepared according to two general routes: (a) via formation of their pyrimidine ring by cyclocondensation of anthranilic acids or their esters with 2-chloro-1,3,4-thiadiazoles (366) (73JPR185; 79FES688; 81FES292), and (b) via formation of their 1,3,4-thiadiazole ring by cyclization of 2-mercapto-3-aminoquinazolines (368) with one-carbon cyclizing agents, such as carbon disulfide, cyanogen bromide, or acid chlorides [70IJC710; 86IJC(B)489; 86IJC(B)957; 87AP166, 87MI3].

(366)   (367)

R¹COCl

(368)   (369), Z = S or NH

The approach of synthesizing this ring system through formation of its fused pyrimidine and 1,3,4-thiadiazole rings was also reported (70IJC-389). Hydrazones (370) derived from anthranilic acid hydrazides and aromatic aldehydes gave 371 upon cyclization with potassium thiocyanate (70IJC389). The pyrimidine and thiadiazole rings of 1,2,4-triazolo[4',3' : 4,5]1,3,4-thiadiazolo[2,3-b]-quinazolin-6-one (374) were also simultaneously formed when isatoic anhydride was condensed with 4-amino-5-mercapto-3-methyl-1,2,4-triazole (372), followed by dehydrative cyclization of 373 (83M339).

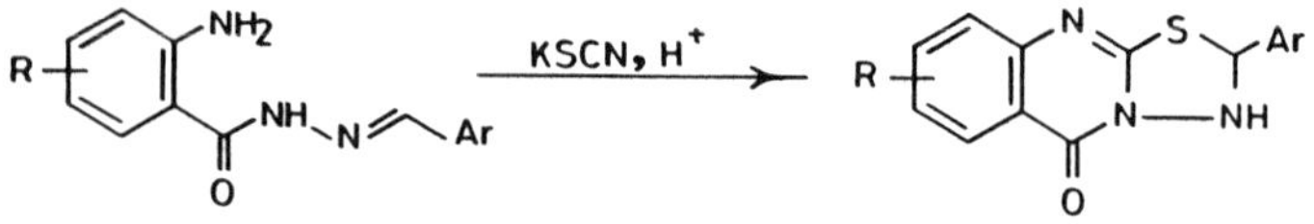

(370) → (371)

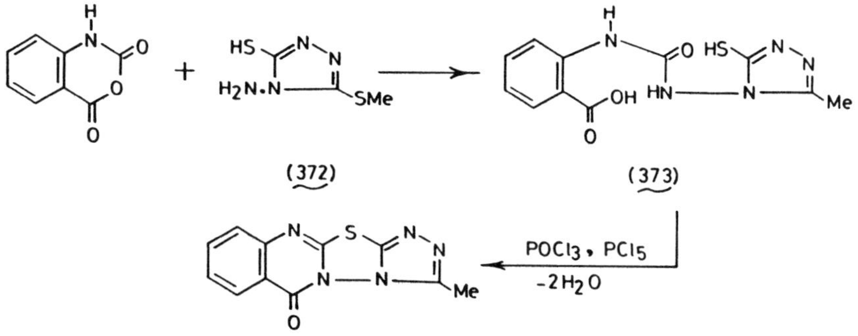

(372) + → (373)

(374)

3. *1,3,4-Thiadiazolo[3,2-c]quinazolines*

Cyclization of 3-aminoquinazoline-4-thiones (**375**) with carbon disulfide gave the zwitterionic 1,3,4-thiadiazolo[3,2-*c*]quinazoline (**376**) [77IJC(B)1110]. Synthesis from properly functionalized 1,3,4-thiadiazole precursors was also reported [83JCS(P1)2011], according to which the 2-(2-nitrophenyl)-1,3,4-thiadiazoles (**377**) were reduced and cyclized to **378**. 1,3,4-Thiadiazoloquinazolines were shown to exhibit amebicidal

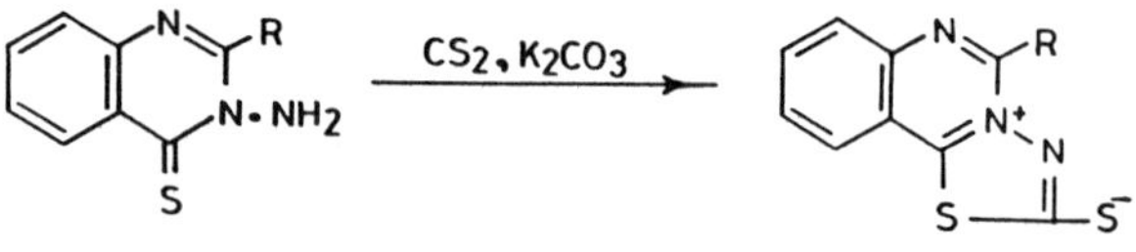

(375) → (376)

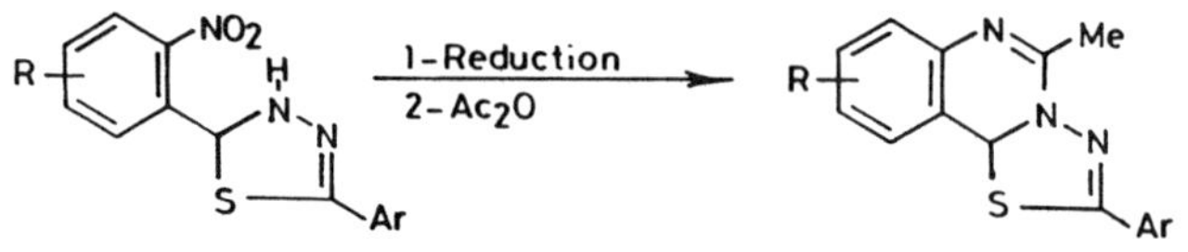

(377) → (378)

(70IJC710), antiinflammatory (81FES292), antifungal [87MI3; 89IJC-(B)200] and hypotensive activities (87AP166).

X. Condensed Selenadiazolo-quinazolines

1,3,4-SELENADIAZOLO-QUINAZOLINES

1,3,4-Selenadiazolo[2,3-a]quinazolines

Similar to their 1,3,4-thiadiazolo analogues, 1,3,4-selenadiazolo[2,3-a]quinazolines (380) were synthesized by the reaction of diazotized anthranilic acid or its esters with either 2-selenocyano-1,3-diketones or 2-chloro-1,3-diketones, followed by treatment with potassium selenocyanate (82G545; 83JHC719).

XI. Condensed Tetrazolo-quinazolines

A. TETRAZOLO[1,5-a]QUINAZOLINES

Two general methods were used for synthesizing these compounds (e.g. 383). (a) Cyclization of 3-substituted 2-hydrazinoquinazolin-4-ones [64ZOB1745; 76GEP(O)2539396; 78JAP(K)7837695, 78PHA507; 80MIP1, 80PHA800; 83PHA25, 83PHA367; 84PHA867] or 4-aryl-2-hydrazino-quinazolines (381) [77ZOR1773; 78JAP(K)7812893; 79MI2] with sodium nitrite in acid media. In the latter case, exclusive cyclization witn *N*-1, rather than N-3, of the quinazoline occurred as a result of its higher nucleophilicity. (b) Reaction of 2-chloroquinazolin-4-ones (63ZOB2334; 78USP4085213) or 4-aryl-2-chloroquinazoline (**382**) [78GEP(O)2805124, 78JAP(K)7812893] with sodium azide, in which cyclization also took place with the more nucleophilic N-1 of the quinazoline.

(381)   (383)   (382)

It was interesting to discover that the reaction of sodium azide with 2,4-dichloroquinazoline (**384**) gave only the 5-azidotetrazolo[1,5-*a*] quinazoline (**385**) and not the 5-azidotetrazolo[1,5-*c*]quinazoline (**386**) or the tetracyclic product (**387**) (33JPR9; 63ZOB2475).

(384)   (385)

(386)   (387)

Synthesis from tetrazole precursors was reported [60LA159; 79JCS(P1)3085], according to which 5-aminotetrazole (**388**) was cyclocondensed with 2-hydroxymethylenecyclohexanone to give tetrazolo [1,5-*a*]quinazoline (**389**) instead of the expected tetrazolo[5,1-*b*] quinazoline (**390**).

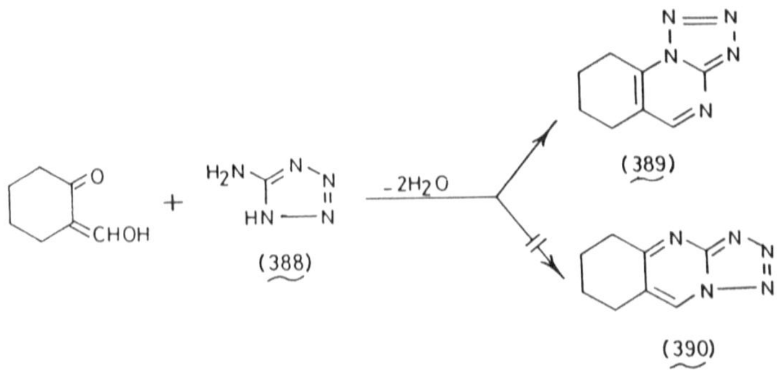

(389)

(388)

(390)

B. Tetrazolo[5,1-b]quinazolines

The title compounds were obtained by cyclization of 3-amino-2-hydroxyquinazolin-4-ones (391) (86MI5; 88MI2) or 1-substituted 2-hydrazinoquinazolin-4-ones (311) [83EGP(D)203545; 84PHA867] with nitrous acid. In the latter case, it was necessary to block N-1 in order to direct the cyclization towards the less nucleophilic N-3 of the quinazoline ring.

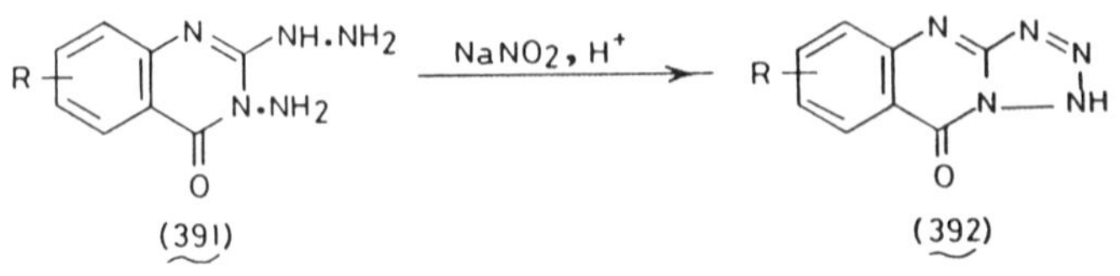

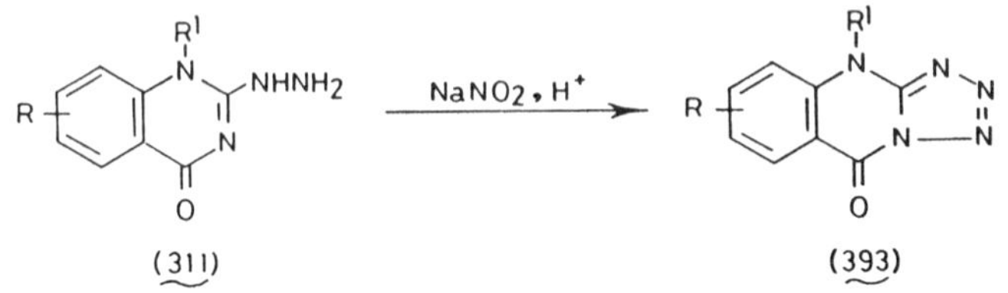

(391)

(392)

(311)

(393)

C. Tetrazolo[1,5-c]quinazolines

Compounds belonging to this ring system (394) were prepared by cyclization of either 4-hydrazinoquinazolines (339) with nitrous acid (63N732, 63ZOB2334; 66KGS130; 67KGS1096; 70KGS100, 70KGS855; 84CJC2570) or 4-haloquinazolines (395) with sodium azide (67KGS1096; 70KGS100, 70KGS855; 90UP1, 90UP2).

(339)   (394)   (395)

5-(2-Aminophenyl)tetrazoles (396) were also cyclized with acetic anhydride (70KGS100), aldehydes (74USP3835137, 74USP3835138; 88CZ135), ketones (74USP3835138; 88CZ135), or phosgene (74USP-3838126) to tetrazolo[1,5-c]quinazolines (397).

(396)   (397)

Various biological applications were reported for tetrazoloquinazolines, including fungicides [78JAP(K)7837695; 83EGP(D)203545; 88MI2], pesticides [76GEP(O)2539396], antiallergics (78USP4085213), antiulcer agents (78USP4085213), bactericides [83EGP(D)203545]; 88MI2], bronchodilators [74USP3835138, 74USP3838126; 83EGP(D)203545], antiinflammatory agents [83EGP(D)203545], analgesics [83EGP(D)203545], antianphylactics [83EGP(D)203545], and antihypertensives (86MI5).

XII. Condensed Azino-quinazolines

A. PYRIDO-QUINAZOLINES

1. *Pyrido[1,2-a]quinazolines*

Synthesis of this ring system from quinazoline precursors was reported by Acheson [75JCS(P1)2322, 75KGS1701; 80H1959]. Reaction of 4-ethoxyquinazoline (398) with dimethyl acetylenedicarboxylate gave the 1 : 2 molar adduct 399 in high yield. The latter underwent rapid rearrangement in the presence of a strong acid (e.g. trifluoroacetic acid) to give the isomer 401.

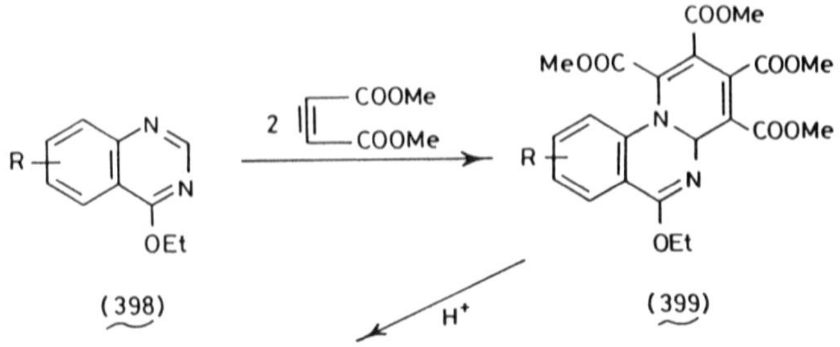

(398)   (399)

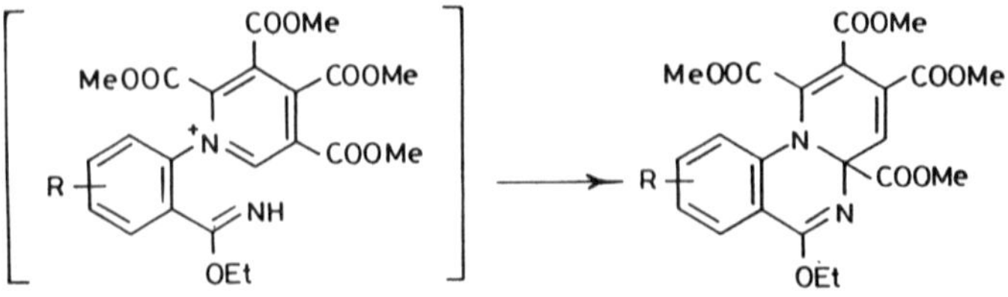

(400)   (401)

Diels–Alder reaction of maleic anhydride or N-phenylmaleimide with 2-styrylquinazolin-4-ones (402) gave the corresponding pyrido[1,2-a]quinazolinones 403 [73MI2; 81IJC(B)290]. 3-Aryl-2-methylquinazolin-4-ones (404) reacted with malonic esters to afford the pyrido [1,2-a]quinazolinones (405) [81ZN(B)252]. Cyclization of properly functionalized quinazolines, such as 2-ethoxycarbonyl-1-(3-ethoxycarbonyl-propyl-4-quinazolinone (83CPB2234) or 1-(4-chlorobutanoyl)-1,2-dihydro-4-phenylquinazoline-2-carbonitrile [86JCS(P1)2295] (406), with sodium hydride gave pyrido[1,2-a]quinazoline (407).

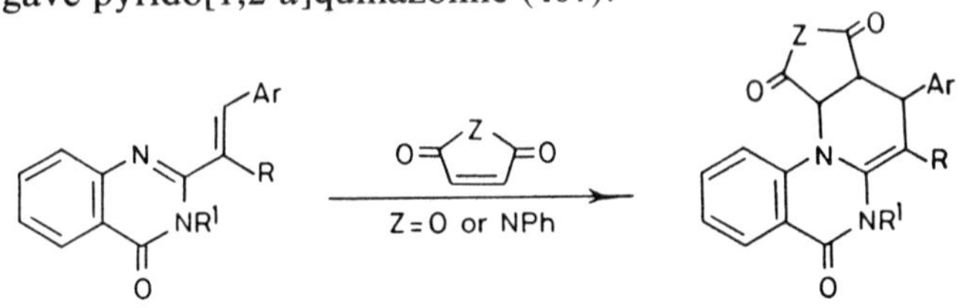

(402)   (403)

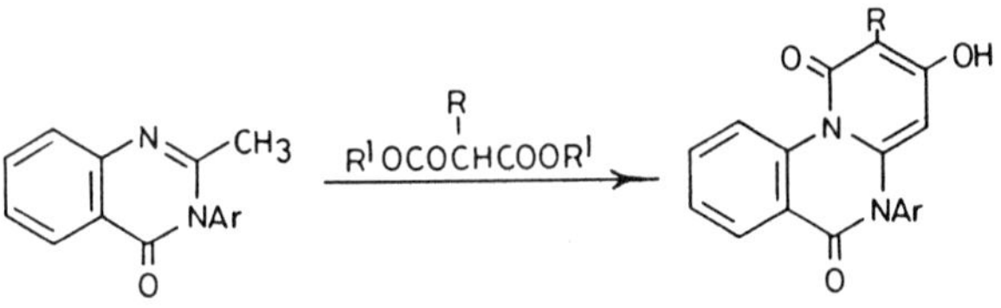

(404)   (405)

(406)   (407)

An alternative synthetic approach for synthesizing pyrido[2,1-*a*]quinazolines involved using pyridine precursors and the formation of the quinazoline nucleus. Thus, reaction of 2-aminopyridines or their salts with methyl 2-chlorobenzoate (68IJC758), 1-chloro-2-formylcyclohexene (73-KGS242), or 2-acetylcyclohexanone (73JHC821) afforded the corresponding pyrido[1,2-*a*]-quinazolines (e.g. (**408**). Dehydrogenative cyclization of 2-hexahydropyridobenzamides (**410**) (78AP586) or 2-hexahydropyridobenzylamines (79AP838) with mercury(II)-EDTA resulted in the formation of the pyrimidine ring of the title compounds (e.g. **411**). Morpholinocyclohexene (**412**) reacted with 2-isothiocyanatopyridine (**413**) to give **414** (81CCC2428).

(408)

(409)   (410)   (411)

(412)   (413)   (414)

A one-step construction of both of heterocyclic rings of the title compounds (**415** and **416**) was accomplished when anthranilamides reacted
with 5-oxoalkanoic acids (68JOC2402; 69USP3441566; 75USP3883524) or
when anthranilic acids reacted with 1-cyano-4-haloalkanes (79KGS1427),
respectively. Cyclization of 2-(2-carboxyethylamino)benzamide (**417**) with
ethyl chloroformylacetate gave 4-ethoxycarbonyl-1,2-dihydropyrido[1,2-
a]quinazoline-3,6-dione (**419**) (83CPB2234).

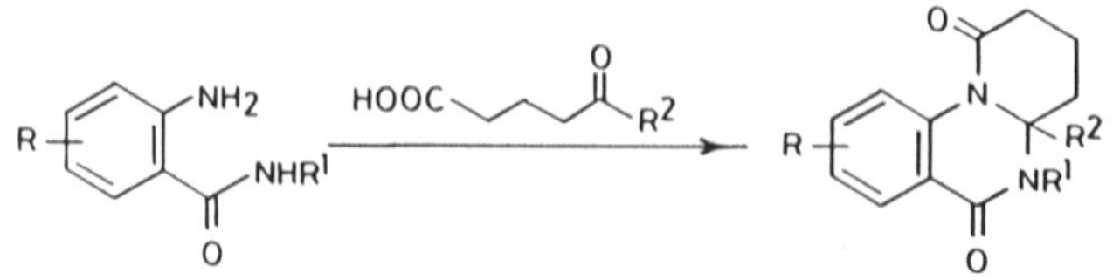

(415)

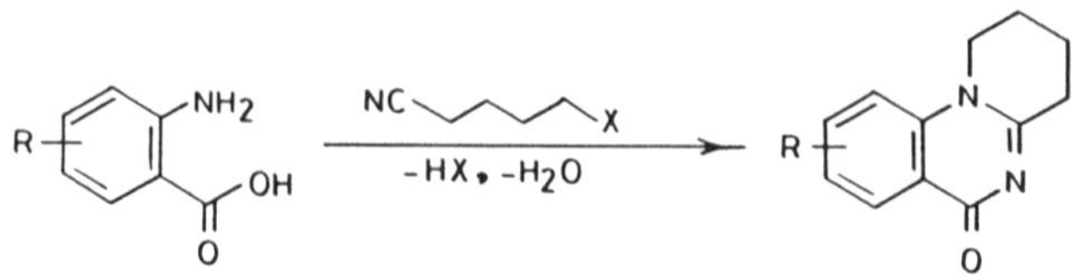

(416)

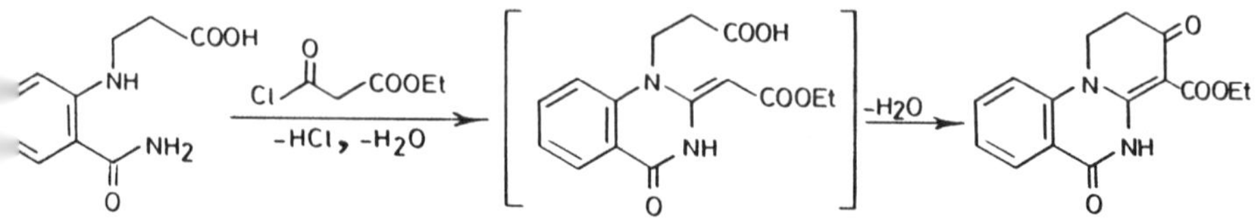

(417)   (418)   (419)

2. Pyrido[2,1-b]quinazolines

The main approach for synthesizing this ring system comprised the
formation of its pyrimidine ring. 2-Aminopyridines provided the two skeletal nitrogens of these compounds when condensed with 2-ethoxy-
carbonylcyclohexanone [70GEP(O)1932885; 71JCS(C)2163; 75JOC2201;
79JHC137, 79MI1; 81MI1], 2,5-dihydroxycyclohexadiene-1,4-dicarboxy-
lic esters [75MI2; 76JAP(K)7643799], 2,5-diaminocyclohexylidene-1,4-
dicarboxylic esters [76JAP(K)7643800], or 2-chlorobenzoic acid [24-
LA311; 83ZN(B)248] to give pyrido[2,1-b]quinazoline (e.g. **420**).

Fusion

(420)

Pyridine derivatives having a leaving group at position 2, such as 2-pyridones, 2-halopyridines, or 2-alkoxypyridines, were also used for synthesizing this system. In this case, the second skeletal nitrogen of the title compounds was provided by the other reacting entity. Thus, reaction of anthranilic acids with 2-pyridones gave the pyrido[2,1-*b*]quinazolin-11-ones (**421**) [62CB2182; 77KPS544; 79JCS(P1)1765; 83NEP8202602]. Isatoic anhydrides [35CB(B)2221; 68JCS(C)1722] or cyclic sulfinamides [76H1487;77JA2306, 77JAP(K)7777093, 77MI2; 78JAP(K)7877075] were also used in place of anthranilic acids.

$-2H_2O$

(421)

Reaction of anthranilic acids with 2-halopyridines (**422**) was the most widely used reaction for synthesizing pyrido[2,1-*b*]quinazolines (**423**) [31JCS2840, 31LA284; 38CB(B)1657; 45JCS927; 55G1210, 55MI1; 56JCS4694; 65NEP641471; 77GEP(O)2645110; 78GEP(O)2812585, 78GEP(O)2812586, 78USP4066767; 79GEP(O)2739020, 79JMC114; 79JMC748, 79SAP7804057; 80GEP(O)2845766, 80JMC92; 81USP4261996, 81ZN(B)-252; 82JMC742, 82MIP1; 83MIP2, 83MIP3, 83MIP4; 84AP824, 84MI4; 85CP1189509; 87MIP1].

Cu or KI

(422)   (423)

2-Alkoxypyridines (**424**) also react with anthranilic acids to give examples of this ring system (**423**) [60GEP1088968; 68ZOB2030; 79JCS(P1)1765; 80MI3; 83JHC93; 85IJC(B)336]. An interesting synthesis of pyrido[2,1-*b*]quinazolines (e.g. **425**) was that in which 2,1-benzoisothiazolin-3-one (84JHC369) or 5-nitroanthranilic acid (72KGS1003, 72KGS1341) reacted with pyridine in the presence of an acid catalyst.

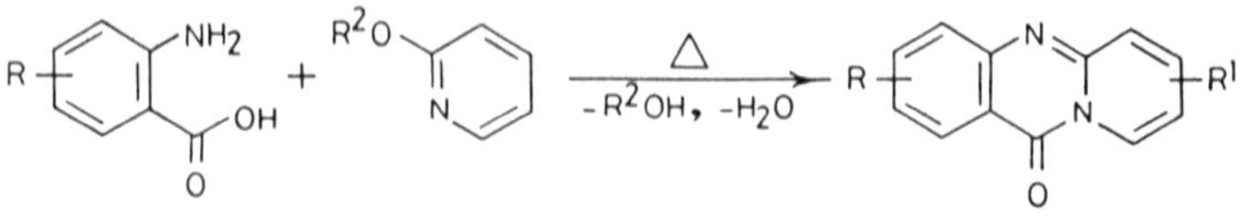

(424)   (423)

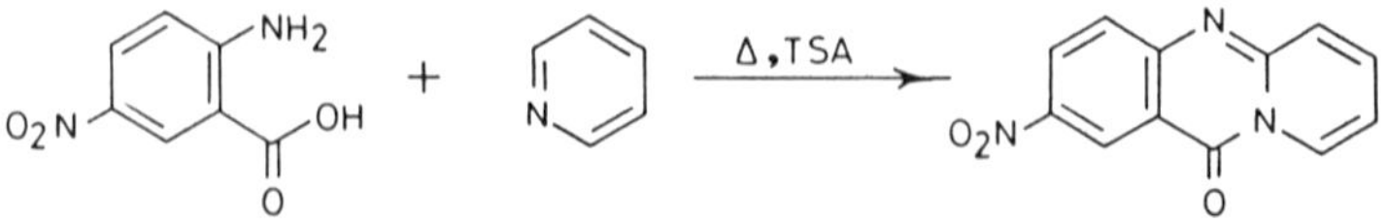

(425)

Condensation of acetanilide with 1-hydroxymethyl-2-piperidone (**426**) in concentrated H_2SO_4 gave a mixture of **427** and its *para*-isomer. Hydrolysis of the *N*-acetyl group of **427** and subsequent dehydrative cyclization gave the pyrido[2,1-*b*]quinazoline **429** (75KPS809). In a series of publications (70TL997; 73AP541; 79CZ266; 80M627, 80PHA389), Moehrle prepared

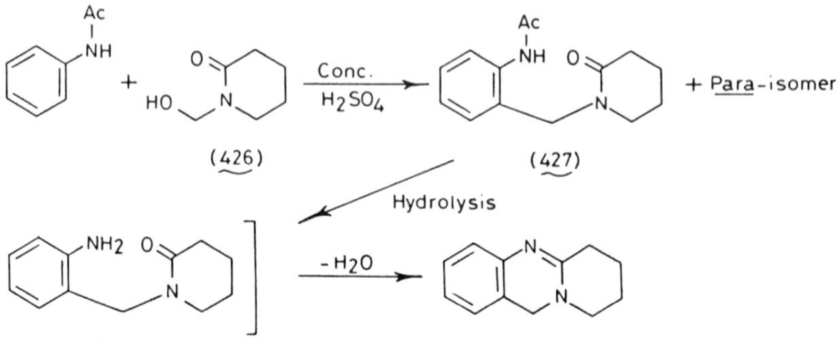

(426)   (427)

Hydrolysis

(428)   (429)

pyrido[2,1-*b*]quinazolines (**431**) by the cyclodehydrogenation of *N*-(2-aminobenzyl)piperidines (**430**) with Hg(II)-EDTA. The same cyclization was also affected by the use of manganese dioxide [68JCS(C)1722].

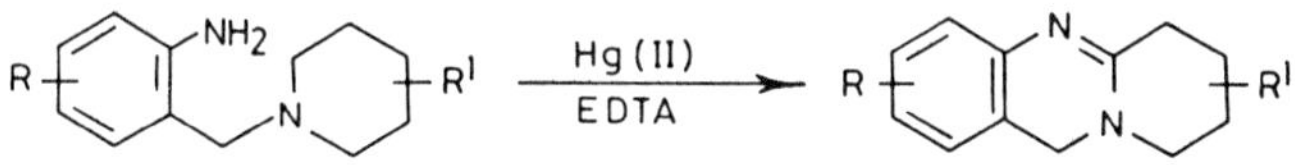

(430)   (431)

Photoirradiation of the mesoionic 1,2,3,4,5,7,8,9,10,11-decahydro-11-oxopyrido[1,2-*b*]cinnolin-6-ium hydroxide inner salt (433) affected clean intramolecular isomerization to the pyrido[2,1-*b*]quinazolin-11-one (436, 78% yield) through the diaziridine and ketene intermediates 434 and 435

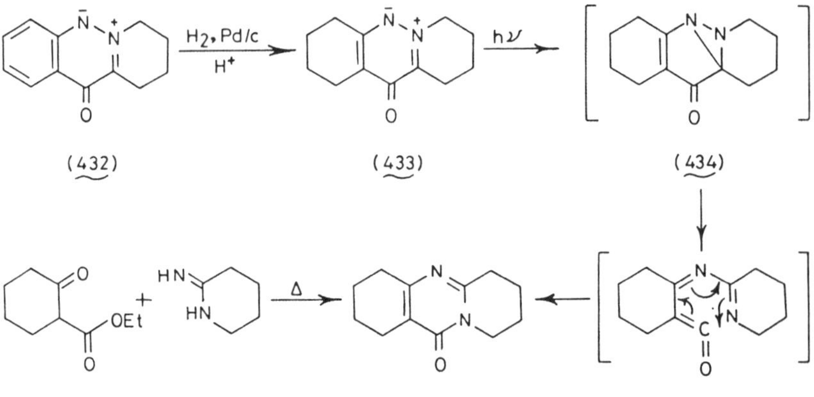

(432)   (433)   (434)

(437)   (436)   (435)

(75JOC2201). The structure of 436 was confirmed by synthesis from 2-iminopiperidine (437) and 2-ethoxycarbonylcyclohexanone (75JOC2201). 3-(Pyrid-2-yl)-1,2,3-benzotriazin-4-one (438) underwnet loss of a nitrogen molecule upon thermolysis, giving pyrido[2,1-*b*]quinazolin-11-one (440) through the β-lactam intermediate (439) [80JCS(P1)633]. A unique synthesis of the title compounds involved the palladium-catalyzed carbonyl insertion into 2-(2-bromophenyl)-aminopyridines (441) to give 423 (87JOC2469).

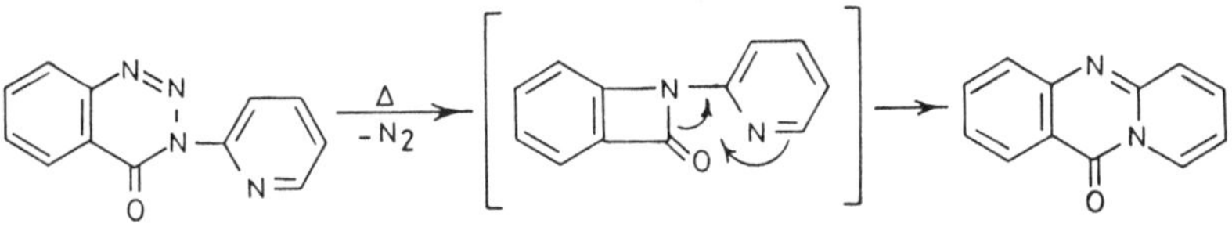

(438)   (439)   (440)

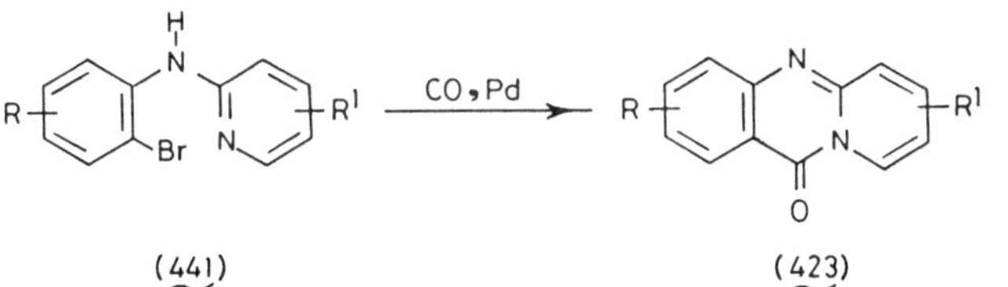

(441)   (423)

Treatment of 4-hydroxy-1-(2-pyridyl)quinolin-2-one (**442**) with sulfuryl chloride gave 3,3-dichloro-1-(2-pyridyl)quinoline-2, 4-dione (**443**) which, upon treatment with sodium carbonate, rearranged to pyrido[2,1-*b*]quinazolin-11-one (**440**) (79CB3424).

Synthesis of pyrido[1,2-*b*]quinazolines through formation of the azole ring from properly substituted quinazoline precursors is also known (61AP556). Thus, the cyclodehydrobromination of the 2-(4-bromo-butyl)quinazolin-4-one (**445**) with alkali was claimed to give only the lin-early annulated 1,2,3,4-tetrahydropyrido[1,2-*b*]quinazolin-4-ones (**446**) (61AP556). Oxidation of the 3-amino-2-(butyn-1-yl)quinazolin-4-ones (**447**) with lead tetraacetate led to intramolecular addition of the produced *N*-nitrene intermediate to the triple bond to give **449** [86JCS(P1)1215].

(447)   (448)

(449)

3. Pyrido[1,2-c]quinazolines

The pyrido[1,2-c]quinazoline (450) was prepared from a quinazoline precursor when 2,4-diphenylquinazoline was reduced with sodium metal in tetrahydrofuran (THF) to the dianion 60, followed by treatment with 1,4-dibromobutane to give 450, together with other products (76JOC497).

(60)   (450)

Cyclization of 2-(pyrid-2-yl)benzenediazonium tetrafluoroborate (451) with alkyl or aryl nitriles afforded the pyrido[1,2-c]quinazolinium salts 452 [72GEP(O)2043665].

(451)   (452)

Formation of both of the azole and diazole rings of 454 was achieved upon cyclization of 5-chloro-2-trichloroacetamidobenzophenone (453) with 1-amino-4-chlorobutane [72GEP(O)2141616].

Many pyridoquinazolines have been reported to possess analgesic [65NEP6414717; 75USP3883524; 79GEP(O)2739020; 80GEP(O)28457-

(453)   (454)

66], antiinflammatory [65NEP6414717; 77JAP(K)7777093; 70GEP(O)2-
739020; 80GEP(O)2845766], antipyretic [75USP3883524; 79GEP(O)2739-
020; 80GEP(O)2845766], antiallergic [77GEP(O)2645110; 78USP4066767;
79GEP(O)2739020; 80GEP(O)2845766; 82USP4332802; 87MIP1], gastric
secretion inhibition (83NEP8202602), antianaphylactic [78GEP(O)2812-
586; 83MIP2, 83MIP3, 83MIP4], passive cutaneous anaphylaxis inhibition
[78GEP(O)2812585; 82MIP1], antiasthmatic (85CP1189509; 87MIP1), antidepressant [77JAP(K)7777093], hypotensive [77JAP(K)7777093], and
platelet activating factor antagonist activities (88JMC466).

B. Quino-quinazolines

1. Quino[2,1-b]quinazolines

The title compounds were synthesized from quinazoline precursors as
exemplified by the dehydrative cyclization of the diformyl derivatives
(456), obtained from 2-methyl-3-phenylquinazoline (455), with phosphoric
acid to give the 6-formylquino[2,1-b]quinazolin-12-one 457 (73IJC532).
Quino[2,1-b]quinazolines were synthesized from quinoline precursors
having a leaving group at position 2 (458, LG = Cl or OR') by cyclocondensation with anthranilic acid or its esters [37JGU2318; 56JCS4173;
79IJC(B)107].

(455)   (456)

(457)

(458), LG = Cl or OR¹   (459)

2. *Quino[1,2-c]quinazolines*

The synthesis of this condensed system by two different approaches was reported in two publications (80JHC1489, 80JHC1665) from the same laboratory. According to the first approach, 1,2,3,4,12,13-hexahydro-1-oxoquino[1,2-c]-quinazolinium perchlorate (**461**) was prepared (80JH-C1489) from the quinazoline precursors (**460**) by acid-catalyzed cyclodehydration.

(460)   (461)

In the second approach, the quino[1,2-c]quinazolinium perchlorates **463** were synthesized (80JHC1489, 80JHC1665) from the 2-(2-aminophenyl)quinolines (**462**) by cyclization with one-carbon cyclizing agents such as acetic anhydride, formic acid, benzoyl chloride, cyanogen bromide, urea, or carbon disulfide. Quinoquinazolines have been reported to show antitumor activity (80JHC1489, 80JHC1665).

(462)   (463)

C. Isoquino-quinazolines

1. *Isoquino[2,1-a]quinazolines*

12-Methyl-6*H*-isoquino[2,1-*a*]quinazolin-5-one (**466**) was synthesized from 2-cyano aniline and 3-methylisocoumarin (84CPB2160).

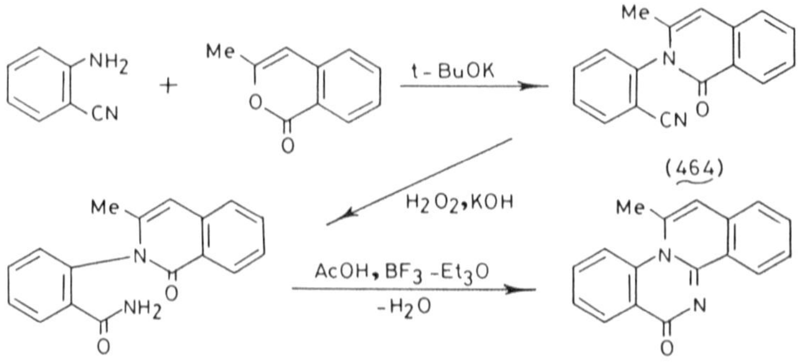

(464)

(465)

(466)

2. Isoquino[2,3-a]quinazolines

5-Phenyl-12H-isoquino[2,3-a]quinazolin-5-one (468) was synthesized by cyclization of 1-(2-chloromethylbenzoyl)-1,2-dihydro-4-phenylquinaz-oline-2-carbonitrile (467) with sodium hydride [86JCS(P1)2295]. Cyclo-condensation of ethyl anthranilate with 2-(bromomethyl)benzyl cyanide (469) gave the 7,12-dihydro-5H-isoquino[2,3-a]quinazolin-5-one (470) (89DOK628).

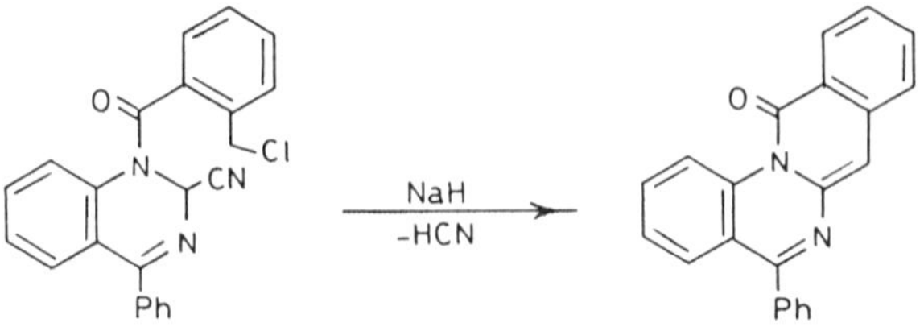

(467)

(468)

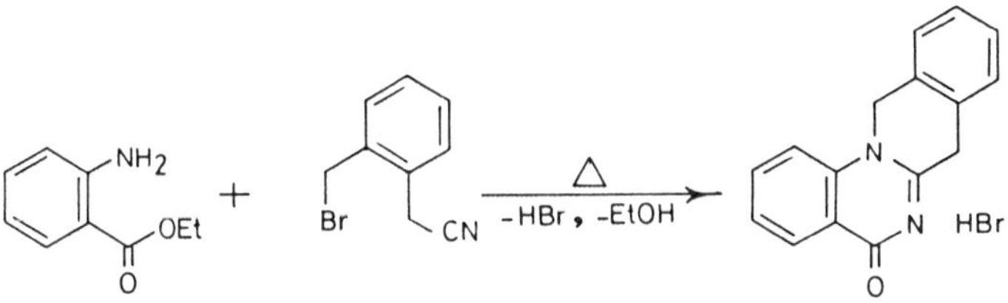

(469)

(470)

3. Isoquino[1,2-b]quinazolines

Isatoic anhydride condensed with 3,4-dihydroisoquinazolines (471) to give the partially reduced title compounds (472) (70USP3497499).

(471)   (472)

The sulfinamide anhydrides (473) underwent regiospecific cyclization with isoquinolines (474) yielding 475 [76JA6186; 79JAP(K)79135799]. 2-Aminobenzaldehydes also condensed with 3,4-dihydroisoquinoline pic-

(473)   (474)   (475)

rate (476) to yield the corresponding isoquino[1,2-*b*]quinazoline picrate (477) (87EUP236251). Molecular rearrangement also took place when anthranilopapaverine (478) was pyrolyzed or photolyzed to give a mixture of the isoquino[1,2-*b*]quinazoline (479) and the mesoionic isoquinolinocinnoline (480) (87H1841).

(476) Y = picrate anion   (477)

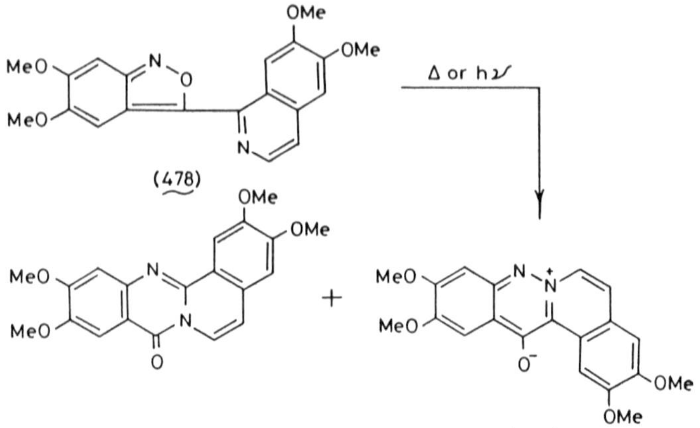

(478)

(479)   +   (480)

4. Isoquino[3,2-b]quinazolines

Molecular rearrangement of 7,12-dihydro-5*H*-isoquino[2,3-*a*]quina-
zolin-5-one hydrobromide (470) by refluxing with *N*-methylpyrrolidone
gave 7,12-dihydro-5*H*-isoquino[3,2-*b*]quinazolin-5-one (481) (89DOK628).

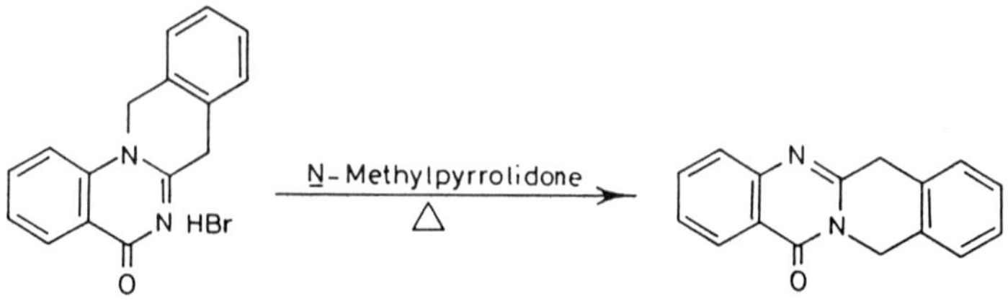

(470)   N-Methylpyrrolidone   (481)

5. Isoquino[2-1-c]quinazolines

Examples (484) of this system were synthesized by cyclization of 1-(2-
aminophenyl)-3,4-dihydroisoquinolines (482) with carbon disulfide
(67USP3297696) or with glyoxalic acid [69CI(L)417]. Some isoquino-
quinazolines have been reported to exhibit hypotensive (70USP3497499)
and central nervous stimulant activities (70USP3497499).

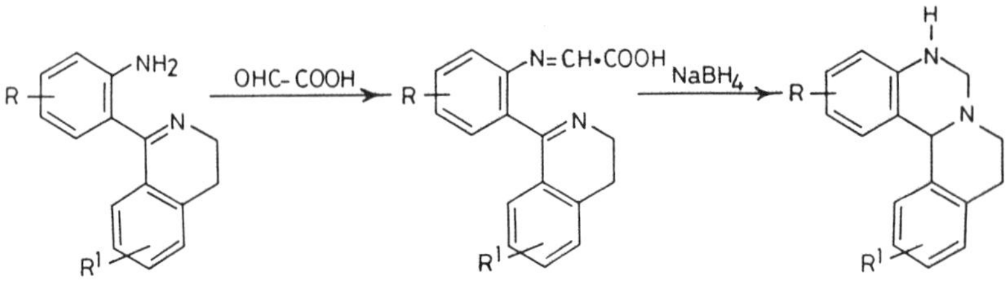

(482)   (483)   (484)

XIII. Condensed Diazino-quinazolines

A. PYRIDAZINO-QUINAZOLINES

1. Pyradazino[1,6-a]quinazolines

The two heterocyclic rings of this system (485) were formed upon coupling salts of diazotized anthranilic acid (87JHC227) or diazotized anthranilic acid ester (88MI1) with butenonitriles.

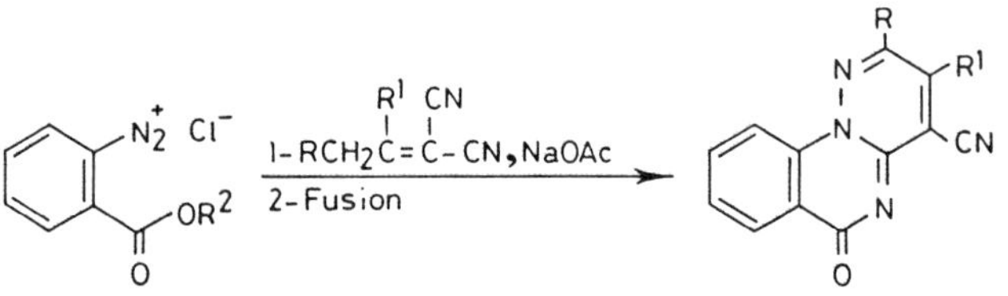

(485)

2. Pyridazino[6,1-b]quinazolines

These compounds (487) were synthesized by cyclocondensation of anthranilic acids or their esters with 3-halopyridazines (486) [61ZC224; 64CB390, 64JHC42; 68CPB972; 771JC(B)250; 81USP4250177; 85MI2].

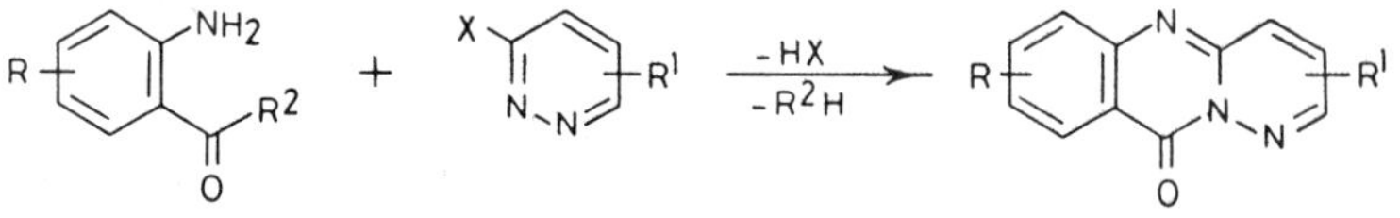

(486)   (487)

A synthesis occurred in which the pyridazine ring of this system was formed, starting with 3-amino-2-(buten-2-yl)quinazolinones (**488**) [84JCS(P1)1905]; oxidative cyclization of this compound with lead tetraacetate gave **490** through the intramolecular trapping of the corresponding *N*-nitrene intermediate (**489**). The pyridazine ring of **492** was formed upon cyclization of 3-(2-chlorocarbonylpyrrol-1-yl)-3-methylquinazolin-4(3*H*)-one (**491**) (84G525).

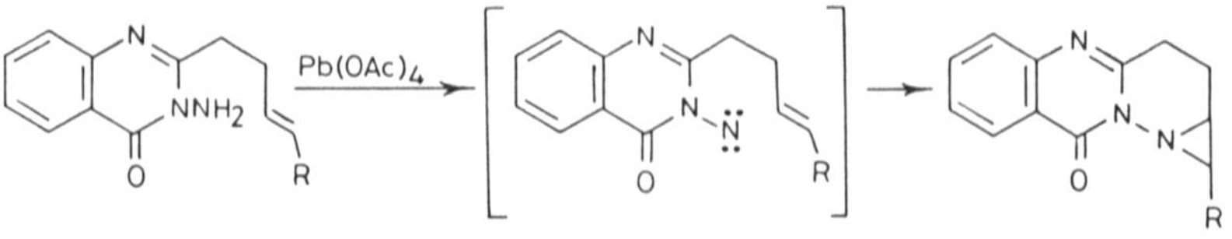

(488)   (489)   (490)

(491)   (492)

The two heterocyclic rings of **493** were formed by cyclocondensation of anthranilic acid hydrazides with levulinic acid (68USP3375250; 74USP3843654; 78MI2). Some pyridazino-quinazolines showed anaphylaxis inhibitory (81USP42550177), antiinflammatory (78MI2), and analgesic activities (78MI2).

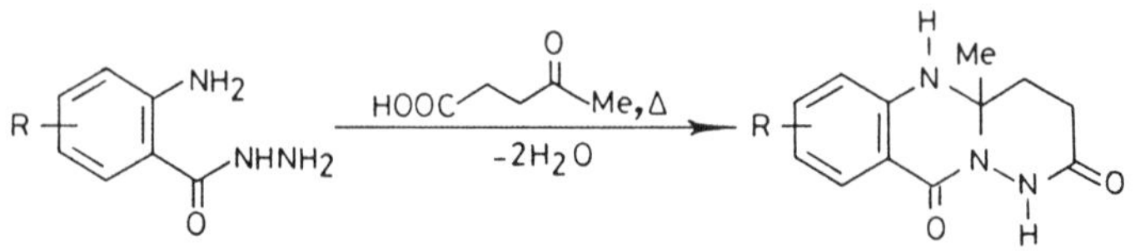

(493)

B. QUINAZOLINO-CINNOLINES

Quinazolino[3,2-b]cinnolines

Oxidation of 3-amino-2-(2,5-dihydroxybenzyl)quinazolin-(3*H*)-one (**494**) with acidic ferric chloride or aqueous sodium hydroxide gave a quinone that spontaneously cyclized to the quinazolino[3,2-*b*]cinnoline-2,7(13*H*)-dione (**495**) [66JCS(C)2190].

(494)   (495)

C. Quinazolino-Phthalazines

Publications on the synthesis of these compounds have almost always referred to these compounds as phthalazino-quinazolines. The title name, however, conforms better with the IUPAC rule B-3.1 on the nomenclature of organic compounds (73MI3).

Quinazolino[2,3-a]phthalazines

Cyclocondensation of anthranilic acid with 1-chlorophthalazines (**496**) gave examples of the title compounds (**497**) [61ZC224; 67B875; 78IJC(B)689].

(496)   (497)

The pyrimidine and pyridazine rings of **498** were built through its synthesis from anthranilic acid hydrazide and phthalic anhydride (67CB875; 78MI2; 82FES719). Reaction of 2-(2-carboxylphenyl)benzoxazinone (**499**) with hydrazine gave the quinazolino[2,3-a]phthalazine (**500**) [86EGP(D)234013]. Quinazolino-phthalazines possess antiinflammatory and analgesic (78MI2) properties.

(498)

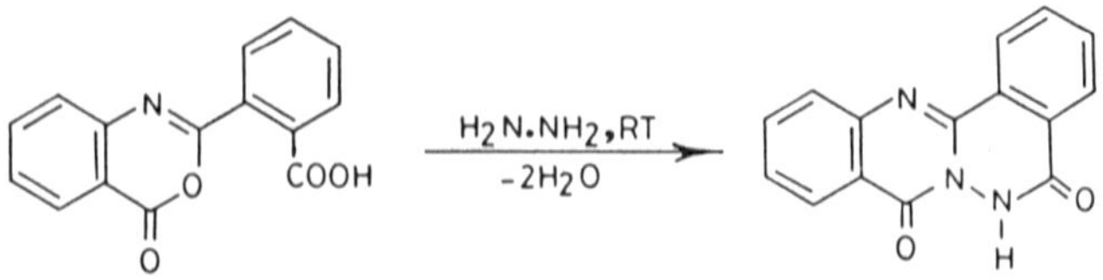

(499)   (500)

D. PYRIMIDO-QUINAZOLINES

1. *Pyrimido[1,2-a]quinazolines*

2-Aminoquinazolines substituted on the amino function with a three-carbon cyclizable moiety, such as 2-(3-hydroxypropyl)amino-4-phenylquinazolines (**501**) (73NKK1944) or 2-(2-chloronicotinoylamino-4-phenylquinazolines (86H3075), gave, upon cyclization, the corresponding pyrimido[1,2-*a*]quinazolines (**502**). Cyclization took place with the quinazoline *N*-1 as a result of its higher nucleophilicity.

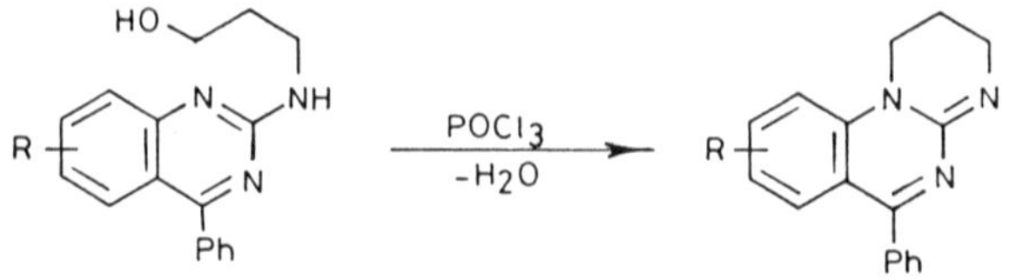

(501)   (502)

N-(Butyn-2-yl)isatoic anhydride (**503**) gave the pyrimido[1,2-*a*]quinazolinone (**504**) on cyclization with 3-allyl-2-methylthiopseudourea (76JOC825).

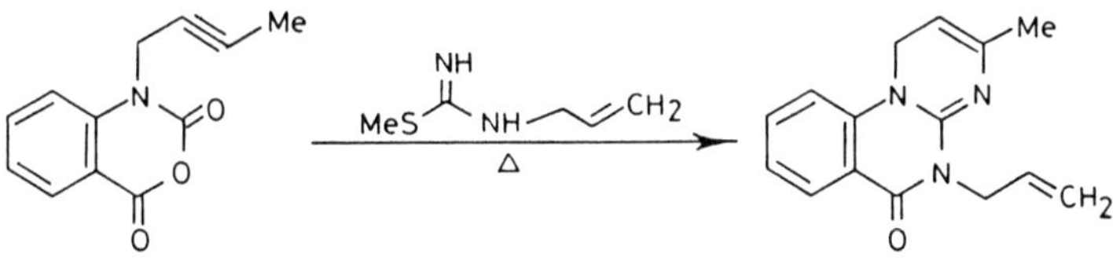

(503)   (504)

2. *Pyrimido[2,1-b]quinazolines*

This linear type of pyrimido-quinazoline was synthesized by (a) formation of the terminal pyrimidine ring, (b) formation of the middle pyrimidine ring, and (c) formation of both pyrimidine rings.

Synthesis according to the first approach requires quinazoline precursors with an appropriate subtituent on position 2 or 3 which, upon
cyclization, forms the terminal pyrimidine ring. Thus, 2-aminoquinazolin-
4(3H)-one **(505)** underwent cyclization with α,β-unsaturated acids
(78KGS105; 80MI1) or their chlorides (87KGS1527) to the pyrimido

(505)   (506)

[2,1-b]quinazolin-2-ones **(506)**. 3-(Dimethylaminopropyl)-6-chloro-3,4-
dihydro-4-phenylquinazolin-2(1H)-one or the corresponding thione **(507)**
was cyclized with phosphoryl chloride to 8-chloro-1,2,3,4-tetrahydro-1-
methyl-6-phenyl-6H-pyrimido[2,1-b]quinazoline **(508)**. The terminal pyrimidine ring of **510** was also formed upon cyclization of 2-chloro-3-(3-
chloropropyl)quinazolin-4-ones **(509)** with primary amines (84USP4451448).

(507), Z= O or S   (508)

(509)   (510)

Synthesis through formation of the middle pyrimidine ring was accomplished by the reaction of anthranilic acids with 2-halopyrimidines **(511)** to
give **512** [59NKZ21181; 68ZC103; 76MI2; 77KGS678; 84BRP2125785,
84GEP(O)3231408; 86URP1235866]. Reaction of isatoic anhydrides with
2-alkylthio-3,4,5,6-tetrahydropyrimidines **(513)** gave the 1,2,3,4-tetrahy-
dropyrimido[2,1-b]quinazolin-6-ones **(514)** [68MI499; 70GEP(O)2025248;

(511)   (512)

(513)   (514)

76USP3969506]. Reductive cyclization of 2-methylthio-1-(2-nitrobenzyl]-6-pyrimidines (515) with stannous chloride gave 6,11-dihydro-4H-pyrimido[2,1-b]quinazolin-4-ones (516) (78JHC77).

(515)   (516)

The two pyrimidine rings of 1,2,3,4-tetrahydro-6H-pyrimido[2,1-b]-quinazolin-6-one (517) were simultaneously formed when ethyl N-(2-methoxy-carbonylphenyl)thiocarbamate was cyclized with 1,3-diaminopropane (82JHC1117). The two heterocyclic rings of 1,2,3,4,5,6-hexahydropyrimido[2,1-b]quinazolin-2-ones (519) were also concomitantly formed when compounds 518 were cyclized with cyanogen bromide [73GEP(O)2305575; 76USP3932407].

(517)

(518)   (519)

3. Pyrimido[6,1-b]quinazolines

The published work on the synthesis of this system involved the reaction of anthranilic acid with a reactive pyrimidine derivative. Thus, 4-chloropyrimidines (520) condense with 2-aminobenzaldehyde (47JCS726) or an anthranilic acid derivative [66T(S)227; 68ZC103; 89MI2] to give the pyrimido-[6,1-*b*]quinazolines (521).

(520)   (521)

4. Pyrimido[1,2-c]quinazolines

Synthesis of the title compounds (523) from quinazoline precursors was carried out by cyclization of 4-(3-hydroxypropylamino)quinazolines (522), obtained from quinazoline-4-thiones and 3-hydroxypropylamine, with phosphoryl chloride (74YZ417; 84MI5). Cyclization of 2-(3-chloropro-pylureido)benzonitrile (524) with ammonia (84JHC1411) or acid-catalyzed cyclization of 2-(3-chloropropylureido)thiobenzamide (526) (89JHC595) afforded the pyrimido[1,2-*c*]-quinazoline (525). The two pyrimidine rings of pyrimido[1,2-*c*]quinazolines (523) were formed upon cyclocondensation of 3,1-benzoxazinones [88EGP(D)258232] or trithioisatoic anhydride (73ZC428; 79PHA390) with 1,3-diaminopropane.

(522)   (523)

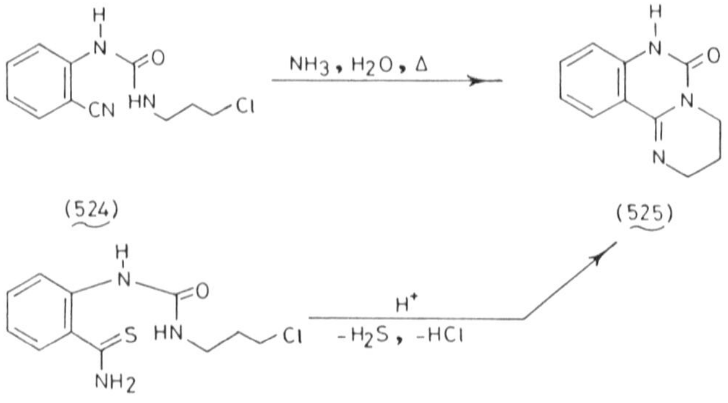

(524)   (525)

(526)

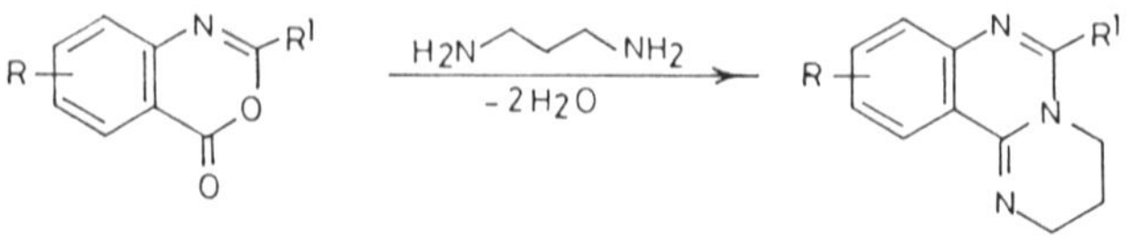

(523)

Syntheses from properly substituted pyrimidine precursors were also described. 2-(2-Aminophenyl-1,4,5,6-tetrahydropyrimidines (527) were cyclized with aldehydes (65NEP6409191; 67USP3309369; 68SZP452537], ketones [76BSF1857), or carbon disulfide (82EUP46446), formic acid or triethyl orthoformate [86JAP(K)6150983] to give pyrimido[1,2-c] quinazolines (528). 2-Trichloroacetamido- or 2-methoxycarbonylamino-benzophenones (529) were cyclized with 1,3-diaminopropane to 1,2,3,4, 6,11b-hexahydro-11b-phenyl-6H-pyrimido[1,2-c]quinazolin-6-one   (530) [67USP3329679; 72GEP(O)2141616; 73GEP(O)2166380]. The two pyrimidine rings of the octahydro-6H-pyrimido[1,2-c]quinazolin-6-thione (531) were formed upon condensation of tetrahydrotrithioisatoic anhydride with 1,3-diaminopropane (79PHA390).

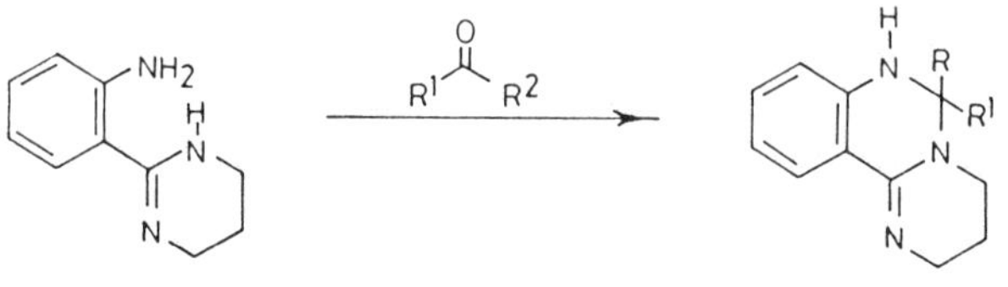

(527)   (528)

(529) R¹= CCl₃ or OMe   (530)

(531)

Various biological activities have been attributed to pyrimidoquinazo-lines; they showed blood sugar level depressant (82EUP46446), analeptic (65NEP6409191; 68SZP452537; 76USP3969506), bronchodilating [70GE-P(O)2025248; 76USP3969506], antipyretic [84GEP(O)3231408], analgesic [84BRP2125785, 84GEP(O)3231408], antiinflammatory (84BRP2125785), adrenomimetic (65NEP6409191; 67USP3309369; 68SZP452537), central nervous system depressant (67USP3329679), fungicidal (80MI1), and hypotensive [70GEP(O)2025248; 76USP3932407] activities.

E. QUINAZOLINO-QUINAZOLINES

One should be aware that, in the literature, some members of the quina-zolino-quinazolines are described by either one of two fusion locants as a result of numbering either of the quinazoline moieties.

1. *Quinazolino[1,2-a]quinazolines*

Amidation of *N*-(2-carboxyphenyl)anthranilic acid (**532**) gave the 2,2′-dicarbamoyldiphenylamines (**533**), which were then cyclized with chloro-acetyl chloride to the quinazolino[1,2-*a*]quinazolines (**534**) (81JOC1571).

(532)   (533)   (534)

Katritzky *et al.* [76CC48; 77JCS(P1)1162] synthesized the quina-
zolin[1,2-*a*]-quinazoline (536) by the reaction of 2,2'-bis(chloroformyl)
diphenylamine (535) with *N,N*-diphenylbenzamidine.

(535)   (536)

2. Quinazolino[3,2-a]quinazolines (or Quinazolino[2,1-b]-quinazolines)

Examples of this ring system (539) were prepared by cyclization of
2-(2-carboxyphenyl)aminoquinazolin-4-one (537) (59JCS1512; 83AP702)
or 2-amino-3-(2-carboxyphenyl)quinazolin-4-one (538) (59JCS1512). Alternatively, cyclodehydration of 3-(2-benzamido)quinazoline-2,4(1*H*,3*H*)-

(537)   (539)   (538)

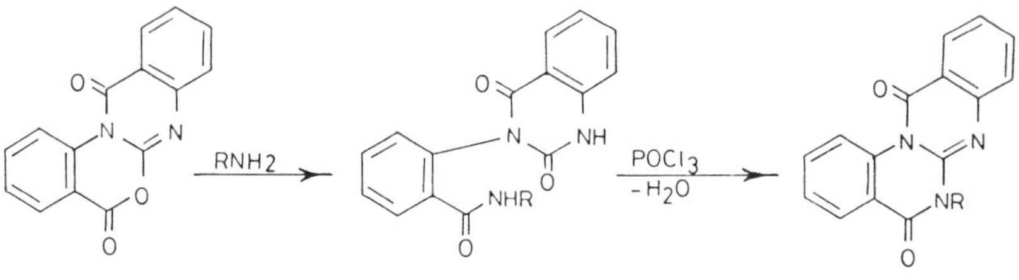

(540)   (541)   (542)

diones (**541**), obtained from **540** and primary amines, gave quinazoline
[3,2-*a*]quinazolines (**542**) (66ACH77; 67MI1). Condensation of 3-
substituted 3-thioxoquinazoline-4-ones (**543**) with anthranilic acid gave the
quinazolino[3,2-*a*]quinazolines (**542**) (71JIC989].

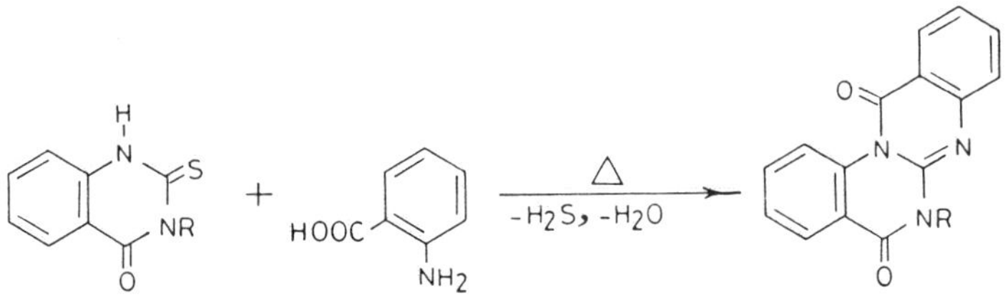

(543)   (542)

It has already been mentioned that the reaction of 2,4-dithioxo-1,3-
benzothiazine (**544**) with anthranilamide gave a complex mixture of products from which benzoisothiazolo[2,3-*a*]quinazolin-5-one (**221**) was isolated (see Section VI.A). The quinazolino[3,2-*a*]quinazoline (**545**) has also
been isolated from this mixture (75JHC1077).

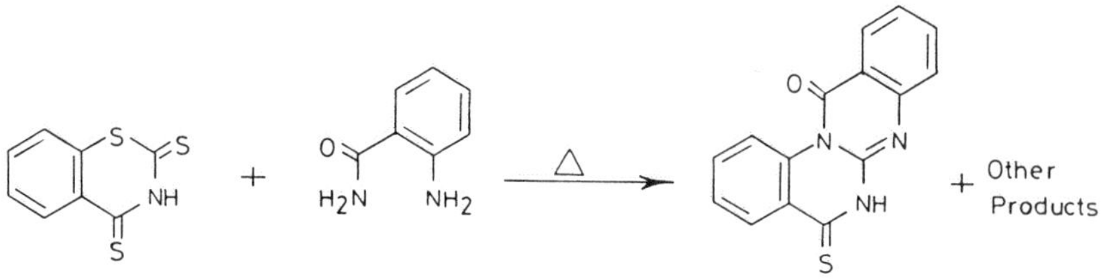

(544)   (545)

3. Quinazolino[3,4-a]quinazolines (or Quinazolino [1,2-c]quinazolines)

The reported examples of this ring system were obtained by the reaction of 2-ureidobenzonitrile (89JPR537) or 2-alkyl-4H-3,1-benzoxazin-4-one (84CPB2160) with anthranilic acid nitrile to give 6-alkyl-13H-quinazolino[3,4-a]quinazolin-13-ones (546).

(546)

4. Quinazolino[2,3-b]quinazolines

Contrary to the cyclization of 2-(2-carboxyphenyl)aminoquinazolin-4-one (537), which gave only the quinazolino[3,2-a]quinazoline (see Section XIII.E.2), cyclization of the corresponding methyl ester (547) gave a mixture of quinazolino[3,2-a]quinazoline (539) and quinazolino[2,3-b] quinazoline (548) (59JCS1512). 5-Chloroanthranilic acid or its amide reacted with thiophosgene to give a mixture of the 2-mercaptoquinazolinone (549) and the quinazolino[2,3-b]-quinazoline (550) (70USP3501473; 75JHC1207).

(547)   (548)

(549)   (550)

5. Quinazolino[4,3-b]quinazolines (or Quinazolino [3,2-c]quinazolines)

Compounds of this class were synthesized by cyclocondensation of 2-(2-aminophenyl)quinazolines (e.g. **551**) with acid anhydrides (56JCS4178; 60JCS4970) or orthoesters [60JCS4970; 64JCS(C)3670] as well as with aldehydes or their Schiff bases [60JCS4970; 79IJC(B)349; 83EGP(D)204095]. Whereas the reaction with acid derivatives gave quinazolino[4,3-b]quinazolines (**552**), the reaction with aldehydes afforded the dihydro derivatives (**553**), which could be dehydrogenated to **552**. One of

the two 1,3-diazine rings of these compounds was also formed by the alternative route of cyclizing anthranilic acid, its methyl ester, or ammonium salt with 2-substituted-4-chlorobenzoquinazolines (**554**) (29JIC723; 56JCS4173; 60JCS4970). Pertinent to this synthesis is the reaction of a mixture of anthranilic acid, quinazolin-4(3H)-ones, and phosphorus trichloride (55MI2). Some biological applications have been reported for quinazolino-quinazoline as being central nervous system depressants (70USP3501473), analgesics (83AP702) and antiinflammatory agents (83AP702).

F. PYRAZINO-QUINAZOLINES

Pyrazino[2,1-b]quinazolines

Synthesis of the pyrazino[2,1-*b*]quinazoline (557) was achieved by oxidative cyclization of the 1-(2-aminobenzyl)-4-methylpiperazine (556) with manganese dioxide [68JCS(C)1722], the pyrimidine ring of 557 was formed during this synthesis. The pyrimidine ring of 559 was also formed upon cyclocondensation of anthranilic acids with the cyclic imidate ester (558), followed by removal of the protective group (66USP3280117).

(556)   (557)

(558)   (559)

Simultaneous formation of the pyrimidine and pyrazine rings of 562 was affected when 2-(*N*-methylglycylamino)-5-chlorobenzophenone (560) was chloroacetylated to 561, followed by cyclization with ammonia (81JOC4489). Deblocking of the *N*-benzyloxycarbonyl protective group of 563 with hydrogen bromide in acetic acid took place with simultaneous dehydrative cyclization to give the pyrazino[2,1-*b*]quinazoline (564) (84KGS983).

(560)   (561)

(562)

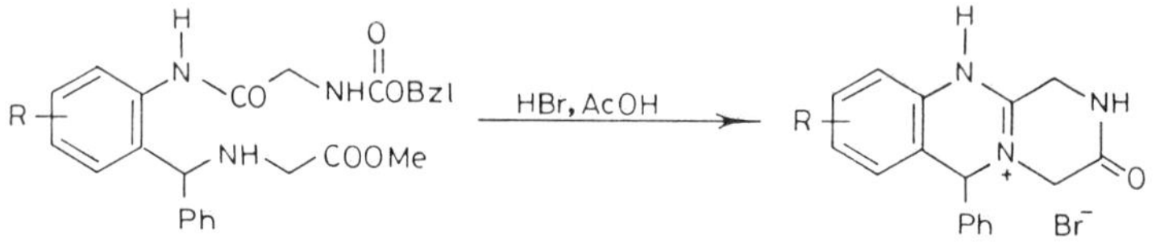

(563)   (564)

Consecutive formation of the two diazine rings was the approach for synthesizing the pyrazino[2,1-*b*]-quinazoline (**567**) from methyl anthranilate and the protected glycylglycine dipeptide (**565**), according to the indicated sequence of reactions (80MI4).

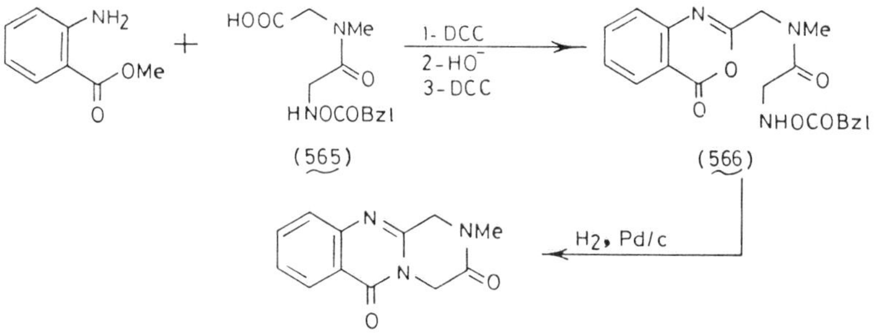

(565)   (566)

(567)

XIV. Condensed Oxazino-quinazolines

A. 1,2-OXAZINO-QUINAZOLINES

1,2-Oxazino[3,2-b]quinazolines

The title compounds (**568**) were prepared by reaction of anthranilic acids with γ-chloroacid chlorides, followed by cyclization with hydroxylamine (77AF766). The title compounds exhibited antiinflammatory, analgesic and antipyretic activities (77AF766).

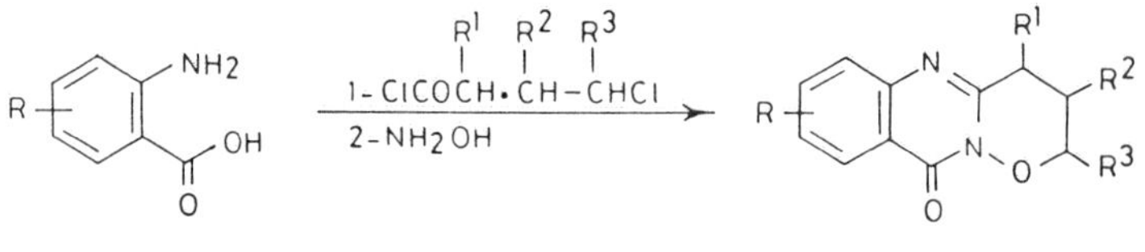

(568)

B. QUINAZOLINO-2,3-BENZOXAZINES

Quinazolino[3,2-c]2,3-benzoxazines

Condensation of anthranilic acid or its methyl ester with the 2,3-benzoxazine imidoyl chloride or imidate ester (569) gave the 5H,8H-quinazolino[3,2-c]2,3-benzoxazin-9-one (570) [74CI(M)492].

(569), LG = Cl or OPh   (570)

C. 1,3-OXAZINO-QUINAZOLINES

1. 1,3-Oxazino[2,3-b]quinazolines

3,4-Dihydro-2H,6H-[1,3]oxazino[2,3-b]quinazolin-6-one (572) was obtained by the base-catalyzed cyclization of the urea derivatives (571) (84JHC1411).

(571)   (572)

2. 1,3-Oxazino[3,2-c]quinazolines

10-Chloro-3,4,7-11b-tetrahydro-2H,6H,11b-phenyl-1, 3-oxazino[3,2-c]-quinazolines (574) were obtained by the reaction of 2-acylamino-5-chlorobenzophenones (573) with 3-aminopropanol [70YZ629; 72GEP-(O)2141616; 80PHA256; 81CPB2135].

(573)   (574)

D. QUINAZOLINO-3,1-BENZOXAZINES

Quinazolino[3,2-a]3,1-benzoxazines

2-Isocyanatobenzoyl chloride reacted with isatoic anhydride (76JOC2728) or nitromethane in benzene [83IJC(B)485] to yield 5*H*,12*H*-quinazolino[3,2-*a*]3,1-benzoxazine-5,12-dione **(540)** (76JOC2728). Cyclodehydration of 3-(2-carboxyphenyl)quinazoline-2,4(1*H*,3*H*)-dione **(575)** (66ACH77;67MI1) also afforded the same compound **(540)**.

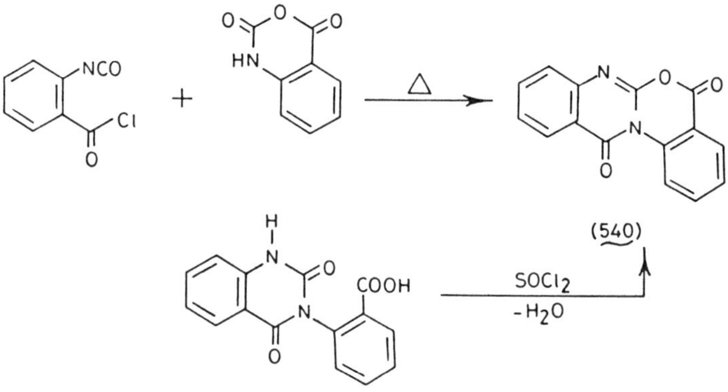

(540)

(575)

E. 1,4-OXAZINO-QUINAZOLINES

1,4-Oxazino[3,4-b]quinazolines

Oxidative cyclization of the 4-(2-aminobenzyl)-1,4-oxazine **(576)** with manganese dioxide gave the 1,4-oxazino[3,4-*b*]quinazoline **(577)** through the formation of its pyrimidine ring [68JCS(C)1722]. Cyclocondensation of anthranilic acids with the thioimidate ester **(578)** gave 3,4-dihydro-1,4-oxazino[3,4-*b*]quinazolin-6(1*H*)-ones **(579)** [79IJC(B)107].

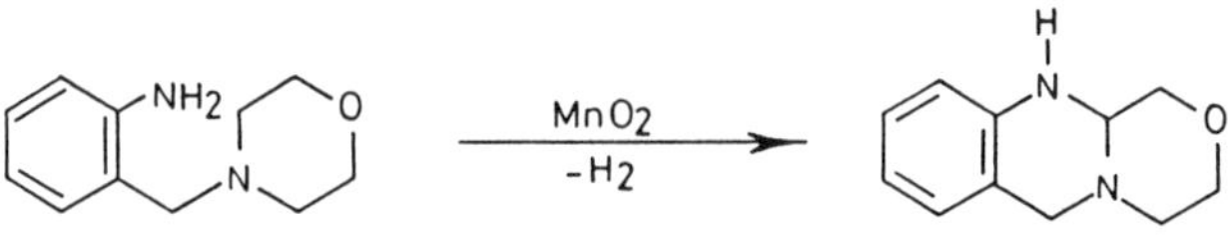

(576)   (577)

(578)   (579)

The title compounds were also synthesized from quinazoline precursors through the formation of their 1,4-oxazine rings. *N*-Methylisatoic anhydride was first condensed with ethanolamine to give *N*-(2-hydroxyethyl)-2-methylaminobenzamide (580). Cyclization of the latter with ethyl pyruvate gave quinazoline derivatives 581 which, upon hydrolysis and dehydrative cyclization with 1-methyl-2-chloropyridinium iodide, afforded the 1,4-oxazines[3,4-*b*]quinazoline (582) (80JHC1163).

(581)   (582)

F. QUINAZOLINO-1,4-BENZOXAZINES

Quinazolino[2,3-c]1,4-benzoxazines

The title compounds (584) were synthesized from 1,4-benzoxazine precursors carrying a good leaving group at position 3 (583, LG = Cl or SMe) by condensation with anthranilic acids [79IJC(B)107; 88S336].

(583), LG= Cl or SMe   (584)

On the other hand, the two heterocyclic rings of the title compounds **(587)** were formed upon synthesis from anthranilic acids and 2-nitrophenoxyacetyl chloride **(585)** followed by reduction and cyclization (84JIC721).

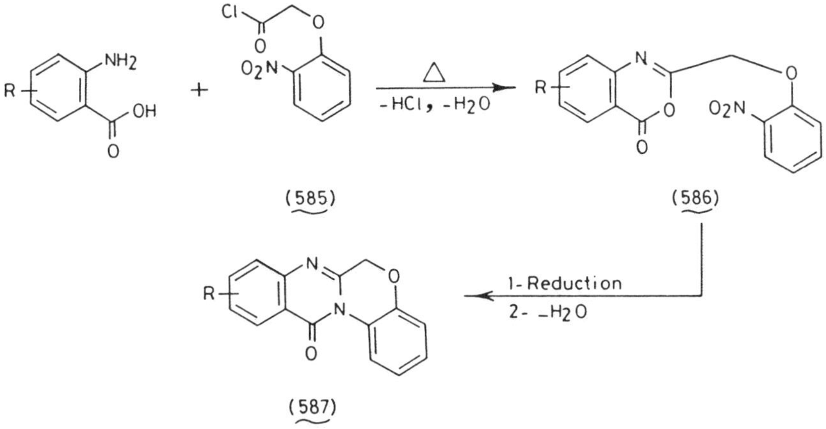

(585)   (586)

1- Reduction
2- $-H_2O$

(587)

XV. Condensed Thiazino-quinazolines

A. 1,3-THIAZINO-QUINAZOLINES

1. *1,3-Thiazino[3,2-a]quinazolines*

3H,6H-1,3-Thiazino[3,2-a]quinazolin-6-ones **(588)** were synthesized by Gakhar (65IJC44) through the formation of both the diazine and thiazine rings by condensation of anthranilic acid with β-thiocyanatoketones.

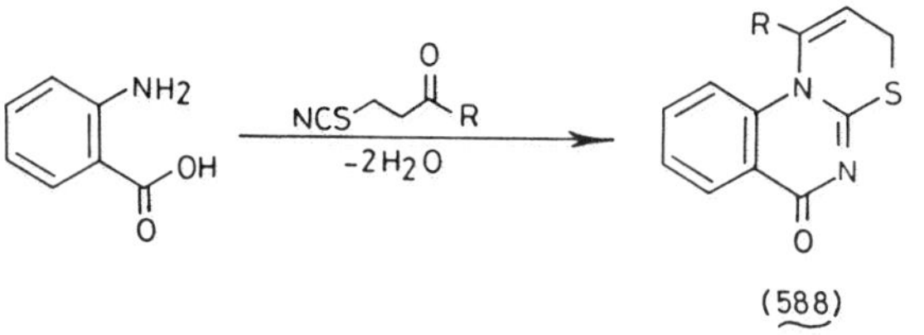

(588)

2. *1,3-Thiazino[2,3-b]quinazolines*

Reaction of methyl isocyanatobenzoate (67HCA1440) or trithioisatoic anhydride (71IJC647; 73JMC633; 77ZC444; 80PHA124) with 3-aminopropanol gave 3,4-dihydro-2H,6H-1,3-thiazino[2,3-b]quinazolines (e.g. **590**). The same compounds (**590**) were also obtained from the reaction of anthranilic acid esters and 3-chloropropylisothiocyanate (79JHC391).

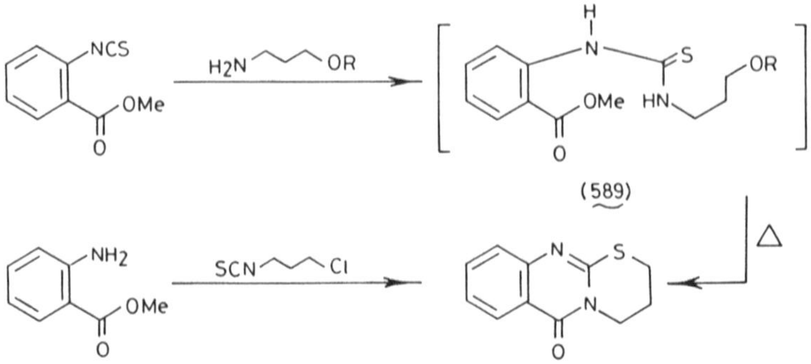

(589)

(590)

2-Mercaptoquinazoline **247** was cyclized with three-carbon cyclizing agents such as 1,3-dibromopropane (73JMC633; 84JIC1050; 85AP502), 3-bromopropanoic acid (78MI1), ethyl 2,4-dibromobutanoate [85JAP(K)6075488], or ethyl 3-chloro-2-butenoate (83AP379) to the corresponding 1,3-thiazino[3,2-*b*]quinazoline (**591**).

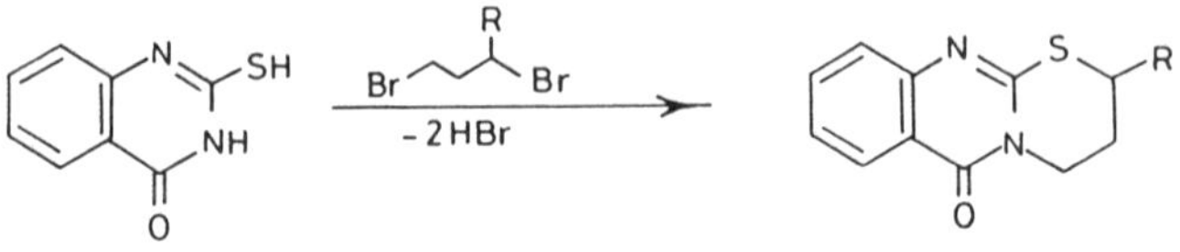

(247)

(591)

A synthesis during which the pyrimidine ring of this system was formed involved the fusion of 2-mercapto-4,4,6-trimethyl-4*H*-1,3-thiazine (**592**) with anthranilic acids to give **593** (62JOC4061).

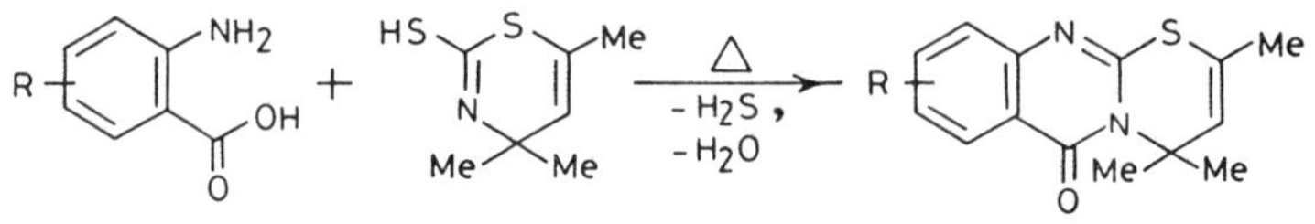

(592)

(593)

3. *1,3-Thiazino[3,2-c]quinazolines*

The tetrahydrothioisatoic anhydride reacted with 3-aminopropanol to give the quinazoline **594** which, upon acid-catalyzed cyclization, gave the two structural isomers 1,3-thiazino[2,3-*b*]quinazoline (**595**) and 1,3-thiazino[3,2-*c*] quinazoline (**596** (78PHA185). Some biological applications have been reported for thiazinoquinazolines such as bactericidal (62JOC4061), and antihypertensive (83AP379; 85AP502).

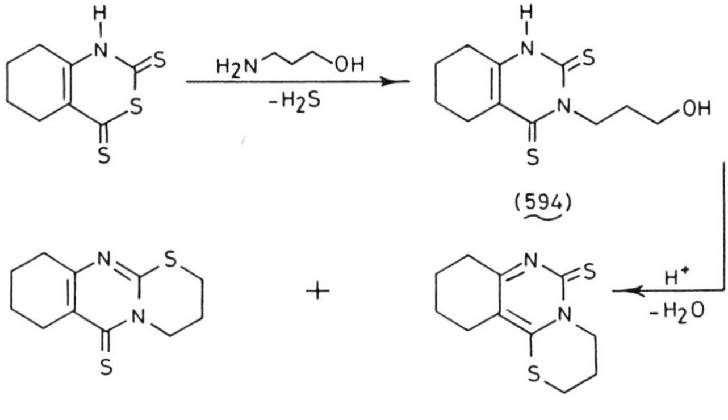

(594)

(595)   +   (596)

B. Quinazolino-3,1-benzothiazines

Quinazolino[2,3-a]3,1-benzothiazines

Reaction of *N*-(2-aminobenzoyl)-2-aminobenzyl alcohol (711JC647) or anthraniloyl anthranilic acids (**597**) (76JIC382) with carbon disulfide gave the 3-(2-carboxyphenyl)-2-mercaptoquinazolin-4-ones (**598**), which underwent cyclodehydration to the title compounds **599** (76JIC382).

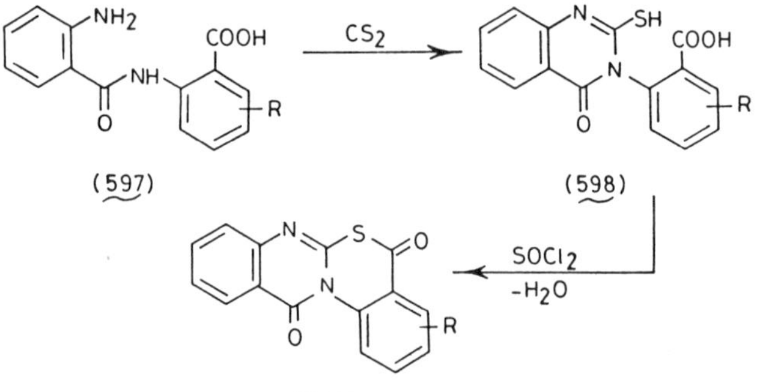

(597)   (598)

(599)

C. 1,4-THIAZINO-QUINAZOLINES

1. 1,4-Thiazino[4,3-a]quinazolines

(Quinazolin-2-ylmethyl)thioacetic acid (600) gave the 1,4-thiazino[4,3-a]-quinazoline (601) when heated with acetic anhydride and pyridine [88IJC(B)578].

(600)   (601)

2. 1,4-Thiazino[3,4-b]quinazolines

3,4-Dihydro-1,4-thiazino[3,4-b]quinazolin-6(1H)-ones (603) were prepared by cyclocondensation of anthranilic acids with 3-methylthio-1,4-thiazine (602) [79IJC(B)107].

(602)   (603)

D. QUINAZOLINO-1,4-BENZOTHIAZINES

Quinazolino[2,3-c]1,4-benzothiazines

The pyrimidine ring of the title compounds (605) was formed when 3-methylthio- [79IJC(B)107] or 3-chloro-1,4-benzothiazines (604) (60JOC853; 88S336) were cyclocondensed with anthranilic acids. Cyclization of 2-(chloroacetamido)benzoic acids (606) with 2-aminothiophenols also gave the same compounds (69IJC881).

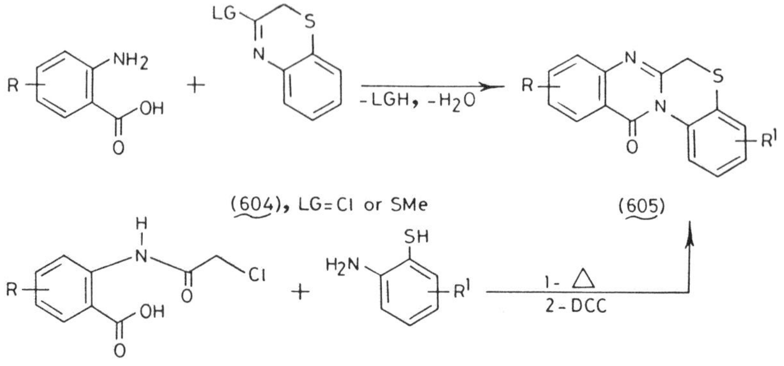

(604), LG = Cl or SMe

(605)

(606)

Isatoic anhydride condensed with the sodium salt of 2-aminothiophenol to give the intermediate benzamide (607), which underwent double ring closure with α-haloketones to quinazolino[2,3-c]1,4-benzothiazoles (608) (72IJC476).

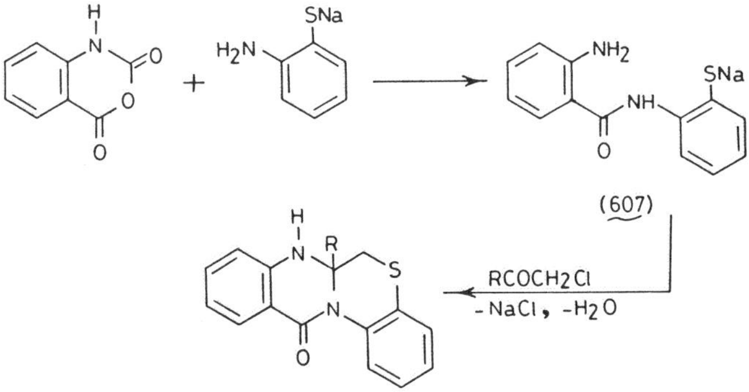

(607)

(608)

XVI. Condensed Triazino-quinazolines

A. QUINAZOLINO-1,2,3-BENZOTRIAZINES

Quinazolino[3,2-c]1,2,3-benzotriazines

Quinazolino[3,2-c]1,2,3-benzotriazin-8(7H)-imine (611) was prepared from 4-amino-2-(2-aminophenyl)quinazoline (609) by diazotization, followed by basification of the resulting diazonium salt (610). The isomeric quinazolino[1,2-*c*]1,2,3-benzotriazine structure (**612**) has not been rigorously excluded [77JCS(P1)107].

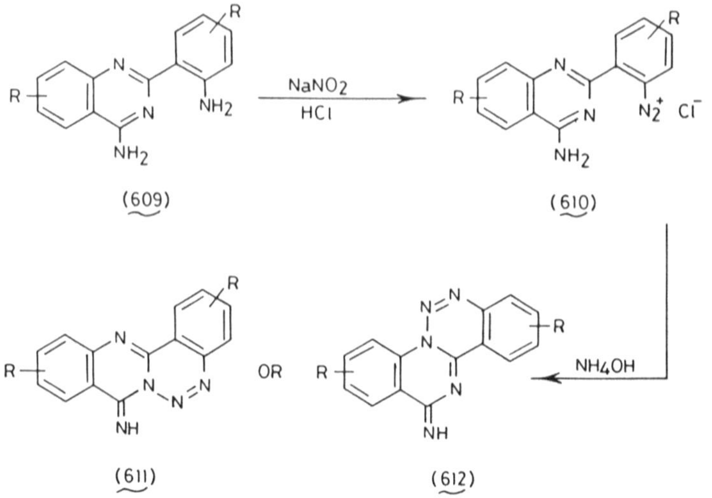
NaNO2
HCl
(609)
(610)
NH4OH
OR
(611)
(612)

Diazotization of anthraniloanthranilamide (**613**) gave 3-(2-carbamoylphenyl)-1,2,3-benzotriazine-4(3*H*)-one (**614**), which underwent base-catalyzed cyclization to the quinazolino[3,2-*c*]1,2,3-benzotriazin-8-one (**615**) (78CJC1616).

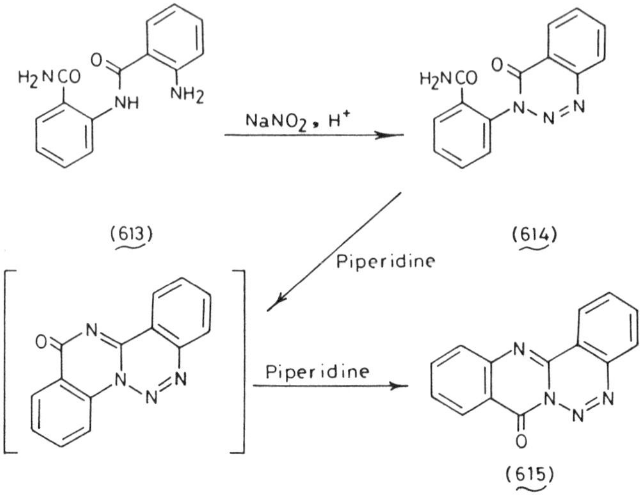
NaNO2, H+
(613)
(614)
Piperidine
Piperidine
(615)

Ring contraction of the triazepine ring of 6-aryl-6,7-dihydroquinolino-[3,2-*d*]1,3,4-benzotriazepin-9(5*H*)-ones (**616**) was affected with lead tetraacetate to give quinazolino[3,2-*c*]1,2,3-benzotriazin-8-one (**615**) [87IJC(B)983].

(616)   (615)

B. 1,2,4-TRIAZINO-QUINAZOLINES

1. *1,2,4-Triazino[2,3-a]quinazolines*

The 1,2,4-triazino[2,3-*a*]quinazoline (**618**) was prepared by thermal cyclization of the 2-(2-carboxamidophenyl)-1,2,4-thiazine precursor (**617**) (74JPR943).

(617)   (618)

2. *1,2,4-Triazino[4,3-a]quinazolines*

Hydrazones (**619**) derived from 3-substituted-2-hydrazinoquinazolin-4-ones and pyruvic acid were cyclized by heating with acetic acid to give the 1,2,4-triazino[4,3-*a*]quinazoline-1,6-diones (**620**) [83EGP(D)160343; 84PHA717; 90JOC344].

(619)   (620)

3. 1,2,4-Triazino[3,2-b]quinazolines

The 1,2,4-triazine ring of 3,4-dihydro-2-methyl-1,2,4-triazino[3,2-b]-quinazoline-3,10-dione (622) was formed when 3-amino-2-(substituted amino)quinazolin-4-one (621) was cyclized with ethyl pyruvate in acetic acid (86JHC833).

(621)   (622)

Synthesis was also achieved from 1,2,4-triazine precursors when 4-amino-3-methylthio-1,2,4-triazinones (623) were cyclocondensed with anthranilic acid to give 624 (84CB1077, 84CB1083).

(623)   (624)

4. 1,2,4-Triazino[3,4-b]quinazolines

When the 3-methylthio-1,2,4-triazine (625) was cyclocondensed with anthranilic acid, it afforded the 1,2,4-triazino[3,4-b]quinazoline (626) (84CB1077); cyclization with N-4 of the 1,2,4-triazine ring was possible as a result of blocking N-2.

(625)   (626)

5. 1,2,4-Triazino[6,1-b]quinazolines

Reaction of 3-isonitrosopyrazolo[5,1-b]quinazoline-2,9-dione with Vilsmeier reagent (627) followed by hydroxylamine gave the 1,2,4-triazino[6,1-b]quinazolin-10-one (628) [77IJC(B)335].

(627)   (628)

6. *1,2,4-Triazino[1,6-c]quinazolines*

The   3-alkylthio-6,7-dihydro-1,2,4-triazino[1,6-*c*]quinazolin-5-ium-1-olates **(630)** were prepared through the formation of their pyrmidine rings by cyclocondensation of 3-alkylthio-6-(2-aminophenyl)-1,2,4-triazin-5(2*H*)ones **(629)** with aldehydes, ketones or acetals (74T3997; 76T1735; 80ACH107; 86SC35).

(629)   (630)

Treatment of the 1,2,4-triazino[5,6-*d*]3,1-benzoxazepine derivatives **(631)** with an excess of methyl iodide affected methylation and intramolecular rearrangement to give the mesoionic 1,2,4-triazolo[1,6-*c*] quinazolinone **(632)** (73MI1).

(631)   (632)

7. *1,2,4-Triazino[2,3-c]quinazolines*

Reaction   of   3-(2-aminophenyl)-1,2,5,6-tetrahydro-1-methyl-1,2,4-triazine **(633)** with phenyl isocyanate gave the urea derivative **(634)**, which was cyclized by heating with polyphosphoric acid to 2,3,4,7-tetrahydro-2-methyl-6*H*-1,2,4-triazino-[2,3-*c*]quinazolin-6-one (**635**) (74JHC747). Cyclization of **634** was also affected with benzoyl chloride (75JHC321).

(633)   (634)   (635)

8. *1,2,4-Triazino[4,3-c]quinazolines*

Two types of precursors were used for synthesizing compounds belonging to this system: (a) quinazoline and (b) 1,2,4-triazine precursors. Synthesis from quinazoline precursors was achieved by cyclocondensation of the (4-thioxoquinazolin-3-yl)acetic esters (**636**) with hydrazine to give the 6-substituted-1,2,4-triazino[4,3-*c*]quinazolin-3-ones (**637**) (82S853; 83BSF226).

(636)   (637)

Cyclization of the urido-1,2,4-trizine derivatives (**634**) to the 1,2,4-triazino[2,3-*c*]quinazoline (**635**) by heating with polyphosphoric acid has already been discussed (see Section XVI.B.7) (74JHC747). However, pyrolytic cyclization of **634** by heating at 200° afforded the 1,2,4-triazino[4,3-*c*]quinazoline (**638**) as a result of eliminative cyclization between the urido function and *N*-4 of the 1,2,4-triazine ring (74JHC747). The structure of **638** was confirmed by an unequivocal synthesis from 2-oxo-4-thioxoquinazoline (**639**) (74JHC747).

(634)   (638)

                                                ↑ KOH

(639)   (640)

Synthesis from 1,2,4-triazine precursors was described in a series of papers and patents by Trepanier (74JHC747; 75JHC321, 75USP3919216, 75USP3919219, 75USP3919220, 75USP3922274), in which 3-(2-aminophenyl)-1-substituted-1,2,4-triazines (**641**) were cyclized with aldehydes, ketones, or acid chlorides. In the hands of other workers (75USP3919215), treatment of **641** with acid chlorides gave the corresponding acylaminophenyltriazines that dehydratively cyclized to a mixture of 1,2,4-triazino[4,3-*c*]quinazolines and 1,2,4-triazino[2,3-*c*]quinazolines.

(641)   (642)

Some medicinal applications were reported for 1,2,4-triazinoquinazolines as being antihistaminics [75USP3919220, 75USP3922274; 83EGP(D)160343], blood platelet aggregation inhibitors [83EGP(D)-160343], inhibitors of reserpine induced ptosis (75USP3919219,

75USP3919220), potentiators of hexobarbital induced sleep (75USP-3919220, 75USP3922274), analgesics (75USP3922274), and antidepressants (75USP3922274).

C. 1,3,5-TRIAZINO-QUINAZOLINES

1. 1,3,5-Triazino[1,2-a]quinazolines

Cyclization of quinazolin-4(3H)-one with an excess of methyl isocyanate gave the 1,3,5-triazino[1,2-a]quinazoline-1,3,6-trione (**643**) (84H501). A one-pot synthesis of 1,3,5-triazino[1,2-a]quinazolines (**645**) was carried out by reacting 2-amino-4-phenylquinazolines (**644**) with chlorocarbonyl isocyanate (85S892).

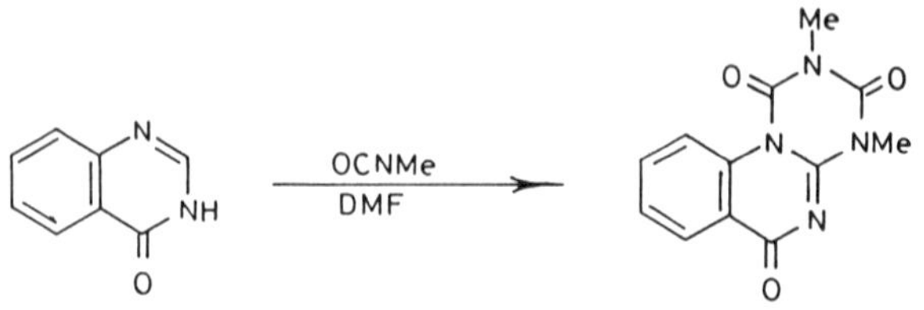

(643)

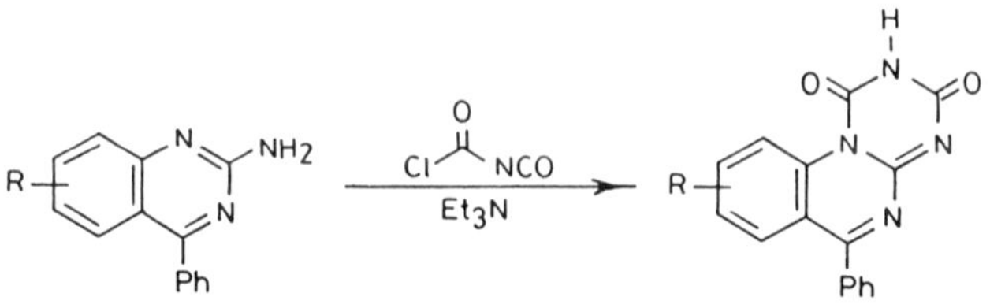

(644)   (645)

2. 1,3,5-Triazino[2,1-b]quinazolines

Methyl anthranilate condensed with **646** to give 2-phenyl-4-imino-1,3,5-triazino[2,1-b]quinazolin-6-one (**647**) [69JCS(D)1040]. Condensation of the cyanuric acid derivative **648** with methyl anthranilate gave **649** which, upon cyclization, afforded a mixture of the two isomeric 4(2)-anilino-4(2)-hydroxy-1,3,5-triazino[2,1-b]quinazolin-6-ones (**650** and **651**) (75USP3887554).

(646)   (647)

(648)   (649)

AcOH, Δ

(650), R = NHPh, R^1 = OH
(651), R = OH, R^1 = NHPh

XVII. Condensed Oxadiazino-quinazolines

1,3,4-OXADIAZINO-QUINAZOLINES

1,3,4-Oxadiazino[2,3-b]quinazolines

3,4-Dihydro-2,2,4-trimethyl-2*H*,6*H*-1,3,4-oxadiazino[2,3-*b*]quinazolines (**654**) were synthesized together with 2,3-dihydro-3-methyl-2-thioxo-6*H*-1,3,4-thiadiazolo[2,3-*b*]quinazolines by the reaction of isatoic anhydrides with 1-[(2-hydroxy-2-methyl)propyl]hydrazine to give the corresponding anthranilic acid hydrazides (**652**), which were cyclized to the quinazoline intermediates **653** and then to **654** (79JHC1339).

XVIII. Condensed Thiadiazino-quinazolines

1,3,4-THIADIAZINO-QUINAZOLINES

1,3,4-Thiadiazino[2,3-b]quinazolines

3,4-Dihydro-4-methyl-2H,6H-1,3,4-thiadiazino[2,3-b]quinazolines (655) were synthesized from isatoic anhydrides and 1-(2-hydroxymethyl)-1-methylhydrazine (79JHC1339). 1,3,4-Thiadiazino[2,3-b]quinazolines (657) were also prepared by the reaction of 3-amino-2-mercaptoquinazolinones (656 with α-haloketones [81IJC(B)14; 82M1145].

XIX. Condensed Tetrazino-quinazolines

1,2,4,5-TETRAZINO-QUINAZOLINES

1,2,4,5-Tetrazino[1,6-c]quinazolines

Condensation of the 3,1-benzothiazine-4-thiones (**658**) with ethoxy-carbonylhydrazine afforded the 3-ethoxycarbonylaminoquinazoline-4-thiones that underwent cyclocondensation with hydrazine to give the 1,2,4,5-tetra-zino[1,6-*c*]quinazolines (**659**) (83BSF226). The reaction of 3-amino-2-phenylquinazolin-4(3*H*)-ones or their thiones (**660**) with isocyanates or isothiocyanates gave the corresponding urea derivatives, which were cyclized with hydrazine to the title compounds (**661**) (88CCC329).

(658)   (659)

(660), Z = O or S   (661)

XX. Condensed Azepino-quinazolines

A. AZEPINO[1,2-*a*]QUINAZOLINES

The title compounds (**662**) were synthesized through the formation of their azepine rings by cycloaddition of two molar equivalents of dimethyl acetylenedicarboxylate to 2-methyl-3-(2-tolyl)quinazolin-4(3*H*)-one (**404**) (72JHC1227).

(404)   (662)

In a different approach, the pyrimidine ring was formed when the 1-(2-aminomethylphenyl)hexahydroazepine (663) was oxidatively cyclized with Hg(II)-EDTA to 664 (79AP838).

(663)   (664)

B. AZEPINO[2,1-b]QUINAZOLINES

Synthesis of the title compounds (666) from quinazoline precursors was reported (76AP542), according to which 2-(5-hydroxypentyl)quinazolin-4(3H)-one (665) was thermally cyclodehydrated. The alternative cyclization product was not obtained.

(665)   (666)

Synthesis through formation of the pyrimidine ring was accomplished by reacting anthranilic acid or its derivatives with a properly functionalized azepine. Thus, cyclocondensation of an anthranilic acid derivative with caprolactam [68JCS(C)1722, 68ZOB2051; 76JGS1564; 77JAP(K)7777093, 77KPS544] or caprolactim esters 667 [59LA166; 60GEP1088968; 68ZOB2030; 76JCS(P1)2182; 80MI3; 85IJC(B)336; 86JHC53] gave 668.

Cyclization of 1-hydroxymethylcaprolactam and acetanilide in the presence of concentrated sulfuric acid also gave the title compounds (75KPS809). The pyrimidine ring of azepino[2,1-*b*]quinazolines (**670**) were also formed when 1-(2-aminobenzyl)hexahydroazepines (**669**) were oxidatively cyclized with manganese dioxide [68JCS(C)1722] or with Hg(II)-EDTA (70TL997; 73AP541; 79CZ266; 80M627, 80PHA389).

(667)   (668)

(669)   (670)

C. Azepino[1,2-*c*]quinazolines

The azepine ring of **671** was constructed through the reaction of 4-methylquinazoline with two molar equivalents of dimethyl acetylenedicarboxylate [68JCS(C)926]. Some azepino-quinazolines were shown to possess antidepressant, antiinflammatory, and hypotensive activities [77JAP(K)7777093].

(671)

XXI. Condensed Diazepino-quinazolines

A. QUINAZOLINO-1,2-BENZODIAZEPINES

Quinazolino[3,2-b]1,2-benzodiazepines

The two isomeric 13,14-dihydro-2- and 4-methoxyquinazolino[3,2-*b*]1,-2-benzodiazepin-7(5*H*)-ones (**674** and **675**, respectively) were obtained by lead tetraacetate oxidation of 3-amino-2-[2-(3-methoxyphenyl)ethyl] quinazolin-4(3*H*)-one (**672**) [81CC160; 82JCS(P1)2407; 85JCS(P1)335]. The nitrene intermediate (**673**) underwent intramolecular trapping to **674** and **675** or loss of nitrogen to give the deaminated product **676.** The oxidation was found to be sensitive to the orientation of the methoxy group, since the *para*-isomer of **673** gave only deaminated product on oxidation.

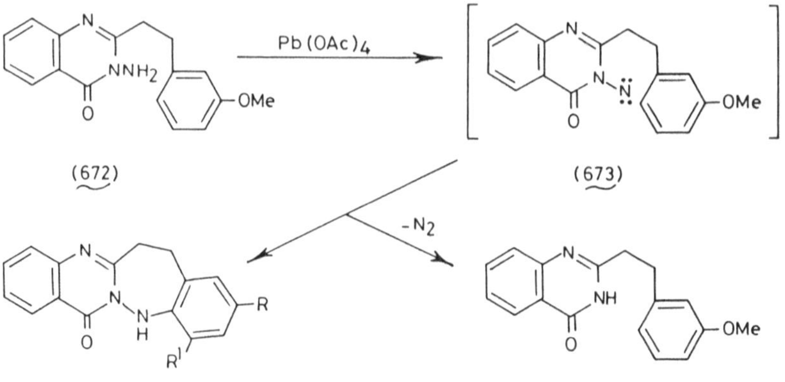

(672)

Pb(OAc)₄ →

(673)

-N₂

(674), R=OMe, R¹=H
(675), R=H, R¹ = OMe

(676)

B. 1,3-DIAZEPINO-QUINAZOLINES

1. *1,3-Diazepino[2,1-b]quinazolines*

Compounds belonging to the title compounds (**678**) were prepared by cyclodehydration of 1-substituted-2-(4-hydroxybutyl)aminoquinazolin-4-ones (**677**) (73SZP532068). Cyclocondensation of the isatoic anhydride with the 2-methylthio-1,3-diazepine **679** gave the 1,3-diazepino[2,1-*b*] quinazoline **680** [70GEP(O)2025248].

(677)   (678)

(679)   (680)

2. *1,3-Diazepino[1,2-c]quinazolines*

Reaction of trithioisatoic anhydride with 1,4-diaminobutane gave the title compound **681** (73ZC428). 1,3-Diazepino-quinazolines possess tranquilizing and bronchodilating properties (73SZP532068).

(681)

C. 1,4-DIAZEPINO QUINAZOLINES

1,4-Diazepino[2,1-b]quinazolines

Members of the title compounds (**683**) were prepared by cyclocondensation of anthranilic acid with the 2-methylthio-1,4-diazepines (**682**) (89MI3).

(682)   (683)

D. QUINAZOLINO-1,4-BENZODIAZEPINES

1. Quinazolino[3,2-a]1,4-benodiazepines

Compounds possessing this structure were prepared by thermal cyclo-condensation of anthranilic acid and the 1,4-benzodiazepine-2-thione (**684**) (77JHC1191). Synthesis of the naturally occuring cholecystokinin antagonist (**687**) followed the same approach; methyl anthranilate was fused with the 2-methylthio-1,4-benzodiazepin-5-one (**686**) to give **687** (85USP-4559338; 86USP4576750, 86USP4594191; 87JOC1644).

Cyclization of the 2-halomethyl-3-[2-(methoxycarbonyl)phenyl]quin-azolin-4(3*H*)-one (**688**) with ammonia or primary amines led to the formation of 6-substituted-6,7-dihydroquinazolino[3,2-*a*]1,4-benzodiaze-pine-5,13(5*H*,13*H*)-diones (**689**) [77JHC1191; 79GEP(O)2758875].

2. *Quinazolino[3,2-d]1,4-benzodiazepines*

Thermal cyclization of the benzoxazinone derivative **690** gave the quinazolino[3,2-*d*]1,4-benzodiazepine-6,9(5*H*,7*H*)-dione (**691**) (85H273, 85USP4554272).

Quinazolino-1,4-benzodiazepines are useful in the treatment of some central nervous system disorders [79GEP(O)2758875] and showed cholecystokinin antagonist activity (85USP4554272, 85USP4559338).

(690)   (691)

XXII. Condensed Oxazepino-quinazolines

1,4-OXAZEPINO-QUINAZOLINES

1,4-Oxazepino[5,4-b]quinazolines

Dehydrative cyclization of 3-(2-hydroxyethyl)-1-methyl-4-oxo-2-quinazolinacetic acid (**692**) yielded the 1,4-oxazepino[5,4-*d*]quinazoline (**693**) (80JHC1163).

(692)   (693)

XXIII. Condensed Thiazepino-quinazolines

QUINAZOLINO-1,5-BENZOTHIAZEPINES

Quinazolino[2,3-a]1,5-benzothiazepines

6,7-Dihydro-13H-quinazolino[2,3-a]1,5-benzothiazepin-13-one   (**695**) was synthesized by cyclocondensation of anthranilic acid with the 4-methylthio-1,5-benzothiazepine (**694**) (88JHC1399).

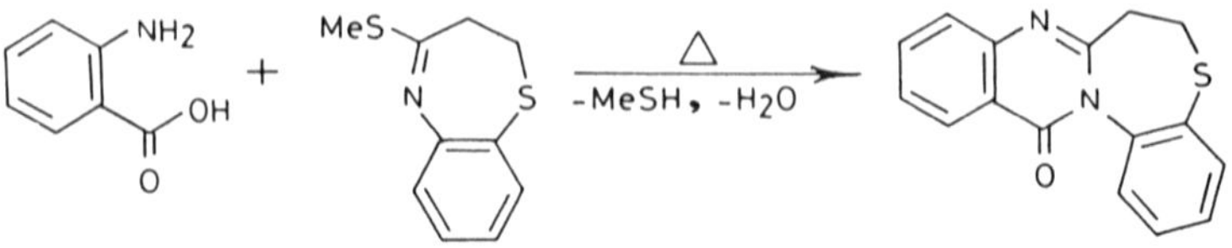

(694)   (695)

XXIV. Condensed Triazepino-quinazolines

QUINAZOLINO-1,3,4-BENZOTRIAZEPINES

Quinazolino[3,2-d]1,3,4-benzotriazepines

Condensative cyclization of 3-amino-3-(2-aminophenyl)quinazolin-4-one (**696**) with aromatic aldehydes gave 6-aryl-6,7-dihydroquinazolino [3,2-d]1,3,4-benzotriazepin-9(5H)-ones (**616**) [87IJC(B)983].

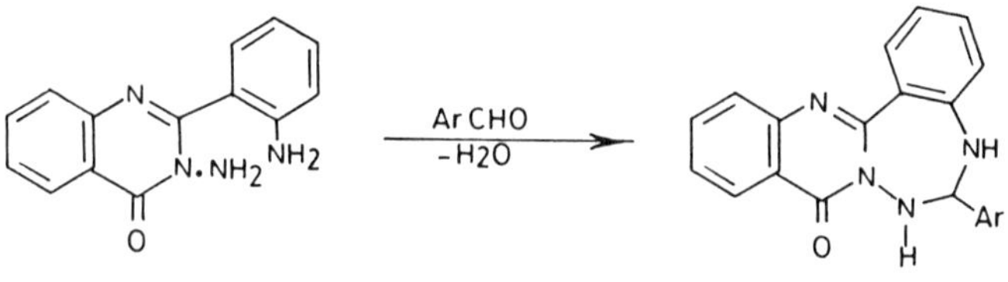

(696)   (616)

XXV. Condensed Tetrazepino-quinazolines

A. 1,2,4,5-Tetrazepino[3,2-*b*]quinazolines

Members of this ring system were prepared by cyclocondensation of 3-amino-2-hydrazinoquinazolin-4-ones **(391)** with benzil to afford the 3,4-diphenyl-1,2,4,5-tetrazepino[3,2-*b*]quinazolin-7(1*H*)-ones **(697)** (78JIC-928; 81H621; 83JIC1071).

(391)   (697)

B. 1,2,4,6-Tetrazepino[1,7-*c*]quinazolines

Cyclization of thiourea derivative **698** by heating with thiourea yielded the title compound **699** (88RRC981).

(698)   (699)

XXVI. Condensed Diazocino-quinazolines

A. 1,4-Diazocino[8,1-*b*]quinazolines

The pyrimidine ring of the title compounds **(701)** were formed upon condensative cyclization of anthranilic acids with the 1-benzyloxy-carbonyl-5-ethoxy-1,4-diazocine **(700)**, followed by removal of the *N*-benzyloxycarbonyl group (66USP3280117).

R—[benzene ring]—NH2 / —OH (C=O) + EtO—[8-membered ring with N, N·C.OBzl, C=O] →[1-MeOH, Δ 2-HBz,AcOH]→ R—[quinazoline fused 8-membered ring with N, N, NH, C=O]

(700)   (701)

B. QUINAZOLINO[3,2-b]1,5-DIBENZODIAZOCINES

Intermolecular cyclocondensation of anthranilic acid by heating with polyphosphoric acid gave a mixture of products from which the trimer (702) having the title structure was isolated (84MI6).

3 [anthranilic acid: benzene ring—NH2, —OH (C=O)] →[PPA, Δ]→ [polycyclic structure]

(702)

XXVII. Condensed Oxazocino-quinazolines

1,4-OXAZOCINO[5,4-b]QUINAZOLINES

Examples of the title compounds (704) were obtained by reacting anthranilic acids with the 5-ethoxy-1,4-oxazocine (703) (66USP3280117).

R—[benzene ring]—NH2 / —OH (C=O) + EtO—[8-membered ring with N, O] →[MeOH, Δ -EtOH, -H2O]→ R—[quinazoline fused 8-membered ring with N, N, O, C=O]

(703)   (704)

XXVIII.　Condensed Thiazocino-quinazolines

1,4-THIAZOCINO[5,4-*b*]QUINAZOLINES

In a synthesis that is similar to their oxygen analogues, the title compounds **706** were prepared by cyclocondensation of anthranilic acids with the 5-ethoxy-1,4-thiazocine (**705**) (66USP3280117).

(705)　　　　　　　　　　　　　　　　　(706)

XXIX.　Conclusion

Taking the number of references as a measure of how extensively the synthesis of the title compounds were studied, one may conclude that great efforts have been directed towards this goal. This may plausibly be rationalized in terms of academic interests as well as diverse biological activities attributed to these compounds. Nevertheless, much remains to be accomplished regarding the synthesis of many of the possible, but unprepared, heterocyclo[n,m-*a,b*, or *c*]quinazolines.

ACKNOWLEDGMENT

The authors would like to thank Dr. B. A. Silwanis for the computer work and proof reading of the manuscript and Mr. A. E. Z. Solaiman for the art work.

References

24LA311　　　　O. Seide, *Justus Liebigs Ann. Chem.* **440,** 311 (1924).
27JCS1708　　　Y. Asahina, R. Helmuth, F. Manske, and R. Robinson, *J. Chem. Soc.,* 1708 (1927).
29JIC723　　　　J. S. Aggarwal and J. N. Ray, *J. Indian Chem. Soc.* **6,** 723 (1929).
30JIC899　　　　T. N. Ghosh and M. V. Betrabet, *J. Indian Chem. Soc.* **7,** 899 (1930).
31JCS2840　　　P. K. Bose and D. C. Sen, *J. Chem. Soc.,* 2840 (1931).
31LA284　　　　A. Binz and C. Räth, *Justus Liebigs Ann. Chem.* **486,** 284 (1931).
33JPR9　　　　　R. Stolle and F. Hanusch, *J. Prakt. Chem.* **136,** 9 (1933).
34JIC463　　　　P. K. Bose and K. B. Pathak, *J. Indian Chem. Soc,* **11,** 463 (1934).
35CB(B)497　　　E. Späth, F. Kuffner, and N. Platzer, *Chem. Ber.* **68B,** 497 (1935).

35CB(B)699	E. Späth, F. Kuffner, and N. Platzer, *Chem. Ber.* **68B**, 699 (1935).
35CB(B)2221	E. Späth and N. Platzer, *Chem. Ber.* **68B**, 2221 (1935).
35JA921	W. E. Hanford and R. Adams, *J. Am. Chem. Soc.* **57**, 921 (1935).
35JA951	R. C. Morris, W. E. Hanford, and R. Adams, *J. Am. Chem. Soc.* **57**, 951 (1935).
35JCS1277	H. R. Juneja, K. S. Narang, and J. N. Ray, *J. Chem. Soc.*, 1277 (1935).
35MI1	K. S. Narang and J. N. Ray, *Curr. Sci.* **3**, 552 (1935) [*CA* **29**, 6238 (1935)].
36CB(B)2052	E. Späth, F. Kuffner, and J. Lintner, *Chem. Ber.* **69B**, 2052 (1936).
36LA1	C. Schöpf and F. Oechler, *Justus Liebigs Ann. Chem.* **523**, 1 (1936).
37JGU2318	O. A. Zeide and G. V. Chelintsev, *J. Gen. Chem. USSR (Engl. Transl.)* **7**, 2318 (1937) [*CA* **32**, 572 (1938)].
37JIC411	T. N. Ghosh, *J. Indian Chem. Soc.* **14**, 411 (1937).
38CB(B)1657	E. Späth and F. Kiffner, *Chem. Ber.* **71B**, 1657 (1938).
45JCS927	V. A. Petrow, *J. Chem. Soc.* 927 (1945).
47JCS726	F. E. King and T. J. King, *J. Chem. Soc.*, 726 (1947).
52MI1	M. C. Khosla, O. P. Vig, I. S. Gupta, and K. S. Narang, *Sci. Cult.* **18**, 43 (1952) [*CA* **47**, 10541 (1953)].
53JA712	L. Katz, *J. Am. Chem. Soc.* **75**, 712 (1953).
53JOC1380	L. Katz, L. S. Karger, W. Schroeder, and M. S. Cohen, *J. Org. Chem.* **18**, 1380 (1953).
53MI1	M. C. Khosla, O. P. Vig, I. S. Gupta, and K. S. Narang, *J. Sci. Ind. Res. Sect. B.* **12**, 466 (1953) [*CA* **49**, 1059 (1955)].
53MI2	M. S. Dhatt and K. S. Narang, *Res. Bull. East Panjab Univ.* **36**, 139 (1953) [*CA* **49**, 4660 (1955)].
54JCS2354	E. F. N. Stephenson, *J. Chem. Soc.*, 2354 (1954).
54JIC848	D. S. Bariana, M. S. Dhatt, H. S. Sachdev, and K. S. Narang, *J. Indian Chem. Soc.* **31**, 848 (1954).
54JOC699	M. L. Sherrill, E. Ortelt, S. Duckworth, and Z. Budenstein, *J. Org. Chem.* **59**, 699 (1954).
54JPS538	J. Sam and J. N. Plampin, *J. Pharm. Sci.* **53**, 538 (1954).
55G1210	S. Carboni and D. Segnini, *Gazz. Chim. Ital.* **85**, 1210 (1955).
55JCS2390	P. Sykes, *J. Chem. Soc.*, 2390 (1955).
55JIC589	G. M. Sharma, I. S. Gupta, and K. S. Narang, *J. Indian Chem. Soc.* **32**, 589 (1955).
55JIC631	H. S. Sachdev and K. S. Narang, *J. Indian Chem. Soc.* **32**, 631 (1955).
55JIC644	D. S. Bariana, H. S. Sachdev, and K. S Narang, *J. Indian Chem. Soc.* **32**, 647 (1955).
55JIC647	D. S. Bariana, H. S. Sachdev, and K. S. Narang, *J. Indian Chem. Soc.* **32**, 644 (1955).
55JOC302	M. S. Dhatt and K. S. Narang, *J. Org. Chem.* **20**, 302 (1955).
55MI1	S. Carboni, *Atti Soc. Toscana Sci. Nat. Pisa Mem.*, *Ser. A* **62**, 261 (1955) [*CA* **50**, 16767 (1956)].
55MI2	M. Indravati, *J. Madras Univ.*, *Sect. B* **25B**, 125 (1955) [*CA* **50**, 11350 (1956)].
56JCS1319	A. K. Kiang, F. G. Mann, A. F. Prior, and A. Topham, *J. Chem. Soc.*, 1319 (1956).
56JCS4173	T. Stephen and H. Stephen, *J. Chem. Soc.*, 4173 (1956).

56JCS4178	T. Stephen and H. Stephen, *J. Chem. Soc.*, 4178 (1956).
56JCS4694	T. Stephen and H. Stephen, *J. Chem. Soc.*, 4694 (1956).
56MI1	G. M. Sharma, H. S. Sachdev, and K. S. Narang, *J. Sci. Ind. Res., Sect. B* **15**, 687 (1956) [*CA* **51**, 8105 (1957)].
57CLY2122	L. Macholan, *Chem. Listy* **51**, 2122 (1957) [*CA* **52**, 5430 (1958)].
57G1191	V. Evdokimov, *Gazz. Chim. Ital.* **87**, 1191 (1957).
57MI1	K. S. Dhami, H. S. Sachev, and K. S. Narang, *J. Sci. Ind. Res., Sect. B* **16**, 311 (1957) [*CA* **52**, 5421 (1958)].
58JA1168	P. L. Southwick and J. Casanova, Jr., *J. Am. Chem. Soc.* **80**, 1168 (1958).
59JCS1512	K. Butler and M. W. Partridge, *J. Chem. Soc.*, 1512 (1959).
59LA166	S. Petersen and E. Tietze, *Justus Liebigs Ann. Chem.* **623**, 166 (1959).
59MI1	V. Evdokimoff, *Rend. Ist. Super. Sanita* (*Engl. Transl.*) **22**, 407 (1959) [*CA* **54**, 1527 (1960)].
59MIP1	K. S. Narang, H. S. Sachdev, and K. S. Dhami, Indian Pat. 58,902 (1959) [*CA* **53**, 20092 (1959)].
59NKZ1181	M. Yanai and T. Kuraishi, *Nippon Kagaku Zasshi* **80**, 1181 (1959) [*CA* **55**, 4515 (1961)].
60GEP1088968	S. Petersen and E. Tietze, Ger. Pat. 1,088,968 (1960) [*CA* **55**, 27381 (1961)].
60GEP1111505	K. H. Menzel, R. Puetter, and G. Wolfrum, Ger. Pat. 1,111,505 (1960) [*CA* **56**, 4904 (1962)].
60JCS3551	R. J. Grout and M. W. Partridge, *J. Chem. Soc.*, 3551 (1960).
60JCS4970	K. Butler, M. W. Partridge, and J. A. Waite, *J. Chem. Soc.*, 4970 (1960).
60JOC853	B. A. Carpenter, J. E. McCarty, and C. A. VanderWerf, *J. Org. Chem.* **25**, 853 (1960).
60LA159	G. Satzinger, *Justus Liebigs Ann. Chem.* **638**, 159 (1960).
60MI1	H. S. Sachdev, N. K. Ralhan, M. S. Atwal, H. S. Garg, and K. S. Narang, *J. Sci. Ind. Res., Sect. B* **19**, 217 (1960) [*CA* **55**, 2670 (1961)].
60MI2	H. S. Sachdev, K. S. Dhami, and K. S. Narang, *J. Sci. Ind. Res., Sect. C* **19**, 11 (1960) [*CA* **54**, 22668 (1960)].
60MI3	H. S. Sachdev and N. K. Ralhan, *J. Sci. Ind. Res., Sect. C* **19**, 109 (1960) [*CA* **55**, 1627 (1961)].
61AP556	H. Boehme and H. Boeing, *Arch. Pharm.* (*Weinheim, Ger.*) **294**, 556 (1961).
61GEP1120455	K. H. Menzel, R. Puetter, and G. Wolfrum, Ger. Pat. 1,120,455 (1961) [*CA* **57**, 4675 (1962)].
61MI1	G. M. Sharma and S. G. Singh, *J. Sci. Ind. Res., Sect. C* **20**, 178 (1961) [*CA* **57**, 12482 (1962)].
61T(14)304	H. S. Sachdev, K. S. Dhami, and M. S. Atwal, *Tetrahedron* **14**, 304 (1961)
61T(15)53	G. M. Sharma, H. S. Sachdev, N. K. Ralhan, H. Singh, G. S. Sandu, K. Ghandi, and K. S. Narang, *Tetrahedron* **15**, 53 (1961).
61ZC224	H. Beyer and C. E. Voelker, *Z. Chem.* **1**, 224 (1961).
62AG249	G. De Stevens and H. Blatter, *Angew. Chem.* **74**, 249 (1962).
62CB2182	G. G. Munoz and R. Madronero, *Chem. Ber.* **95**, 2182 (1962).

62GEP1139123 R. Puetter, G. Wolfrum, and K. H. Menzel, Ger. Pat. 1,139,123 (1962) [CA **58**, 9095 (1963)].

62JCS945 M. Davis and F. G. Mann, *J. Chem. Soc.*, 945 (1962).

62JOC3701 J. C. Howard and G. Klein, *J. Org. Chem.* **27**, 3701 (1962).

62JOC4061 N. K. Rathan, K. K. Narula, and H. S. Sachdev, *J. Org. Chem.* **27**, 4061 (1962).

62T1019 G. M. Sharma, K. K. Soni, and K. S. Narang, *Tetrahedron* **18**, 1019 (1962).

63ACH457 K. Lempert and G. Doleschall, *Acta Chim. Acad. Sci. Hung.* **37**, 457 (1963) [CA **59**, 12809 (1963)].

63BSB365 A. L. L. Poot, J. F. Willems, and F. C. Heugebaert, *Bull. Soc. Chim. Belg.* **72**, 365 (1963).

63CB1271 K. Lempert and G. Doleschall, *Chem. Ber.* **96**, 1271 (1963).

63CI(M)709 A. Alemagna and T. Bacchetti, *Chim. Ind.* (Milan) **45**, 709 (1963).

63IJC318 A. S. Narang and K. S. Narang, *Indian J. Chem.* **1**, 318 (1963).

63JIC545 H. Singh and K. S. Narang, *J. Indian Chem. Soc.* **40**, 545 (1963).

63JMC450 K. S. Dhami, S. S. Arora, and K. S. Narang, *J. Med. Chem.* **6**, 450 (1963).

63N732 G. S. Sidhu, G. Thyagarajan, and N. Rao, *Naturwissenschaften* **50**, 732 (1963).

63T1587 M. S. Gibson, *Tetrahedron* **19**, 1587 (1963).

63ZOB2334 I. Ya. Postovskii and I. N. Goncharova, *Zh. Obshch. Khim.* **33**, 2334 (1963) [CA **59**, 13987 (1963)].

63ZOB2475 I. N. Goncharova and I. Ya. Postovskii, *Zh. Obshch. Khim.* **33**, 2475 (1963) [CA **60**, 1743 (1964)].

64CB390 H. Beyer and C. E. Voelcker, *Chem. Ber.* **97**, 390 (1964).

64IJC285 M. S. Puar, H. S. Sachdev, and N. K. Ralhan, *Indian J. Chem.* **2**, 285 (1964).

64JCS(C)3670 M. W. Partridge, S. A. Slorach, and H. J. Vipond, *J. Chem. Soc. C*, 3670 (1964).

64JHC42 T. Kuraishi and R. N. Castle, *J. Heterocycl. Chem.* **1**, 42 (1964).

64JIC591 H. S. Suri, G. M. Sharma, and K. S. Narang, *J. Indian Chem. Soc.* **41**, 591 (1964).

64JIC715 H. Singh, K. S. Bhandari, and K. S. Narang, *J. Indian Chem. Soc.* **41**, 715 (1964).

64JIC855 H. Singh, N. Kaur, and K. S. Narang, *J. Indian Chem. Soc.* **41**, 855 (1964).

64MI1 Z. Kolodynska and S. Biniecki, *Acta Pol. Pharm.* (*Engl. Transl.*) **21**, 225 (1964). [CA **62**, 13144 (1965)].

64ZOB1745 N. N. Vereshchagina and I. Ya. Postovskii, *Zh. Obshch. Khim.* **34**, 1745 (1964) [CA **61**, 8307 (1964)].

65ACH357 G. Doleschall and K. Lempert, *Acta Chim. Acad. Sci. Hung.* **45**, 357 (1965).

65IJC44 H. K. Gakhar, *Indian J. Chem.* **3**, 44 (1965).

65IJC284 A. P. Bhaduri, N. M. Khanna, and M. L. Dhar, *Indian J. Chem.* **3**, 284 (1965).

65JIC155 A. S. Narang and S. Singh, *J. Indian Chem. Soc.* **42**, 155 (1965).

65JIC220 A. S. Narang, S. Singh, K. K. Sharma, and K. S. Narang, J. Indian Chem. Soc. **42**, 220 (1965).

65NEP6409191   A. Wander A. G., Neth. Pat. 6,409,191 (1965) [*CA* **63**, 2948 (1965)].
65NEP6414717   Laboratoires U.P.S.A., Neth. Pat. 6,414,717 (1965) [*CA* **64**, 712 (1966)].
65ZOR1154   N. N. Vereshchagina, I. Ya. Postovskii, and S. L. Mertsalov, *Zh. Org. Khim.* **1**, 1154 (1965) [*CA* **63**, 13256 (1965)].
66ACH77   G. Doleschall and K. Lempert, *Acta Chim. Acad. Sci. Hung.* **48**, 77 (1966).
66CB1532   G. Doleschall and K. Lempert, *Chem. Ber.* **99**, 1532 (1966).
66IJC527   H. Singh, M. S. Saini, and K. S. Narang, *Indian J. Chem.* **4**, 527 (1966).
66JCS(C)2190   M. J. Kort and M. Lamchen, *J. Chem. Soc. C,* 2190 (1966).
66JCS(C)2290   G. Tennant, *J. Chem. Soc. C,* 2290 (1966).
66KGS130   I. Ya. Postovskii, N. N. Vereschagina, and S. L. Mertsalov, *Khim. Geterotsikl. Soedin.,* 130 (1966) [*CA* **65**,710 (1966)].
66T(S)227   C. A. R. Hurt and H. Stephen, *Tetrahedron*, Suppl. **7**, 227 (1966).
66TL2609   G. F. Field, W. J. Zally, and L. H. Sternbach, *Tetrahedron Lett.,* 2609 (196).
66USP3271400   J. Bernstein and E. R. Spitzmiller, U.S. Pat. 3,271,400 (1966) [*CA* **65**, 18601 (1966)].
66USP3280117   R. G. Griot, U.S. Pat. 3,280,117 (1966) [*CA* **66**, 2584 (1967)].
67AG(E)878   E. C. Taylor and F. Yoneda, *Angew. Chem., Int. Ed. Engl.* **6**, 878 (1967).
67CB875   C. E. Voelcker, J. Marth, and H. Beyer, *Chem. Ber.* **100**, 875 (1967).
67CR(C)265   C. Foulcher, J. Barrans, and H. Goncalves, *C. R. Hebd. Seances Acad. Sci., Ser. C* **265**, 407 (1967) [*CA* **68**, 12948 (1968)].
67HCA1440   E. Cherbuliez, B. Willhalm, O. Espejo, S. Jaccard, and J. Rabinowitz, *Helv. Chim. Acta* **50**, 1440 (1967) [*CA* **67**, 64354 (1967)].
67KGS1096   N. N. Vereshchagina, I. Ya. Postovskii, and S. L. Mertsalov, *Khim. Geterotsikl. Soedin.,* 1096 (1967) [*CA* **69**, 59188 (1968)].
67MI1   G. Doleschall and K. Lempert, *Kem. Kozl.* **27**, 73 (1967) [*CA* **67**, 54051 (1967)].
67TL2701   A. M. Shkrob, Yu. I. Krylova, V. K. Antonov, and M. M. Shemyakin, *Tetrahedron Lett.,* 2701 (1967).
67USP3297696   H. Ott, U.S. Pat. 3,297,696 (1967) [*CA* **66**, 65505 (1967)].
67USP3309369   O. Schindler, U.S. Pat. 3,309,369 (1967) [*CA* **67**, 73617 (1967)].
67USP3313815   R. T. Wolfe and A. R. Surrey, U.S. Pat. 3,313,815 (1967) [*CA* **67**, 64428 (1967)].
67USP3329679   T. S. Sulkowski and S. J. Childress, U.S. Pat. 3,329, 679 (1967) [*CA* **68**, 49646 (1968)].
67ZC456   G. Westphal and H. H. Stroh, *Z. Chem.* **7**, 456 (1967).
68CB2106   W. Ried and J. Valentin, *Chem. Ber.* **101**, 2106 (1968).
68CPB972   M. Yanai, T. Kinoshita, and S. Takeda, *Chem. Pharm. Bull.* **16**, 972 (1968).
68IJC758   C. M. Gupta, A. P. Bhaduri, and N. M. Khanna, *Indian J. Chem.* **6**, 758 (1968).
68JCS(C)926   R. M. Acheson, M. W. Foxton, and J. K. Stubbs, *J. Chem. Soc. C,* 926 (1968).
68JCS(C)1722   O. M. Cohn and H. Suschitzky, *J. Chem. Soc. C,* 1722 (1968).
68JHC179   S. C. Bell and G. Conklin, *J. Heterocycl. Chem.* **5**, 179 (1968).

68JHC185   S. C. Bell and P. H. L. Wei, *J. Heterocycl. Chem.* **5**, 185 (1968).
68JOC1359   E. G. Garcia, J. G. Riley, and R. I. Fryer, *J. Org. Chem.* **33**, 1359
                (1968).
68JOC1719   E. Taylor and Y. Shvo, *J. Org. Chem.* **33**, 1719 (1968).
68JOC2402   P. Aeberli and W. J. Houlihan, *J. Org. Chem.* **33**, 2402 (1968).
68JPS1445   J. Sam, J. L. Valentine, and C. W. Richmond, *J. Pharm. Sci.* **57**,
                1445 (1968).
68M1499   E. Ziegler, W. Steiger, and T. Kappe, *Monatsh. Chem.* **99**, 1499
                (1968).
68P307   A. Chatterjee and M. Ganguly, *Phytochemistry* **7**, 307 (1968).
68SZP452537   O. Schindler, Swiss Pat. 452,537 (1968) [*CA* **69**, 96766 (1968)].
68USP3375250   F. K. Kirchner and A. W. W. Zalay, U.S. Pat. 3,375,250 (1968) [*CA*
                69, 52170 (1968)].
68ZC103   C. E. Voelcker, M. Schoenfeld, and H. Beyer, *Z. Chem.* **8**, 103
                (1968).
68ZOB2030   A. M. Shkrob, Yu. I. Krylova, V. K. Antonov, and M. M. Shem-
                yakin, *Zh. Obshch. Khim.* **38**, 2030 (1968) [*CA* **70**, 47397
                (1969)].
68ZOB2051   A. M. Shkrob, Yu. I. Krylova, V. K. Antonov, and M. M. Shemya-
                kin, *Zh. Obshch. Khim.* **38**, 2051 (1968) [*CA* **70**, 77902 (1969)].
69CB1480   L. Cpauano and W. Ebner, *Chem. Ber.* **103**, 1480 (1969).
69CI(L)417   D. C. Bishop and M. J. Tucker, *Chem. Ind.* (*London*), 417 (1969).
69G715   F. Gatta and V. R. Landi, *Gazz. Chim. Ital.* **99**, 715 (1969).
69IJC444   A. N. Kaushal, A. P. Taneja, and K. S. Narang, *Indian J. Chem.* **7**,
                444 (1969).
69IJC765   H. Singh, S. L. Jain, V. K. Sharma, and K. S. Narang, *Indian J.
                Chem.* **7**, 765 (1969).
69IJC881   A. Singh, A. S. Uppal, and K. S. Narang, *Indian J. Chem.* **7**, 881
                (1969).
69IJC1191   A. S. Narang, A. N. Kaushal, H. Singh, and K. S. Narang, *Indian
                J. Chem.* **7**, 1191 (1969).
69JCS(D)423   D. R. Sutherland and G. Tennant, *J. Chem. Soc. D*, 423 (1969).
69JCS(D)1040   J. T. Shaw and J. Ballentine, *J. Chem. Soc. D*, 1040 (1969).
69JHC947   J. B. Wright, *J. Heterocycl. Chem.* **6**, 947 (1969).
69JOC2123   M. Kurihara, *J. Org. Chem.* **34**, 2123 (1969).
69MI1   E. Ajello, *Atti Accad. Sci. Lett. Arti Palermo, Parte 1* **30**, 237
                (1969–1970) [*CA* **77**, 152110 (1972)].
69USP3441566   W. J. Houlihan, U.S. Pat. 3,441,566 (1969) [*CA* **71**, 70630 (1969)].
69USP3459754   S. C. Bell, U.S. Pat. 3, 459, 754 (1969) [*CA* **71**, 81407 (1969)].
69USP3471497   S. C. Bell and P. H. L. Wei, U.S. Pat. 3,471,497 (1969) [*CA* **72**,
                31836 (1970)].
69USP3475432   S. C. Bell and P. H. L. Wei, U.S. Pat. 3,475,432 (1969) [*CA* **71**,
                124487 (1969)].
70AJC781   S. Beveridge and J. L. Huppatz, *Aust. J. Chem.* **23**, 781 (1970).
70BSF2615   H. Goncalves, C. Foulcher, and F. Mathis, *Bull. Soc. Chim. Fr.*,
                2615 (1970).
70CB82   L. Capuano, W. Ebner, and J. Schrepfer, *Chem. Ber.* **103**, 82
                (1970).
70GEP(O)1932885 M. H. Sherlock, Ger. Pat. Offen. 1,932,885 (1970) [*CA* **72**, 132770
                (1970)].

70GEP(O)1946188   G. E. Hardtmann and H. Ott, Ger. Pat. Offen. 1,946,188 (1970) [*CA* **72,** 132774 (1970)].

70GEP(O)2025248   G. E. Hardtmann, Ger. Pat. Offen. 2,025,248 (1970) [*CA* **74,** 42373 (1971)].

70IJC389   S. K. Modi, H. K. Gakhar, and K. S. Narang, *Indian J. Chem.* **8,** 389 (1970).

70IJC710   S. K. Modi, V. Kumar, and K. S. Narang, *Indian J. Chem.* **8,** 710 (1970).

70IJC793   S. K. Modi, H. K. Gakhar, and K. S. Narang, *Indian J. Chem.* **8,** 793 (1970).

70IJC1065   A. S. Narang, S. L. Jain, and K. S. Narang, *Indian J. Chem.* **8,** 1065 (1970).

70JIC758   B. D. Singh and D. N. Chaudhury, *J. Indian Chem. Soc.* **47,** 758 (1970).

70JOC3448   K. T. Potts and E. G. Brugel, *J. Org. Chem.* **35,** 3448 (1970).

70KGS100   I. Ya. Postovskii and B. V. Golomolzin, *Khim. Geterotsikl. Soedin.,* 100 (1970) [*CA* **72,** 121484 (1970)].

70KGS428   V. I. Shvedov, I. A. Kharizomenova, L. B. Altukhova, and A. N. Grinev, *Khim. Geterotsikl. Soedin.,* 428 (1970) [*CA* **73,** 25403 (1970)].

70KGS855   B. V. Golomolzin and I. Ya. Postovskii, *Khim. Geterotsikl. Soedin.,* 855 (1970) [*CA* **73,** 120586 (1970)].

70MI1   A. S. Narang, A. N. Kaushal, B. K. Bahl, and K. S. Narang, *Indian J. Appl. Chem.* **33,** 228 (1970).

70MI2   H. K. Gakhar, S. D. Sharma, and K. S. Narang, *Indian J. Appl. Chem.* **33,** 387 (1970).

70MI3   V. G. Zubenko and M. K. Mikhalevich, *Khim. Issled. Farm.* 26 (1970) [*CA* **76,** 46138 (1972)].

70MI4   D. Brutane, A. Ya. Strakov, and I. A. Strakova, *Latv. PSR Zinat. Akad. Vestis, Kim. Ser.,* 485 (1970) [*CA* **74,** 13089 (1971)].

70TL997   H. Moehrle and P. Gundlach, *Tetrahedron Lett.,* 997 (1970).

70TL3249   H. Moehrle and P. Gundlach, *Tetrahedron Lett.,* 3249 (1970.

70USP3497499   W. J. Houlihan and R. E. Manning, U.S. Pat. 3,497,499 (1970) [*CA* **72,** 121573 (1970)].

70USP3501473   S. C. Bell, U.S. Pat. 3,501,473 (1970) [*CA* **72,** 121569 (1970)].

70USP3505315   S. C. Bell, U.S. Pat. 3, 505, 315 (1970)[*CA* **72,** 132808 (1970)].

70USP3505333   S. C. Bell, U. S. Pat. 3.505,333 (1970) [*CA* **73,** 25507 (1970)].

70USP3506663   S. C. Bell, U. S. Pat. 3,506,663 (1970) [*CA* **73,** 25511 (1970)].

70USP3509147   W. J. Houlihan, U. S. Pat. 3,509,147 (1970) [*CA* **73,** 25510 (1970)].

70USP3531482   H. Ott, U.S. Pat. 3,531,482 (1970)[*CA* **74,** 31771 (1971)].

70YZ629   Y. Sato, T. Tanaka, and T. Nagasaki, *Yakugaku Zasshi* **90,** 629 (1970) [*CA* **73,** 25393 (1970)].

71BRP1242863   Aktieselskabet Pharmacia, Br. Pat. 1,242,863 (1971) [*CA* **75,** 129827 (1971)].

71CB2789   F. P. Woerner, H. Reimlinger, and R. Merenyi, *Chem. Ber.* **104,** 2789 (1971).

71GEP(O)2051961   R. L. Duncan, Jr., Ger. Pat. Offen. 2,051,961 (1971) [*CA* **75,** 36115 (1971)].

71GEP(O)2058185   W. H. W. Lunn, Ger. Pat. Offen. 2,058,185 (1971) [*CA* **75,** 63815 (1971)].

71IJC647   S. Singh, A. N. Kaushal, A. S. Narang, and K. S. Narang, *Indian J. Chem.* **9**, 647 (1971).
71JCS(C)2163   T. H. Brown and K. Bowden, *J. Chem. Soc. C*, 2163, (1971).
71JHC141   W. H. W. Lunn and R. W. Harper, *J. Heterocycl. Chem.* **8**, 141 (1971).
71JHC1071   T. Kametani, K. Nyu, T. Yamanaka, and S. Takano, *J. Heterocycl. Chem.* **8**, 1071 (1971).
71JIC395   B. Singh and D. N. Chaudhury, *J. Indian Chem. Soc.* **48**, 395 (1971).
71JIC443   B. Singh and D. N. Chaudhury, *J. Indian Chem. Soc.* **48**, 443 (1971).
71JIC743   B. D. Singh and S. K. P. Sinha, *J. Indian Chem.* **48**, 743 (1971).
71JIC953   H. K. Gakhar, G. S. Gill, and J. S. Multani, *J. Indian Chem. Soc.* **48**, 953 (1971).
71JIC989   S. K. P. Sinha, *J. Indian Chem. Soc.* **48**, 989 (1971).
71TL4387   T. Onaka, *Tetrahedron Lett.*, 4387 (1971).
71USP3595861   S. C. Bell and P. H. L. Wei, U.S. Pat. 3,595,861 (1971) [*CA* **75**, 98585 (1971)].
71USP3609139   W. J. Houlihan, U.S. Pat, 3,609,139 (1971) [*CA* **75**, 151851 (1971)].
71USP3621025   T. Y. W. Jen and B. Loev, U.S. Pat. 3,621,025 (1971) [*CA* **76**, 34288 (1972)].
72BRP1279842   P. D. Sorrentino, Br. Pat. 1,279,842 (1972) [*CA* **77**, 114442 (1972)].
72CI(L)255   H. Singh, S. Singh, and K. B. Lal, *Chem. Ind.* (*London*), 255 (1972) [*CA* **76**, 140707 (1972)].
72GEP(O)2043665   R. R. Schmidt, Ger. Pat. Offen. 2,043,665 (1972) [*CA* **77**, 34478 (1972)].
72GEP(O)2141616   S. Inaba, M. Yamamoto, K. Ishizumi, K. Mori, M. Koshiba, and H. Yamamoto, Ger. Pat. Offen. 2,141,616 (1972) [*CA* **76**, 140859 (1972)].
72GEP(O)2146076   F. G. Kathawala and G. E. Hardtmann, Ger. Pat. Offen. 2,146,076 (1972) [*CA* **77**, 48501 (1972)].
72IJC476   A. N. Kaushal, S. Singh, A. P. Taneja, and S. Narang, *Indian J. Chem.* **10**, 476 (1972).
72IJC605   S. K. Modi, S. Singh, and K. S. Narang, *Indian J. Chem.* **10**, 605 (1972).
72JAP(K)7242750   T. Kametani, Jpn. Kokai 72/42,750 (1972) [*CA* **78**, 72192 (1973)].
72JHC1227   J. B. Taylor, D. R. Harrison, and F. Freid, *J. Heterocycl. Chem.* **9**, 1227 (1972).
72JIC1185   M. P. Thakur and S. K. P. Sinha, *J. Indian Chem. Soc.* **49**, 1185 (1972).
72JMC727   T. Jen, P. Dienel, H. Bownan, J. Petta, A. Helt, and P. Loev, *J. Med. Chem.* **15**, 727 (1972).
72KGS1003   M. V. Loseva, B. M. Bolotin, and Yu. S. Ryabokobylko, *Khim. Geterotsikl. Soedin.*, 1003 (1972) [*CA* **77**, 139958 (1972)].
72KGS1341   M. V. Loseva and B. M. Bolotion, *Khim. Geterotiskl. Soedin.*, 1341 (1972) [*CA* **78**, 43392 (1973)].
72MI1   B. Stanovnik and M. Tisler, *Croat. Chem. Acta* **44**, 415 (1972) [*CA* **78**, 43405 (1973)].
72MI2   H. K. Gakhar and P. Singh, *Res. Bull. Panjab Univ., Sci.* **23**, 271 (1972) [*CA* **87**, 39403 (1977)].
72URP334219   V. V. Korshak, A. L. Rusanov, and Ts. G. Iremashvili, U.S.S.R. Pat. 334,219 (1972) [*CA* **77**, 48504 (1972)].

72USP3707468   S. C. Bell and G. T. Conklin, U.S. Pat. 3,707,468 (1972) [*CA* **78**, 72188 (1973)].

72YZ1184   T. Kametani, K. Nyu, and T. Yamanaka, *Yakugaku Zasshi* **92**, 1184 (1972) [*CA* **78**, 4208 (1973)].

72ZC289   S. Leistner, G. Wagner, and E. Iffland, *Z. Chem.* **12**, 289 (1972).

73AP541   H. Moehrle and P. Gundlach, *Arch. Pharm. (Weinheim, Ger.)* **306**, 541 (1973).

73GEP(O)2166380   S. Inaba, M. Yamamoto, K. Ishizumi, K. Mori, M. Koshiba, and H. Yamamoto, Ger. Pat. Offen. 2,166,380 (1973) [*CA* **80**, 27279 (1974)].

73GEP(O)2212371   P. Thieme, H. Koeing, and A. Amann, Ger. Pat. Offen. 2,212,371 (1973) [*CA* **79**, 137184 (1973)].

73GEP(O)2228259   P. Thieme, H. Koeing, and A. Amann, Ger. Pat. Offen. 2,228,259 (1973) [*CA* **80**, 83034 (1974)].

73GEP(O)2234174   G. E. Hardtmann, Ger. Pat. Offen. 2,234,174 (1973) [*CA* **78**, 111383 (1973)].

73GEP(O)2257376   G. E. Hardtmann, Ger. Pat. Offen. 2,257,376 (1973) [*CA* **79**, 42540 (1973)].

73GEP(O)2261095   G. E. Hardtmann and F. G. Kathawala, Ger. Pat. Offen. 2,261,095 (1973) [*CA* **79**, 66385 (1973)].

73GEP(O)2305172   P. H. Eerdekens, Ger. Pat. Offen. 2,305,172 (1973) [*CA* **79**, 137934 (1973)].

73GEP(O)2305575   W. N. Beverung, Jr. and R. A. Partyka, Ger. Pat. Offen. 2,305,575 (1973) [*CA* **79**, 115614 (1973)].

73GEP(O)2319851   G. E. Hardtmann, Ger. Pat. Offen. 2,319,851 (1973) [*CA* **80**, 27284 (1974)].

73IJC500   M. P. Thakur and S. K. P. Sinha, *Indian J. Chem.* **11**, 500 (1973).

73IJC532   R. S. Pandit and S. Seshadri, *Indian J. Chem.* **11**, 532 (1973).

73IJC750   H. Singh, K. B. Lal, and S. Singh, *Indian J. Chem.* **11**, 750 (1973).

73IJC959   H. Singh and K. B. Lal, *Indian J. Chem.* **11**, 959 (1973).

73JHC821   K. T. Potts, R. Dugas, and C. R. Surapaneni, *J. Heterocycl. Chem.* **10**, 821 (1973).

73JMC633   T. Jen, B. Dienel, F. Dowalo, H. VanHoeven, P. Bender, and B. Loev, *J. Med. Chem.* **16**, 633 (1973).

73JPR185   T. R. Vakula, V. R. Rao, and V. R. Srinivasan, *J. Prakt. Chem.* **315**, 185 (1973).

73KGS242   V. A. Chuiguk and V. V. Oksanich, *Khim. Geterotsilk. Soedin.*, 242 (1973) [*CA* **79**, 6745 (1973)].

73KGS1285   M. K. Pordeli, V. V. Oksanich, and V. A. Chuiguk, *Khim Geterotsikl. Soedin.*, 1285 (1973) [*CA* **80**, 3442 (1974)].

73MI1   G. Doleschall and K. Lempert, *Acta Chim. Acad. Sci. Hung.* **77**, 345 (1973) [*CA* **79**, 105206 (1973)].

73MI2   A. Sammour, T. Zimaity, and M. A. Abdo, *Egypt. J. Chem.* **16**, 215 (1973) [*CA* **81**, 152159 (1974)].

73MI3   S. Coffey, ed., "Rodd's Chemistry of Carbon Compounds," 2nd Ed., Vol. IVA, p. 562, Elsevier, Amsterdam.

73NKK1944   S. Kwon, F. Ikeda, and K. Isagawa, *Nippon Kagaku Kaishi*, 1944 (1973) [*CA* **80**, 37067 (1974)].

73S426   P. Thieme and H. Koening, *Synthesis*, 426 (1973).

73SZP532068   G. E. Hardtmann, Swiss Pat. 532,068 (1973) [*CA* **78**, 111345 (1973)].

73TL1417   T. Yamazaki and H. Schechter, *Tetrahedron Lett.*, 1417 (1973).
73TL1643   R. A. Bowie, D. A. Thomason, and J. A. J. Jarvis, *Tetrahedron Lett.*, 1643 (1973).
73URP405895   P. M. Kochergin, I. A. Mazur, and A. F. Vlasenko, U.S.S.R. Pat. 405,895 (1973) [*CA* **81**, 37566 (1974)].
73USP3745216   T. Y. Jen and B. Loev, U.S. Pat. 3,745,216 (1973) [*CA* **79**, 92273 (1973)].
73ZC428   S. Leistner and G. Wagner, *Z. Chem.* **13**, 428 (1973).
74CI(M)492   G. Pifferi, L. F. Zerilli, G. Tuan, and P. Consonni, *Chim. Ind. (Milan)* **56**, 492 (1974) [*CA* **82**, 31302 (1975)].
74CPB601   Y. Yamada, T. Oine, and I. Inoue, *Chem. Pharm. Bull.* **22**, 601 (1974).
74FES579   G. DeMartino, S. Massa, and G. Valitutti, *Farmaco, Ed. Sci.* **29**, 579 (1974) [*CA* **81**, 135867 (1974)].
74GEP(O)2252122 K. D. Kampe, Ger. Pat. Offen. 2,252,122 (1974) [*CA* **81**, 13543 (1974)].
74GEP(O)2402454 G. E. Hardtmann, Ger. Pat. Offen. 2,402,454 (1974) [*CA* **81**, 120678 (1974)].
74JAP(K)7431687 T. Oine, Y. Yamada, I. Inoue, and M. Miyoshi, Jpn. Kokai 74/31,687 (1974) [*CA* **81**, 63695 (1974)].
74JAP(K)7431697 T. Oine, Y. Yamada, I. Inoue, and M. Miyoshi, Jpn. Kokai 74/31,697 (1974) [*CA* **81**, 91564 (1974)].
74JCS(P1)534   D. R. Sutherland and G. Tennant. *J. C. S. Perkin 1*, 534 (1974).
74JHC737   A. J. Hubert, *J. Heterocycl. Chem.* **11**, 737 (1974).
74JHC747   D. L. Trepanier, S. Sunder, and W. H. Braun, *J. Heterocycl. Chem.* **11**, 747 (1974).
74JIC453   S. K. P. Sinha and M. P. Thakur, *J. Indian Chem. Soc.* **51**, 453 (1974).
74JIC457   S. K. P. Sinha and M. P. Thakur, *J. Indian Chem. Soc.* **51**, 457 (1974).
74JOC3508   E. Claudi, P. Franchetti, M. Grifantini, and S. Martelli, *J. Org. Chem.* **39**, 3508 (1974).
74JOC3599   G. E. Hardtmann and H. Ott, *J. Org. Chem.* **39**, 3599 (1974).
74JOC3828   S. C. Pakrashi and A. K. Chakravarty, *J. Org. Chem.* **39**, 3828 (1974).
74JPR943   J. Slouka and V. Bekarek, *J. Prakt. Chem.* **316**, 943 (1974).
74KGS1405   V. V. Korshak, A. L. Rusanov, S. N. Loent'eva, L. Kh. Plieva, Yu. E. Doroshenko, and I. Batirov, *Khim. Geterotsikl. Soedin.*, 1405 (1974) [*CA* **82**, 43314 (1975)].
74KPS681   Kh. Shakhidoyatov, A. Irisbaev, and Ch. Sh. Kadyrov, *Khim. Prir. Soedin.*, 681 (1974) [*CA* **82**, 86470 (1975)].
74SAP7302111   G. E. Hardtmann, S. Afr. Pat. 73/02,111 (1974) [*CA* **83**, 97352 (1975)].
74T3997   G. Doleschall and K. Lempert, *Tetrahedron* **30**, 3997 (1974).
74URP433149   P. M. Kochergin, I. A. Mazur, and R. S. Cinyak, U.S.S.R. Pat. 433,149 (1974) [*CA* **81**, 105555 (1974)].
74URP445665   P. M. Kochergin, I. A. Mazur, and A. F. Vlasenko, U.S.S.R. Pat. 445,665 (1974) [*CA* **82**, 112099 (1975)].
74URP452563   V. I. Shvedov, A. N. Grinev, G. N. Kurilo, and A. A. Cherkasova, U.S.S.R. Pat. 452,563 (1974) [*CA* **82**, 140170 (1975)].

74USP3835137   E. R. Wagner, U.S. Pat. 3,835,137 (1974) [*CA* **81**, 152258 (1974)].

74USP3835138   E. R. Wagner, U.S. Pat. 3,835,138 (1974) [*CA* **81**, 152259 (1974)].

74USP3838126   E. R. Wagner, U.S. Pat. 3,838,126 (1974) [*CA* **82**, 4298 (1975)].

74USP3843654   F. K. Krichner and A. W. Zalay, U.S. Pat. 3,843,654 (1974) [*CA* **82**, 112098 (1975)].

74USP3850932   F. G. Kathawala, U.S. Pat. 3,850,932 (1974) [*CA* **82**, 140175 (1975)].

74USP3853858   R. P. Ryan, U.S. Pat. 3,853,858 (1974) [*CA* **82**, 171014 (1975)].

74YZ417   T. Yoshikaw and K. Shitago, *Yakugaku Zasshi* **94**, 417 (1974) [*CA* **81**, 120561 (1974)].

75BSF1411   L. Legrand and N. Lozac'h, *Bull. Soc. Chim. Fr.*, 1411 (1975).

75BSF2118   L. Legrand and N. Lozac'h, *Bull. Soc. Chim. Fr.*, 2118 (1975).

75FES536   M. Cardellini, P. Franchetti, M. Grifantini, S. Martelli, and F. Petrelli, *Farmaco, Ed. Sci.* **30**, 536 (1975) [*CA* **83**, 188196 (1975)].

75GEP(O)2508333   M. Yamamoto, S. Morooka, M. Koshiba, S. Inaba, and H. Yamamoto, Ger. Pat. Offen. 2,508,333 (1975) [*CA* **83**, 206322 (1975)].

75GEP(O)2508543   M. Yamamoto, S. Morooka, M. Koshiba, S. Inaba, and H. Yamamoto, Ger. Pat. Offen. 2,508,543 (1975) [*CA* **84**, 44118 (1976)].

75JAP(K)7514699   Y. Kishida, Y. Iwano, and K. Hirai, Jpn. Kokai 75/14, 699 (1975)[*CA* **83**, 114460 (1975)].

75JCS(P1)2182   D. J. Brown and K. Ienage, *J. C. S. Perkin 1*, 2182 (1975).

75JCS(P1)2322   P. J. Abbott, R. M. Acheson, M. Y. Kornilov, and J. K. Stabbs, *J. C. S. Perkin 1*, 2322 (1975).

75JHC321   D. L. Trepanier and S. Sunder, *J. Heterocycl. Chem.* **12**, 321 (1975).

75JHC1077   S. Palazzo, L. I. Giannola, and M. Neri, *J. Heterocycl. Chem.* **12**, 1077 (1975).

75JHC1207   S. C. Bell, C. Gochman, and P. H. L. Wei, *J. Heterocycl. Chem.* **12**, 1207 (1975).

75JIC886   S. N. Sawhney, S. P. Singh, and O. P. Bansal, *J. Indian Chem. Soc.* **52**, 886 (1975).

75JMC224   W. N. Beverung and R. A. Partyka, *J. Med. Chem.* **18**, 224 (1975).

75JMC447   G. E. Hardtmann, G. Koletar, O. R. Pfister, J. H. Gogerty, and L. C. Iorio, *J. Med. Chem.* **18**, 447 (1975).

75JOC2201   R. Y. Ning, J. F. Blount, W. Y. Chen, and P. B. Madan, *J. Org. Chem.* **40**, 2201 (1975).

75KGS1096   V. L. Shvedov, G. N. Kurilo, A. A. Cherkasova, and A. N. Grinev, *Khim. Geterotsikl. Soedin.*, 1096 (1975) [*CA* **83**, 206200 (1975)].

75KGS1701   R. M. Acheson, P. J. Abbott, J. K. Stabbs, and M. Yu. Kornilov, *Khim. Geterotsikl. Soedin.*, 1701 (1975) [*CA* **84**, 150043 (1976)].

75KPS435   A. Irisbaev, Kh. M. Shakhidoyatov, and Ch. Sh. Kadyrov, *Khim. Prir. Soedin.* **11**, 435 (1975) [*CA* **84**, 105879 (1976)].

75KPS809   A. Irisbaev, Kh. M. Shakhidoyatov, and Ch. Sh. Kadyrov, *Khim. Prir. Soedin.* **11**, 809 (1975) [*CA* **84**, 150823 (1976)].

75MI1   R. S. Sinyak and I. A .Mazur, *Farm. Zh. (Kiev)* **30**, 29 (1975) [*CA* **83**, 97198 (1975)].

75MI2   Y. Yokoyama, K. Shibata, O. Fujii, and E. Iwamoto, *Toyo Soda Kenkyu Hokoku* **19**, 71 (1975) [*CA* **85**, 125771 (1976)].

75URP481613   V. I. Shvedov, A. N. Grinev, G. N. Kurilo, and A. A. Cherkasova, U.S.S.R. Pat. 481,613 (1975) [*CA* **84**, 17409 (1976)].

75UPS3875160   G. H. Hardtmann, U.S. Pat. 3,875,160 (1975) [*CA* **83**, 79274 (1975)].

75USP3883524   E. H. Wolf and B. J. Duffy, U.S. Pat. 3,883,524 (1975) [*CA* **83**, 131624 (1975)].

75USP3887554   R. A. E. Winter and T. J. Villani, U.S. Pat. 3,887,554 (1975) [*CA* **83**, 147505 (1975)].

75USP3897434   A. S. Katner, U.S. Pat. 3,897,434 (1975)[*CA* **84**, 44113 (1976)].

75USP3919210   G. E. Hardtmann, U.S. Pat. 3,919,210 (1975) [*CA* **84**, 59548 (1976)].

75USP3919215   D. L. Trepanier and S. Sunder, U.S. Pat. 3,919,215 (1975) [*CA* **84**, 121906 (1976)].

75USP3919216   D. L. Trepanier and S. Sunder, U.S. Pat. 3,919,216 (1975) [*CA* **84**, 121905 (1976)].

75USP3919219   D. L. Trepanier and S. Sunder, U.S. Pat. 3,919,219 (1975) [*CA* **84**, 74316 (1976)].

75USP3919220   D. L. Trepanier and S. Sunder, U.S. Pat. 3.919,220 (1975) [*CA* **84**, 121907 (1976)].

75USP3922274   D. L. Trepanier and S. Sunder, U.S. Pat. 3,922,274 (1975) [*CA* **84**, 59582 (1976)].

76AP542   H. Moehrle and C. M. Seidel, *Arch. Pharm. (Weinheim, Ger.)* **309**, 542 (1976).

76BSF1857   L. Legrand, *Bull. Soc. Chim. Fr.*, 1857 (1976) [*CA* **87**, 5904 (1977)].

76CC48   J. C. Cass and A. R. Katritzky, *J. C. S. Chem. Commun.*, 48 (1976).

76GEP(O)2539396 R. Bowie, J. M. Cox, G. M. Farrel, and M. C. Shephard, Ger. Pat. Offen. 2,539,396 (1976) [*CA* **85**, 5681 (1976)].

76H1487   T. Kametani, C. V. Loc, T. Higa, M. Koizumi, M. Ihara, and K. Fukumoto, *Heterocycles* **4**, 1487 (1976).

76IJC(B)354   G. Devi, R. S. Kapil, and S. P. Popli, *Indian J. Chem., Sect. B* **14**, 354 (1976).

76IJC(B)685   H. Singh and K. B. Lal, *Indian J. Chem., Sect. B* **14**, 685 (1976).

76IJC(B)705   C. S. Rao, A. D. Pandya, P. N. Mody, and M. P. Dave, *Indian J. Chem., Sect. B* **14**, 705 (1976).

76IJC(B)879   V. P. Arya, K. G. Dave, V. G. Khadse, and S. J. Shenoy, *Indian J. Chem., Sect B* **14**, 879 (1976).

76JA6186   T. Kametani, T. Higa, C. V. Loc, M. Ihara, M. Koizumi, and K. Fukumoto, *J. Am. Chem. Soc.* **98**, 6186 (1976).

76JAP(K)7643799 T. Yokoyama, K. Shibata, O. Fujii, and E. Iwamoto, Jpn. Kokai 76/43,799 (1976) [*CA* **85**, 95806 (1976)].

76JAP(K)7643800 T. Yokoyama, K. Shibate, O. Fujii, and E. Iwamoto, Jpn. Kokai 76/43,800 (1976) [*CA* **85**, 178985 (1976)].

76JAP(K)76100098 M. Yamamoto, S. Morooka, M. Koshiba, S. Inaba, and H. Yamamoto, *Jpn. Kokai* 76/100,098 (1976) [*CA* **86**, 121364 (1977)].

76JCS(P1)653   R. Hull and M. L. Swain, *J. C. S. Perkin 1*, 653 (1976).

76JHC421   C. F. Beam, N. D. Heindel, M. Chun, and A. Stefanski, *J. Heterocycl. Chem.* **13**, 421 (1976).

76JIC382   S. Singh, S. Kaur, and A. N. Kaushal, *J. Indian Chem. Soc.* **53**, 382 (1976).

76JOC497   J. G. Smith, J. M. Sheepy, and E. M. Levi, *J. Org. Chem.* **41**, 497 (1976).

76JOC825   G. M. Coppola, G. E. Hardtmann, and O. R. Pfister, *J. Org. Chem.* **41**, 825 (1976).

76JOC2728   N. P. Peet, S. Sunder, and W. H. Braun, *J. Org. Chem.* **41**, 2728 (1976).

76KFZ60   I. A. Mazur, A. F. Vlasenko, B. A. Samura, V. I. Linenko, and
                 P. M. Kochergin, *Khim. Farm. Zh.* **10**, 60 (1976) [*CA* **86**, 155597
                 (1977)].
76KGS834   A. F. Vlasenko, B. E. Mandrichenko, G. K. Rogul'chenko, R. S.
                 Sinyak, I. A. Mazur, and P. M. Kochergin, *Khim. Geterotsikl.
                 Soedin.*, 834 (1976) [*CA* **85**, 123861 (1976)].
76KGS1268   I. A. Mazur, R. S. Sinyak, R. I. Katkevich, and P. M. Kochergin,
                 Khim. Geterotsikl. Soedin., 1268 (1976) [*CA* **86**, 29752 (1977)].
76KGS1564   Kh. M. Shakhidoyatov, A. Irisbaev, L. M. Yun, E. Oripov, and
                 Ch. Sh. Kadyrov, *Khim. Geterotsikl. Soedin.*, 1564 (1976) [*CA*
                 86, 106517 (1977)].
76MI1   V. V. Korshak, A. L. Rusanov, M. K. Kereselidze, and T. K.
                 Dzhashiashvili, *Izv. Akad. Nauk. Gruz. SSR, Ser. Khim.* **2**, 189
                 (1976) [*CA* **86**, 43648 (1977)].
76MI2   V. F. Sedova and V. P. Mamaev, *Izv. Sib. Otd. Akad. Nauk SSSR,
                 Ser. Khim. Nauk*, 97 (1976) [*CA* **86**, 155598 (1977)].
76S469   K. D. Kampe, *Synthesis*, 469 (1976).
76T1735   G. Doleschall and K. Lempert, *Tetrahedron* **32**, 1735 (1976).
76TL1935   H. Breuer, *Tetrahedron Lett.*, 1935 (1976).
76URP514836   V. I. Shvedov, G. N. Kurilo, A. A. Cherkasova, and A. N. Grinev,
                 U.S.S.R. Pat. 514,836 (1976) [*CA* **85**, 143132 (1976)].
76URP527423   F. S. Babichev, F. P. Trinus, L. A. Gromov, V. A. Kovtunenko,
                 V. A. Portnyagina, T. F. Troyan, and A. K. Tyltin, U.S.S.R. Pat.
                 527,423 (1976) [*CA* **86**, 5491 (1977)].
76URP539885   V. I. Shvedov, G. N. Kurilo, A. A. Cherkasova, and A. N. Grinev,
                 U.S.S.R. Pat. 539,885 (1976) [*CA* **86**, 155692 (1977)].
76USP3932407   W. N. Beverung, Jr. and A. Partyka, U.S. Pat. 3,932,407 (1976) [*CA*
                 84, 105646 (1976)].
76USP3963720   G. E. Hardtmann, U.S. Pat. 3,963,720 (1976) [*CA* **85**, 160146
                 (1976)].
76USP3969506   G. E. Hardtmann, U.S. Pat. 3,969,506 (1976) [*CA* **86**, 5485 (1977)].
76USP3978059   G. E. Hardtmann, U.S. Pat. 3,978,059 (1976) [*CA* **86**, 29867 (1977)].
76USP3982000   G. E. Hardtmann, U.S. Pat. 3,982,000 (1976) [*CA* **86**, 5489 (1977)].
76USP3983119   W. N. Beverung, Jr., R. A. Partyka, and T. A. Jenks, U.S. Pat.
                 3,983,119 (1976) [*CA* **86**, 89901 (1977)].
76USP3983120   W. N. Beverung, Jr. and R. A. Partyka, U.S. Pat. 3,983,120 (1976)
                 [*CA* **86**, 72688 (1977)].
76USP3984556   G. E. Hardtmann, U.S. Pat. 3,984,556 (1976) [*CA* **86**, 16701 (1977)].
76USP3988340   R. A. Partyka and W. N. Beverung, Jr., U.S. Pat. 3,988,340 (1976)
                 [*CA* **86**, 106634 (1977)].
76USP4000275   W. H. W. Lunn, U.S. Pat. 4,000,275 (1976) [*CA* **86**, 140082 (1977)].
77AF766   D. B. Reisner, B. J. Ludwig, E. Simon, T. Dejneka, and R. D.
                 Sofia, *Arzneim.-Forsch.* **27**, 766 (1977).
77FZ37   I. A. Mazur, *Farm. Zh. (Kiev)*, 37 (1977) [*CA* **88**, 104849 (1978)].
77FZ84   R. S. Sinyak, I. A. Mazur, P. M. Steblyuk, and P. M. Kochergin,
                 Farm. Zh. (Kiev), 84 (1977) [*CA* **87**, 102264 (1977)].
77FZ88   A. F. Vlasenko, I. A. Mazur, and P. M. Kochergin, *Farm. Zh.
                 (Kiev)*, 88 (1977) [*CA* **86**, 189845 (1977)].
77GEP(O)2557425 K. Schromm, A. Mentrup, E. O. Renth, and A. Fuegner, Ger. Pat.
                 Offen. 2,557,425 (1977) [*CA* **87**, 117898 (1977)].

77GEP(O)2645110   C. F. Schwender and B. R. Sunday, Ger. Pat. Offen. 2,645,110
                    (1977) [CA 87, 23327 (1977)].
77IJC(B)41   P. B. Talukdar, S. K. Sengupta, A. K. Datta, and T. K. Roy, Indian
                    J. Chem., Sect. B 15, 41 (1977).
77IJC(B)250   H. Jahine, A. Sayed, H. A. Zaher, and O. Sherif, Indain J. Chem.,
                    Sect B 15, 250 (1977).
77IJC(B)335   V. Purnaprajna and S. Seshadri, Indian J. Chem., Sect. B 15, 335
                    (1977).
77IJC(B)751   A. Singh, Indian J. Chem., Sect. B 15, 751 (1977).
77IJC(B)1100   V. B. Rao and C. V. Ratnam, Indian J. Chem., Sect. B 15, 1100
                    (1977).
77IJC(B)1110   P. B. Talukdar, S. K. Sengupta, and A. K. Datta, Indian J. Chem.,
                    Sect. B 15, 1110 (1977).
77JA2306   T. Kametani, C. V. Loc, T. Higa, M. Koizumi, M. Ihara, and K.
                    Fukumoto, J. Am. Chem. Soc. 99, 2306 (1977).
77JAP(K)77144697   K. Noda, A. Nakagawa, S. Yamazaki, K. Noguchi, T. Hachitani,
                    and H. Ide, Jpn. Kokai 77/144,697 (1977) [CA 88, 190879 (1978)].
77JAP(K)7777093   M. Koizumi, I. Matsuura, and Y. Murakami, Jpn. Kokai 77/77,093
                    (1977) [CA 88, 6930 (1978)].
77JCS(P1)107   A. Gescher, M. F. G. Stevens, and C. P. Turnbull, J. C. S. Perkin 1,
                    107 (1977).
77JCS(P1)1162   A. Banerji, J. C. Cass, and A. R. Katritzky, J. C. S. Perkin 1, 1162
                    (1977).
77JHC1191   D. R. Harrison, P. D. Kennewell, and J. B. Taylor, J. Heterocycl.
                    Chem. 14, 1191 (1977).
77JPR919   S. Johne, B. Jung, D. Groeger, and R. Radeglia, J. Prakt. Chem. 319,
                    919 (1977).
77KGS377   V. I. Shvedov, G. N. Kurilo, A. A. Cherkasova, and A. N. Grinev,
                    Khim. Geterotsikl. Soedin., 377 (1977) [CA 87, 39334 (1977)].
77KGS678   V. F. Sedova, V. I. Mamatyuk, V. A. Samsonov, and V. P.
                    Mamaev, Khim. Geterotsikl. Soedin., 678 (1977) [CA 87, 84938
                    (1977)].
77KPS544   Kh. M. Shakhidoyatov and Ch. Sh. Kadyrov, Khim. Prir. Soedin.,
                    544 (1977) [CA 88, 6830 (1978)].
77MI1   V. V. Korshak, A. L. Rusanov, I. Batirov, I. Ya. Kalontarov. F. F.
                    Niyazi, and L. Kh. Plieva, Dokl. Akad. Nauk Tadzh SSR 20, 37
                    (1977) [CA 88, 152537 (1978)].
77MI2   T. Kametani, K. Fukumoto, M. Ihara, and C. V. Loc, Symp. Het-
                    erocycl. [Pap.], 243 (1977) [CA 89, 163805 (1978)].
77UKZ711   F. S. Babichev and Yu. M. Volovenko, Ukr. Khim. Zh. (Russ. Ed.)
                    43, 711 (1977) [CA 87, 184451 (1977)].
77USP4020062   G. E. Hardtmann, U.S. Pat. 4,020,062 (1977) [CA 87, 85039 (1977)].
77USP4025511   G. E. Hardtmann, U.S. Pat. 4,025,511 (1977) [CA 87, 85044 (1977)].
77ZC444   S. Leistner, A. P. Giro, and G. Wagner, Z. Chem. 17, 444 (1977).
77ZN(B)94   M. I. Ali, H. A. Hammouda, and A. E. M. Abd-Elfattah, Z. Natur-
                    forsch., B: Anorg. Chem., Org. Chem. 32B, 94 (1977).
77ZOR1773   A. V. Bogatskii, S. A. Andronati, Z. I. Zhilina, and N. I. Danilina,
                    Zh. Org. Khim. 13, 1773 (1977) [CA 88, 6829 (1978)].
78AP586   H. Moehrle and H. J. Hemmerling, Arch. Pharm. (Weinheim, Ger).
                    311, 586 (1978).

78CJC1616	F. D. Eddy, K. Vaughan, and M. F. G. Stevens, *Can. J. Chem.* **56,** 1616 (1978).
78FES271	S. Massa and G. DeMartino, *Farmaco, Ed. Sci.,* **33,** 271 (1978) [*CA* **89,** 43355 (1978)].
78FRP2393001	G. Ferrand, J. P. Maffrand, and J. M. Pereillo, Fr. Demande Pat. 2,393,001 (1978) [*CA* **91,** 175390 (1979)].
78GEP(O)2726389	B. R. Vogt, Ger. Offen. Pat. 2,726,389 (1978) [*CA* **88,** 121240 (1978)].
78GEP(O)2805124	M. Yamamoto, M. Koshiba, and H. Yamamoto, Ger. Pat. Offen. 2,805,124 (1978) [*CA* **89,** 197593 (1978)].
78GEP(O)2812585	R. W. Kierstead and J. W. Tilley, Ger. Pat. Offen. 2,812,585 (1978) [*CA* **90,** 87500 (1979)].
78GEP(O)2812586	R. W. Kierstead and J. W. Tilley, Ger. Pat. Offen. 2,812,586 (1978) [*CA* **90,** 38953 (1979)].
78H1375	T. Nagasaka, F. Hamaguchi, N. Ozawa, and S. Ohki, *Heterocycles* **9,** 1375 (1978).
78H1729	T. Kurihara and Y. Sakamoto, *Heterocycles* **9,** 1729 (1978).
78IJC(B)331	H. Singh and C. S. Gandhi, *Indian J. Chem., Sect. B* **16,** 331 (1978).
78IJC(B)537	R. P. Gupta, M. L. Sachdeva, R. N. Handa, and H. K. Pujari, *Indian J. Chem., Sect. B* **16,** 537 (1978).
78IJC(B)689	H. Jahine, H. A. Zaher, Y. Akhnookh, and Z. El-Gendy, *Indian J. Chem., Sect. B.* **16,** 689 (1978).
78JAP(K)7812893	M. Yamamoto, M. Koshiba, and H. Yamamoto, Jpn. Kokai 78/ 12,893 (1978) [*CA* **89,** 43477 (1978)].
78JAP(K)7823997	M. Yamamoto, M. Koshiba, and H. Yamamoto, Jpn. Kokai 78/ 23,997 (1978) [*CA* **89,** 43488 (1978)].
78JAP(K)7837695	Imperial Chemical Industries Ltd., Jpn. Kokai 78/37,695 (1978) [*CA* **89,** 146922 (1979)].
78JAP(K)7844592	F. Ishikawa, A. Kosasayama, and K. Abiko, Jpn. Kokai 78/44,592 (1978) [*CA* **89,** 109566 (1978)].
78JAP(K)7844593	F. Ishikawa, A. Kosasayama, and K. Abiko, Jpn. Kokai 78/44,593 (1978) [*CA* **89,** 197584 (1978)].
78JAP(K)7877075	T. Kametani, Jpn. Kokai 78/77,075 (1978) [*CA* **89,** 180037 (1978)].
78JHC77	R. Friary, *J. Heterocycl. Chem.* **15,** 77 (1978).
78JIC928	V. K. Singh and K. C. Joshi, *J. Indian Chem. Soc.* **55,** 928 (1978).
78KGS105	Kh. M. Shakhidoyatov, L. M. Yun, and Ch. Sh. Kadyrov, *Khim. Geterotsikl. Soedin,* 105 (1978) [*CA* **88,** 121104 (1978)].
78MI1	M. I. Ali and A. E. G. Hammam, *J. Chem. Eng. Data* **23,** 351 (1978).
78MI2	V. Pestellini, M. Ghelardoni, G. Volterra, and P. DelSoldata, *Eur. J. Med. Chem.—Chim. Ther.* **13,** 296 (1978).
78PHA124	K. Kottke and H. Kuehmstedt, *Pharmazie* **33,** 124 (1978).
78PHA125	K. Kottke and H. Kuehmstedt, *Pharmazie* **33,** 125 (1978).
78PHA185	S. Leistner, A. P. Giro, and G. Wagner, *Pharmazie* **33,** 185 (1978).
78PHA462	K. Kottke and H. Kuehmstedt, *Pharmazie* **33,** 462 (1978).
78PHA507	K. Kottke and H. Kuehmstedt, *Pharmazie* **33,** 507 (1978).
78TL3007	H. Karpf and H. Junek, *Tetrahedron Lett.,* 3007 (1978).
78USP4066767	C. F. Schwender and B. R. Sunday, U.S. Pat. 4,066,767 (1978) [*CA* **88,** 136658 (1978)].
78USP4085213	J. S. Bindar, U.S. Pat. 4,085,213 (1978) [*CA* **89,** 197582 (1978)].
78USP4105766	E. J. Alexander, U.S. Pat. 4,105,766 (1976) [*CA* **90,** 87505 (1979)].

78USP4110452   G. C. Rovnyak and B. R. Vogt, U.S. Pat. 4,110,452 (1978) [*CA* **90**, 87512 (1979)].
78USP4112096   B. R. Vogt, U.S. Pat. 4,112,096 (1978) [*CA* **92**, 146801 (1980)].
78USP4112098   B. R. Vogt, U.S. Pat. 4,112,098 (1978) [*CA* **90**, 137848 (1979)].
78USP4128644   B. R. Vogt, U.S. Pat. 4,128,644 (1978) [*CA* **90**, 104003 (1979)].
78USP4129653   V. T. Bandurco and S. Levine, U.S. Pat. 4,129,653 (1978) [*CA* **90**, 137861 (1979)].
79AP552   W. Zimmerman and K. Eger, *Arch. Pharm. (Weinheim, Ger.)* **312**, 552, (1979).
79AP838   H. Moehrle and J. Gerloff, *Arch. Pharm. (Weinheim, Ger.)* **312**, 838 (1979).
79CB3424   T. Kappe and W. Lube, *Chem. Ber.* **112**, 3424 (1979).
79CP1057752   M. Yamamoto, S. Morooka, M. Koshiba, S. Inaba, and H. Yamamoto, Can. Pat. 1,057,752 (1979) [*CA* **91**, 211436 (1979)].
79CZ266   H. Moehrle and J. Herbke, *Chem.-Ztg.* **103**, 266 (1979).
79FES688   F. Russo, M. Santagati, and A. Santagati, *Farmaco, Ed. Sci.* **34**, 688 (1979) [*CA* **91**, 193251 (1979)].
79GEP(O)2739020   H. Biere and J. F. Kapp, Ger. Pat. Offen. 2,739,020 (1979) [*CA* **90**, 204131 (1979)].
79GEP(O)2758875   K. H. Mayer, H. Heitzer, F. Hoffmeister, A. Heise, and S. Kazda, Ger. Pat. Offen. 2,758,875 (1979) [*CA* **91**, 211448 (1979)].
79GEP(O)2832138   M. S. Chodnekar and A. Kaiser, Ger. Pat. Offen. 2,832,138 (1979) [*CA* **90**, 186967 (1979)].
79GEP(O)2916992   B. R. Vogt, and L. M. Simpkins, Ger. Pat. Offen. 2,916,992 (1979) [*CA* **92**, 111052 (1980)].
79IJC(B)39   P. B. Talukdar, S. K. Sengupta, and A. K. Datta, *Indian J. Chem., Sect. B* **18**, 39 (1979).
79IJC(B)107   Mrs. K. Bhandari, V. Virmani, V. A. Murti, P. C. Jain, and N. Anand, *Indian J. Chem., Sect. B* **17**, 107 (1979).
79IJC(B)125   Y. P. Reddy, G. S. Reddy, and K. K. Reddy, *Indian J. Chem., Sect. B* **18**, 125 (1979).
79IJC(B)349   P. Hanumanthu, and C. V. Ratnam, *Indian J. Chem., Sect. B* **17**, 349 (1979).
79IJC(B)632   S. M. Sondhi, M. P. Mahajan, and N. K. Ralhan, *Indian J. Chem., Sect. B* **17**, 632 (1979).
79JAP(K)79135799   T. Kametani, *Jpn. Kokai* 79/135,799 (1977) [*CA* **92**, 163996 (1980)].
79JCS(P1)1765   G. Bernàth, G. Toth, F. Fülöp, G. Göndös, and L. Gera, *J. C. S. Perkin 1*, 1765 (1979).
79JCS(P1)3085   J. S. Bajwa, and P. J. Sykes, *J. C. S. Perkin 1*, 3085 (1979).
79JCS(P2)1708   R. A. Bowie, P. N. Edwards, S. N. J. Taylor, and D. A. Thomason, *J. C. S. Perkin 2*, 1708 (1979).
79JHC137   G. Bernàth, F. Fülöp, I. Hermecz, Z. Meszàros, and G. Toth, *J. Heterocycl. Chem.* **16**, 137 (1979).
79JHC391   L. G. Payne, J. Prztycki, A. A. Patchett, and M. T. Wu, *J. Heterocycl. Chem.* **16**, 391 (1979).
79JHC623   M. L. Cotter, V. Bandurco, E. Wong, and Z. G. Hajos, *J. Heterocycl. Chem.* **16**, 623 (1979).
79JHC957   C. F. Beam, J. Brown, D. R. Dawkins, W. P. Fives, and N. D. Heindel, *J. Heterocycl. Chem.* **16**, 957 (1979).

79JHC1339	S. Sunder, and N. P. Peet, *J. Heterocycl. Chem.* **16**, 1339 (1979).
79JHC1497	M. L. Cotter, V. Bandurco, and E. Wong, *J. Heterocycl. Chem.* **16**, 1497 (1979).
79JMC114	C. F. Schwender, B. R. Sunday, and D. J. Herzig, *J. Med. Chem.* **22**, 114 (1979).
79JMC748	C. F. Schwender, B. R. Sunday, D. J. Herzig, E. K. Kusner, P. R. Schumann, and D. L. Gawlak, *J. Med. Chem.* **22**, 748 (1979).
79KGS832	G. N. Kurilo, S. Yu. Ryabova, and A. N. Grinev, *Khim. Geterotsikl. Soedin.*, 832 (1979) [*CA* **91**, 175292 (1979)].
79KGS1427	V. A. Kovtunenko, A. K. Tyltin, and L. V. Soloshonok, *Khim. Geterotiskl. Soedin.*, 1427 (1979) [*CA* **92**, 110954 (1980)].
79MI1	F. Fülöp, G. Bernàth, I. Hermecz, and Z. Meszaros, *Acta. Phys. Chem.* **25**, 79 (1979) [*CA* **92**, 146712 (1979)].
79MI2	A. V. Bogatskii, Z. I. Zhilina, S. a. Andronati, and N. I. Danilina, *Khim. Prom-St., Ser.: Reakt. Osobo Chist. Veshchesteva* 33 (1979) [*CA* **91**, 123703 (1979)].
79MI3	R. R. Crenshaw and T. A. Montzka, *Res. Discl.* **183**, 360 (1979) [*CA* **91**, 140812 (1979)].
79MIP1	G. N. Kurilo, S. Yu. Ryabova, and A. N. Grinev, U. S. S. R. Pat., 690,017 (1979) [*CA* **92**, 76545 (1980)].
79PHA138	G. Wagner, and E. Bunk, *Pharmazie* **34**, 131 (1979).
79PHA390	S. Leistner, A. P. Giro, and G. Wagner, *Pharmazie* **34**, 390 (1979).
79SAP77804057	C. F. Schwender, and B. R. Sunday, S. Afr. Pat., 78/04,057 (1979) [*CA* **92**, 111050 (1980)].
79USP4168380	R. A. LeMahieu, U.S. Pat., 4,168,380 (1979) [*CA* **92**, 76540 (1980)].
80ACH107	G. Doleschall, and K. Simon-Ormai, *Acta. Chim. Acad. Sci. Hung.* **104**, 107 (1980).
80AP729	J. Troschuetz, *Arch. Pharm., (Weinheim, Ger.)* **313**, 729 (1980).
80CC808	D. E. Davies, D. L. R. Reeves, and R. Storr, *J. C. S. C. Commun.*, 808 (1980).
80CPB702	K. Ozaki, Y. Yamada, and T. Oine, *Chem. Pharm. Bull.* **28**, 702 (1980).
80CPB2024	F. Ishikawa, A. Kosasayama, and K. Higashi, *Chem. Pharm. Bull.* **28**, 2024 (1980).
80EGP(D)139715	K. Kottke, H. Kuehmstedt, V. Hagen, H. Renner, and S. Schnitzler, Ger. Pat. (East) DD 139,715 (1980) [*CA* **96**, 69016 (1982)].
80EGP(D)142337	M. Suesse and S. Johne, Ger. Pat. (East) DD 142,337 (1980) [*CA* **94**, 175162 (1981)].
80EUP15065	J. C. Sircar, T. Capiris, and S. J. Kesten, Eur. Pat. 15,065 (1980) [*CA* **94**, 103413 (1981)].
80GEP(O)2845766	H. Biere, J. F. Kapp, and I. Boettcher, Ger. Pat. Offen. 2,845,766 (1980) [*CA* **93**, 186397 (1980)].
80GEP(O)3019019	E. Berenyi, E. Szirt, P. Gorog, L. Petocz, I. Kosoczky, A. Kovacs, and G. Urmos, Ger. Pat. Offen. 3,019,019 (1980) [*CA* **94**, 156963 (1981)].
80H3	H. Singh, C. S. Gandhi, and M. S. Bal, *Heterocycles* **14**, 3 (1980).
80H1959	K. Matsumoto, S. Nakamura, and R. M. Acheson, *Heterocycles* **14**, 1959 (1980).
80HCA1	R. Heckendorn and T. Winkler, *Helv. Chim. Acta* **63**, 1 (1980).

80JAP(K)8047684   Takeda Chemical industries, Ltd., Jpn. Kokai 80/47, 684 (1980) [*CA* **93**, 186400 (1980)].
80JAP(K)8055188   Sumitomo Chemical Co., Ltd., Jpn. Kokai 80/55,188 (1980) [*CA* **94**, 65713 (1981)].
80JCS(P1)633   T. McC. Paterson, R. K. Smalley, H. Suschitzky, and (in part) A. J. Barker, *J. C. S. Perkin 1*, 633 (1980).
80JHC155   I. Antonini, G. Gristalli, P. Franchetti, M. Grifantini, and S. Martelli, *J. Heterocycl. Chem.* **17**, 155 (1980).
80JHC945   T. Kurihara, T. Tani, S. Maeyama, and Y. Sakamoto, *J. Heterocycl. Chem.* **17**, 945 (1980).
80JHC1163   G. M. Coppola and M. J. Shapiro, *J. Heterocycl. Chem.* **17**, 1163 (1980).
80JHC1489   S. D. Phillips and R. N. Castle, *J. Heterocycl. Chem.* **17**, 1489 (1980).
80JHC1553   E. P. Papadopoulos, *J. Heterocycl. Chem.* **17**, 1553 (1980).
80JHC1665   S. D. Phillips and R. N. Castle, *J. Heterocycl. Chem.* **17**, 1665 (1980).
80JMC92   J. W. Tilley, R. A. LeMahieu, M. Carson, R. W. Kierstead, H. W. Baruth, and B. Yaremko, *J. Med. Chem.* **23**, 92 (1980).
80M627   H. Moehrle and J. Herbke, *Monatsh. Chem.* **111**, 627 (1980).
80MI1   Kh. M. Shakhidoyatov, E. O. Eripov, L. M. Yun, M. Ya. Yamankulov, and Ch. Sh. Kadyrov, *Fungitsidy*, 66 (1980) [*CA* **94**, 192253 (1981)].
80MI2   V. K. Srivastava, B. R. Pandy, R. C. Gupta, J. P. Barthwal, and K. Kishor, *Indian J. Pharm. Sci.* **2**, 29 (1980) [*CA* **94**, 208790 (1981)].
80MI3   G. Bernàth, G. Toth, F. Fülöp, G. Göndös, and L. Gera, *Magy. Kem. Foly.* **86**, 232 (1980) [*CA* **94**, 15667 (1981)].
80MI4   B. Liberek and J. Zarebski, *Pept., Proc. Eur. Pept. Symp., 16th, 1980*, 236 (1981) [*CA* **97**, 216690 (1982)].
80MIP1   J. H. E. Marsden and N. Harrison, Braz. Pedido PI 79/04,117 (1980) [*CA* **93**, 239455 (1980)].
80PHA124   S. Leistner and G. Wagner, *Pharmazie* **35**, 124 (1980).
80PHA256   F. Bahr and G. Dietz, *Pharmazie* **35**, 256 (1980).
80PHA293   S. Leistner, G. Wagner, and T. Strohscheidt, *Pharmazie* **35**, 293 (1980).
80PHA389   H. Moehrle and J. Herbke, *Pharmazie* **35**, 389 (1980).
80PHA582   S. Leistner and G. Wagner, *Pharmazie* **35**, 582 (1980).
80PHA800   K. Kottke and H. Kuehmstedt, *Pharmazie* **35**, 800 (1980).
80USP4208521   R. R. Crenshaw and T. A. Montzka, U.S. Pat. 4,208,521 (1980) [*CA* **93**, 204685 (1980)].
81AP271   R. Domanig, *Arch. Pharm. (Weinheim, Ger.)* **314**, 271 (1981).
81CC160   R. S. Atkinson, J. R. Malpass, and K. L. Woodthorpe, *J. C. S. Chem. Commun.*, 160 (1981).
81CCC2428   M. Marchalin, J. Svetlik, and A. Martvon, *Collect. Czech. Chem. Commun.* **46**, 2428 (1981).
81CPB2135   M. Yamamoto and H. Yamamoto, *Chem. Pharm. Bull.* **29**, 2135 (1981).
81EUP23773   R. A. Conley, M. M. Lam, and L. B. High, Eur. Pat. 23,773 (1981) [*CA* **95**, 25115 (1981)].
81EUP27268   R. A. Lemahieu, Eur. Pat. 27,268 (1981) [*CA* **95**, 81019 (1981)].
81EUP29559   H. Ott, Eur. Pat. 29,559 (1981) [*CA* **95**, 150698 (1981)].

81EUP34529   R. Westwood, W. R. Tully, and R. Murdoch, Eur. Pat. 34,529 (1981) [*CA* **96**, 20114 (1982)].

81FES292   F. Russo, M. Santagati, A. Santagati, M. Amico-Roxas, R. Bitetti, and A. Russo, *Farmaco, Ed. Sci.* **36**, 292 (1981) [*CA* **95**, 80829 (1981)].

81H621   M. R. Chaurasia and S. K. Sharma, *Heterocycles* **16**, 621 (1981).

81IJC(B)14   H. K. Gakhar, S. C. Gupta, and N. Kumar, *Indian J. Chem., Sect. B* **20**, 14 (1981).

81IJC(B)290   M. A. Elkasaby and A. N. Noureldin, *Indian J. Chem., Sect. B* **20**, 290 (1981).

81IJC(B)579   R. Rastogi and S. Sharma, *Indian J. Chem., Sect. B* **20**, 579 (1981).

81JHC117   J. C. Sircar, T. Capiris, and S. J. Kesten, *J. Heterocycl. Chem.* **18**, 117 (1981).

81JHC287   D. H. Kim, *J. Heterocycl. Chem.* **18**, 287 (1981).

81JHC515   E. P. Papadopoulos, *J. Heterocycl. Chem.* **18**, 515 (1981).

81JHC801   D. H. Kim, *J. Heterocycl. Chem.* **18**, 801 (1981).

81JMC735   J. C. Sircar, T. Capiris, S. J. Kesten, and D. J. Herzig, *J. Med. Chem.* **24**, 735 (1981).

81JMC1455   V. T. Bandurco, E. M. Wong, S. D. Levine, and Z. G. Hajos, *J. Med. Chem.* **24**, 1455 (1981).

81JOC1571   K. Ozaki, Y. Yamada, and T. Oine, *J. Org. Chem.* **46**, 1571 (1981).

81JOC4489   K. Hirai, T. Ishiba, H. Sugimoto, and T. Fujishita, *J. Org. Chem.* **46**, 4489 (1981).

81JPR647   M. Suesse and S. Johne, *J. Prakt. Chem.* **233**, 647 (1981).

81KGS844   V. G. Zabrodnyaya, Yu. N. Portnov, A. N. Kost, and V. G. Voronin, *Khim. Geterotsikl. Soedin.*, 844 (1981) [*CA* **95**, 150581 (1981)].

81KGS1264   N. V. Abbakumova, Yu. G. Putsykin, Yu. A. Baskakov, and Yu. A. Kondrat'ev, *Khim. Geterotsikl. Soedin.*, 1264 (1981) [*CA* **96**, 20064 (1982)].

81MI1   I. Hermecz, J. Kokosi, L. Vasvari-Debreczy, A. Horvath, B. Podanyi, G. Szasz, and Z. Meszaros, *Int. Conf. Chem. Biotechnol. Biol. Act. Nat. Prod. [Proc.], 1st, 1981,* Vol. 3, p. 69 (1981) [*CA* **97**, 127863 (1982)].

81PHA62   M. El Enany and S. Botros, *Pharmazie* **36**, 62 (1981).

81RRC109   E. Georgescu, I. Druta, and M. Petrovanu, *Rev. Roum. Chim.* **26**, 109 (1981) [*CA* **94**, 208802 (1981)].

81USP4250177   C. F. Schwender and B. R. Sunday, U.S. Pat. 4,250,177 (1981) [*CA* **95**, 43158 (1981)].

81USP4261996   J. C. Sircar and T. Capiris, U.S. Pat. 4,261,996 (1981) [*CA* **95**, 115584 (1981)].

81USP4261997   J. C. Sircar and S. J. Kesten, U.S. Pat. 4,261, 997 (1981) [*CA* **95**, 97838 (1981)].

81USP4282226   B. R. Vogt and L. G. Magbanua, U.S. Pat. 4,282,226 (1981) [*CA* **95**, 150702 (1981)].

81USP4302585   P. H. L. Wei and S. C. Bell, U.S. Pat. 4,302,585 (1981) [*CA* **96**, 104227 (1982)].

81ZN(B)252   F. S. G. Soliman, W. Stadlbauer, and T. Kappe, *Z. Naturforsch., B: Anorg. Chem., Org. Chem.* **36B**, 252 (1981).

81ZN(B)366   H. A. Daboun, A. M. Abd-Elfattah, M. M. Hussein, and A. F. A.

	Shalaby, Z. Naturforsch., B: Anorg. Chem., Org. Chem. **36B**, 366 (1981).
82AP866	S. Abuzar and S. Sharma, Arch. Pharm. (Weinheim, Ger.) **315**, 866 (1982).
82BRP2086903	W. R. Tully, Br. Pat. 2,086,903 (1982) [CA **97**, 144869 (1982)].
82CP1137474	T. A. Jenks, W. N. Beverung, Jr., and R. A. Partyka, Can. Pat. 1,137,474 (1982) [CA **98**, 179403 (1983)].
82CPB1036	M. Yamato and Y. Takeuchi, Chem. Pharm. Bull. **30**, 1036 (1982).
82EUP46267	M. S. Chodnekar, A. Kaiser, and F. Kienzle, Eur. Pat. 46,267 (1982) [CA **97**, 23812 (1982)].
82EUP46446	J. Frei and E. Schweizer, Eur. Pat. 46,446 (1982) [CA **96**, 217872 (1982)].
82EUP53767	B. Junge, S. Goldmann, G. Thomas, and B. Garthoff, Eur. Pat. 53,767 (1982) [CA **97**, 163014 (1982)].
82EUP54180	M. S. Chodnekar and F. Kienzle, Eur. Pat. 54,180 (1982) [CA **97**, 216205 (1982)].
82FES719	F. Russo, A. Santagati, and M. Santagati, Farmaco, Ed. Sci. **37**, 719 (1982) [CA **98**, 89287 (1983)].
82G545	H. M. Hassaneen, A. O. Abdelhamid, A. Shetta, and A. S. Shawali, Gazz. Chim. Ital. **112**, 545 (1982).
82H249	J. Horiuchi, M. Yamato, N. Katagiri, and T. Kato, Heterocycles **19**, 249 (1982).
82H1375	H. A. Daboun, M. A. Abdel Aziz, and F. E. A. Yousif, Heterocycles **19**, 1375 (1982).
82JCS(P1)2407	R. S. Atkinson, J. R. Malpass, and K. L. Woodthorpe, J. C. S. Perkin 1, 2407 (1982).
82JHC73	A. S. Shawali, A. O. Abdelhamid, H. M. Hassaneen, and A. Shetta, J. Heterocycl. Chem. **19**, 73 (1982).
82JHC1117	W. D. Dean and E. P. Papadopoulos, J. Heterocycl. Chem. **19**, 1117 (1982).
82JIC666	H. K. Gakhar, S. Kiran, and S. B. Gupta, J. Indian Chem. Soc. **59**, 666 (1982).
82JMC742	C. F. Schwender, B. R. Sunday, and V. L. Decker, J. Med. Chem. **25**, 742 (1982).
82KPS498	A. Karimov, M. V. Telezhenetskaya, and S. Yu. Yunusov, Khim. Prir. Soedin., 498 (1982) [CA **98**, 72516 (1983)].
82M1145	H. K. Gakhar, S. Kiran, and S. B. Gupta, Monatsh. Chem. **113**, 1145 (1982).
82MI1	H. A. Daboun and M. A. Abdel Aziz, Egypt J. Chem. **26**, 401 (1982) [CA **101**, 211081 (1984)].
82MIP1	M. Izquierdo Sanjose, I. Fernandez Fernandez, C. Fuentes Manso, and L. Lucero de Pablo, Span. Pat. 499,375 (1982) [CA **97**, 55830 (1982)].
82PHA605	S. Leistner, K. Hentschel, and G. Wagner, Pharmazie **37**, 605 (1982).
82S853	K. D. Deodhar, A. D. D'Sa, S. R. Pednekar, and D. S. Kanekar, Synthesis, 853 (1982).
82T1527	A. O. Abdelhamid, H. M. Hassaneen, M. Abbas, and A. S. Shawali, Tetrahedron **38**, 1527 (1982).

82USP4332802	K. Schromm, A. Mentruep, E. O. Renth, and A. Fuegner, U.S. Pat. 4,332,802 (1982) [*CA* **97**, 144870 (1982)].
83AP379	K. C. Liu, M. H. Yen, J. W. Chern, and Y. O. Lin, *Arch. Pharm.* (*Weinheim, Ger.*) **316**, 379 (1983).
83AP394	H. A. Daboun and M. A. Abdel Aziz, *Arch. Pharm.* (*Weinheim, Ger.*) **316**, 394 (1983).
83AP569	K. C. Liu, J. W. Chern, M. H. Yen, and Y. O. Lin, *Arch. Pharm.* (*Weinheim, Ger.*) **316**, 569 (1983).
83AP702	A. Kamal and P. B. Sattur, *Arch. Pharm.* (*Weinheim, Ger.*) **316**, 702 (1983).
83BCJ1227	H. A. El-Sherief. A. E. Abdel-Rahman, G. M. El-Naggar, and A. M. Mahmoud, *Bull. Chem. Soc. Jpn.* **56**, 1227 (1983).
83BSF226	L. Legrand and N. Lozach, *Bull. Soc. Chim. Fr.,* 266 (1983).
83CPB2234	K. Ozaki, Y. Yamada, and T. Oine, *Chem. Pharm. Bull.,* **31**, 2234 (1983).
83EGP(D)158549	K. Kottke, H. Kuehmstedt, V. Hagen, H. Renner, and S. Schnitzler, Ger. Pat. (East) DD 158,549 (1983) [*CA* **99**, 70757 (1983)].
83EGP(D)160343	K. Kottke, H. Kuehmstedt, H. Landmann, and H. Wehlan, Ger. Pat. (East)DD 160,343 (1983) [*CA* **100**, 6560 (1984)].
83EGP(D)203545	K. Kottke, H. Kuehmstedt, H. Landmann, and H. Wehlan, Ger. Pat. (East)DD 203,545 (1983) [*CA* **100**, 103388 (1984)].
83EGP(D)204095	G. Kempter, M. Plesse, H. U. Lehm, R. Stoeckel, and A. Jumar, Ger. Pat. (East)DD 204,095 (1983) [*CA* **101**, 7182 (1984)].
83EUP73060	F. Kienzle, Eur. Pat. 73,060 (1983) [*CA* **99**, 38484 (1983)].
83EUP76199	W. R. Tully, R. Westwood, D. A. Rowlands, and S. Clements-Jewery, Eur. Pat. 76,199 (1983) [*CA* **99**, 194989 (1983)].
83EUP80176	R. Schlecker, L. Friedrich, and D. Lenke, Eur. Pat. 80,176 (1983) [*CA* **99**, 175790 (1983)].
83GEP(O)3142727	G. Seybold and B. Wuerzer, Ger. Pat. Offen. 3,142,727 (1983) [*CA* **99**, 53778 (1983)].
83GEP(O)3220438	W. Opitz, H. Jacobi, and B. Pelster, Ger. Pat. Offen. 3,220,438 (1983) [*CA* **100**, 103383 (1984)].
83GEP(O)3233766	G. E. Hardmann and W. J. Houlihan, Ger. Pat. Offen. 3,233,766 (1983) [*CA* **99**, 5642 (1983)].
83H1549	M. R. Chaurasia and A. K. Sharma, *Heterocycles* **20**, 1549 (1983).
83HCA148	F. Kienzle, A. Kaiser, and R. E. Minder, *Helv. Chim. Acta* **66**, 148 (1983).
83IJC(B)485	B. K. Misra, Y. R. Rao, and S. N. Mahapatra, *Indian J. Chem., Sect. B* **22**, 485 (1983).
83IJC(B)496	S. K. Phadtare, S. K. Kamat, and G. T. Panse, *Indian J. Chem., Sect. B* **22**, 496 (1983).
83JCS(P1)2011	M. Lempert-Sreter, K. Lempert, and J. Moeller, *J. C. S. Perkin 1,* 2011 (1983).
83JHC93	I. Hermecz, B. Podànyi, and Z. Meszàros, *J. Heterocycl. Chem.* **20**, 93 (1983).
83JHC719	A. O. Abdelhamid, H. M. Hassaneen, and A. S. Shawali, *J. Heterocycl. Chem.* **20**, 719 (1983).
83JIC1071	M. R. Chaurasia and A. K. Sharma, *J. Indian Chem. Soc.* **60**, 1071 (1983).
83JMC107	R. A. LeMahieu, M. Carson, A. F. Welton, H. W. Baruth, and B. Yaremko, *J. Med. Chem.* **26**, 107 (1983).

83JPR88	C. Bischoff and E. Schroeder, *J. Prakt. Chem.* **325**, 88 (1983).
83M339	H. K. Gakhar, A. Jain, J. K. Gill, and S. B. Gupta, *Monatsh. Chem.* **114**, 339 (1983).
83MI1	V. K. Pandey, A. K. Agarwal, and H. C. Lohani, *Biol. Membr.* **8**, 74 (1983) [*CA* **103**, 37444 (1985)].
83MI2	A. G. Hammam, A. S. Ali, and N. M. Youssif, *Egypt. J. Chem.* **26**, 461 (1983) [*CA* **101**, 230459 (1984)].
83MIP1	A. Blade Font and J. Baratau Boixader, Span. Pat. 510,915 (1983) [*CA* **99**, 105275 (1983)].
83MIP2	M. I. Fernandez Fernandez, C. Fuentes Manso, M. Izquierdo Sanjose, and M. L. Lucero de Pablo, Span. Pat. 511,163 (1983) [*CA* **100**, 139136 (1984)].
83MIP3	M. I. Fernandez Fernandez, C. Fuentes Manso, M. Izquierdo Sanjose, and M. L. Lucero de Pablo, Span. Pat. 511,866 (1983) [*CA* **100**, 68317 (1984)].
83MIP4	M. I. Fernandez Fernandez, C. Fuentes Manso, M. Izquierdi Sanjose, A. Mosqueira Toribio, and M. L. Lucero de Pablo, Span. Pat. 513,504 (1983) [*CA* **103**, 123510 (1985)].
83NEP8202602	Farmitalia Carlo Erba S. P. A., Neth. Pat. 8,202,602 (1983) [*CA* **99**, 38481 (1983)].
83PHA25	K. Kottke, H. Kuehmstedt, and D. Knoke, *Pharmazie* **38**, 25 (1983).
83PHA367	K. Kottke, H. Kuehmstedt, and G. Griesner, *Pharmazie* **38**, 367 (1983).
83ZC215	S. Leistner, K. Hentschel, and G. Wagner, *Z. Chem.* **23**, 215 (1983).
83ZN(B)248	R. Wintersteiger and O. S. Wolfbeis, *Z. Naturforsch., B: Anorg. Chem., Org. Chem.* **38B**, 248 (1983).
84AP824	J. Reisch, A. S. M. El-Moghazy, and I. Mester, *Arch. Pharm. (Weinheim, Ger.)* **317**, 824 (1984).
84BCJ1138	H. A. El-Sherief, A. M. Mahmoud, and A. A. Esmaiel, *Bull. Chem. Soc. Jpn.* **57**, 1138 (1984).
84BRP2125785	V. A. Portnyagina, V. K. Karp, I. S. Barkova, F. P. Trinus, N. A. Mokhort, and A. G. Fadeicheva, Br. Pat. 2,125,785 (1984) [*CA* **101**, 90964 (1984)].
84CB1077	S. A. L. Abdel-Hady, M. A. Badawy, Y. A. Ibrahim, and W. Pfleiderer, *Chem. Ber.* **117**, 1077 (1984).
84CB1083	M. A. Badawy, S. A. L. Abdel-Hady, M. M. Eid, and Y. A. Ibrahim, *Chem. Ber.* **117**, 1083 (1984).
84CC1348	K. T. Potts and P. Murphy, *J. C. S. Chem. Commun.*, 1348 (1984).
84CCC1795	S. Stankovsky and M. Sokyrova, *Collect. Czech. Chem. Commun.* **49**, 1795 (1984).
84CJC2570	E. S. Hand and D. C. Barker, *Can. J. Chem.* **62**, 2570 (1984).
84CPB2160	K. Ozaki, Y. Yamada, and T. Oine, *Chem. Pharm. Bull.* **32**, 2160 (1984).
84EGP(D)206555	K. Kottke, H. Kuehmstedt, H. Wehlan, and H. Landmann, Ger. Pat. (East)DD 206,555 (1984) [*CA* **101**, 90970 (1984)].
84EGP(D)206996	K. Kottke, H. Kuehmstedt, H. Landmann, and H. Wehlan, Ger. Pat. (East)DD 206,996 (1984) [*CA* **102**, 149281 (1985)].
84EUP116948	G. H. Jones, M. C. Venuti, R. Alvarez, and J. J. Bruno, Eur. Pat. 116,948 (1984) [*CA* **102**, 78902 (1985)].
84EUP129258	F. Ishikawa and S. Ashida, Eur. Pat. 129,258 (1984) [*CA* **102**, 166773 (1985)].

148   MOHAMMED A. E. SHABAN *et al.*   [Refs.

84G525

A. K. El-Shafei and H. Abdel-Ghany, *Gazz. Chim. Ital.* **114**, 525 (1984).

84GEP(O)3231408

V. A. Portnyagina, V. K. Karp, I. S. Barkova, F. P. Trinus, N. A. MoKhort, A. G. Fadeicheva, G. I. Kozhushko, T. K. Ryabukha, A. G. Panteleimonov, *et al.*, Ger. Pat. Offen. 3,231,408 (1984) [*CA* **101**, 23,491 (1984)].

84H501

M. Sawada, Y. Furukawa, Y. Takai, and T. Hanafusa, *Heterocycles*, **22**, 501 (1984).

84IJC(B)161

S. B. Barnela and S. Seshadri, *Indian J. Chem., Sect. B* **23**, 161 (1984).

84IJC(B)1293

A. Kamal and P. B. Sattur, *Indian J. Chem., Sect. B* **23**, 1293 (1984).

84JCS(P1)1143

M. Lempert-Sreter, K. Lempert, and J. Moeller, *J. C. S. Perkin 1*, 1143 (1984).

84JCS(P1)1905

R. S. Atkinson, J. R. Malpass, K. L. Skinner, and K. L. Woodthorpe, *J. C. S. Perkin 1*, 1905 (1984).

84JHC369

M. Davis, R. J. Hook, and W. Y. Wu, *J. Heterocycl. Chem.* **21**, 369 (1984).

84JHC1049

A. O. Abdelhamid and C. Parkanyi, *J. Heterocycl. Chem.* **21**, 1049 (1984).

84JHC1411

E. P. Papadoupoulos, *J. Heterocycl. Chem.* **21**, 1411 (1984).

84JIC436

H. Singh, L. D. S. Yadav, and B. K. Bhattacharya, *J. Indian Chem. Soc.* **61**, 436 (1984).

84JIC721

Y. D. Kulkarni and S. H. R. Abdi, *J. Indian Chem. Soc.* **61**, 721 (1984).

84JIC1050

G. D. Gupta and H. K. Pujari, *J. Indian Chem. Soc.* **61**, 1050 (1984).

84JOC1964

E. E. Schweizer and K. J. Lee, *J. Org. Chem.* **49**, 1964 (1984).

84KGS983

O. P. Rudenko, M. M. Botoshanskii, Yu. A. Simonov, A. V. Bogatskii, and O. P. Povolotskaya, *Khim. Geterotsikl. Soedin.*, 983 (1984) [*CA* **102**, 6410 (1985)].

84MI1

M. P. Pandey, *Acta Cienc. Indica [Ser.] Chem.* **10**, 178 (1984) [*CA* **104**, 50842 (1986)].

84MI2

E. Pomarnacka, S. Angielski, and A. Hoppe, *Acta Pol. Pharm.* **41**, 141 (1984) [*CA* **102**, 166689 (1985)].

84MI3

J. Kalinowska-Torz, *Acta Pol. Pharm.* **41**, 161 (1984) [*CA* **102**, 185043 (1985)].

84MI4

P. Corti, E. Lencioni, C. Aprea, L. Micheli, and C. Murratzu, *Boll. Chim. Farm.* **123**, 95 (1984) [*CA* **101**, 90892 (1984)].

84MI5

S. Stankovsky and A. Filip, *Chem. Zvesti* **38**, 677 (1984) [*CA* **102**, 113417 (1985)].

84MI6

Y. Abe, K. Nishino, K. Nakamura, M. Kondo, K. Imagawa, and Y. Ikutani, *Osaka Kyoiku Daigaku Kiyo, Dai-3-bumon* **33**, 47 (1984) [*CA* **103**, 178223 (1985)].

84PHA717

K. Kottke and H. Kuehmstedt, *Pharmazie* **39**, 717, (1984).

84PHA867

K. Kottke and H. Kuehmstedt, *Pharmazie* **39**, 867 (1984)].

84S881

P. Molina, A. Arques, I. Cartagena, and M. V. Valcarcel, *Synthesis*, 881 (1984).

84TL4309

W. Verboom, M. R. J. Hamzink, D. N. Reinhoudt, and R. Visser, *Tetrahedron Lett.* **25**, 4309 (1984).

84USP4451448

G. E. Hardtmann and W. J. Houlihan, U. S. Pat 4,451,448 (1984) [*CA* **101**, 130708 (1984)].

84USP4451464   G. E. Hardtmann and W. J. Houlihan, U. S. Pat. 4,451,464 (1984)
                  [CA 101, 130707 (1984)].
84USP4452787   G. E. Hardtmann and W. J. Houlihan, U. S. Pat. 4,452,787 (1984)
                  [CA 101, 130709 (1984)].
85AP502   K. C. Liu and L. Y. Hsu, Arch. Pharm. (Weinheim Ger.) 318, 502
                  (1985).
85CC544   R. S. Atkinson, J. Fawcett, M. J. Grimshire, and D. R. Russell, J. C.
                  S. Chem. Commun., 544 (1985).
85CP1189509   R. W. Kierstead and J. W. Tilley, Can. Pat. 1,189,509 (1985) [CA
                  104, 50890 (1986)].
85CPB950   T. Higashino, H. Kokubo, and E. Hayashi, Chem. Pharm. Bull. 33,
                  950 (1985).
85CPB3336   F. Ishikawa, H. Yamaguchi, J. Saegusa, K. Inamura, T. Mimura, T.
                  Nishi, K. Sakuma, and S. Ashida, Chem. Pharm. Bull. 33, 3336
                  (1985).
85EUP133234   F. Ishikawa, Eur. Pat. 133,234 (1985) [CA 102, 203983 (1985)].
85EUP142057   M. Carson, R. A. Lemahieu, and J. W. Tilley, Eur. Pat. 142,057
                  (1985) [CA 103, 160530 (1985)].
85H273   M. G. Bock, R. M. DiPardo, D. W. Cochran, C. H. Homnick, and
                  R. M. Freidinger, Heterocycles 23, 273 (1985).
85H623   D. Ranganathan, S. Bamezai, and P. V. Ramachandran, Hetero-
                  cycles 23, 623 (1985).
85H2357   P. Molina, A. Arques, I. Cartagena, and M. V. Valcarcel, Hetero-
                  cycles 23, 2357 (1985).
85IJC(B)336   M. P. Jain, V. N. Gupta, and C. K. Atal, Indian J. Chem., Sect. B 24,
                  336 (1985).
85IJC(B)789   B. K. Chowdhury, E. O. Afolabi, G. Osuide, and E. N. Sokomba,
                  Indian J. Chem., Sect. B 24, 789 (1985).
85IJC(B)873   S. B. Barnela, S. Padmanabhan, and S. Seshadri, Indian J. Chem.,
                  Sect. B. 24, 873 (1985).
85IJC(B)1035   S. K. P. Sinha, M. P. Singh, Y. N. Singh, C. S. P. Singh, M. P. Shahi,
                  B. D. Singh, and P. Kumar, Indian J. Chem., Sect. B 24, 1035
                  (1985).
85JAP(K)60152416 H. Ogawa and N. Tanaka, Jpn. Kokai 60/152,416 (1985) [CA 104,
                  62069 (1986)].
85JAP(K)6075488   S. S. Pharmaceutical Co., Ltd., Jpn. Kokai 60/75,488 (1985) [CA
                  103, 142003 (1985)].
85JCS(P1)335   R. S. Atkinson and N. A. Gawad, J. C. S. Perkin 1, 335 (1985).
85JMC1387   F. Ishikawa, J. Saegusa, K. Inamura, K. Sakuma, and S. Ashida,
                  J. Med. Chem. 28, 1387 (1985).
85JOC1666   K. T. Potts, K. G. Bordeaux, W. R. Kuehnling, and R. L. Salsbury,
                  J. Org. Chem. 50, 1666 (1985).
85KGS1368   V. A. Kovtunenko, V. V. Ishchenko, A. K. Tyltin, A. V. Turov, and
                  F. S. Babichev, Khim. Geterotsikl. Soedin., 1368 (1985) [CA 104,
                  168435 (1986)].
85MI1   J. Kalinowska-Torz, Acta Pol. Pharm. 42, 112 (1985) [CA 105,
                  133838 (1986)].
85MI2   G. H. Sayed, M. Y. El-Kady, I. Abd-Elmawgoud, and M. Hamdy,
                  J. Chem. Soc. Pak. 7, 263 (1985) [CA 106, 18467 (1987)].
85PHA55   K. Kottke and H. Kuehmstedt, Pharmazie 40, 55 (1985).

85RRC611   E. I. Georgescu, F. Georgescu, M. Gheorghiu, E. G. Georgescu, and
                M. Petrovanu, *Rev. Roum. Chim.* **30**, 611 (1985) [*CA* **105**, 153019
                (1986)].
85S892   A. Kamal and P. B. Sattur, *Synthesis,* 892 (1985).
85URP1182043   I. I. Popov, S. L. Boroshko, and B. A. Tertov, U. S. S. R. Pat. Su
                1,182,043 (1985) [*CA* **105**, 226611 (1988)].
85USP4554272   M. G. Bock and R. M. Freidinger, U. S. Pat. 4,554,272 (1985) [*CA*
                104, 148923 (1986)].
85USP4559338   M. G. Bock, R. M. Freindinger, and B. E. Evans, U. S. Pat.
                4,559,338 (1985) [*CA* **105**, 42854 (1986)].
86AP188   K. C. Liu and M. K. Hu, *Arch. Pharm. (Weinheim, Ger.)* **319**, 188
                (1986).
86EGP(D)234013 M. Suesse, F. Adler, and S. Johne, Ger. Pat. (East)DD 234,013
                (1986) [*CA* **106**, 5074 (1987)].
86EUP181282   J. E. Francis and K. O. Gelotte, Eur. Pat. 181,282 (1986) [*CA* **105**,
                153069 (1986)].
86FES852   S. Vomero, M. Anzini, A. Segre, and E. Rossi, *Farmaco, Ed. Sci.*
                41, 852 (1986) [*CA* **107**, 23310 (1987)].
86H3075   A. Kamal, M. V. Rao, and P. B. Sattur, *Heterocycles* **24**, 3075
                (1986).
86IJC(B)489   U. S. Pathak, M. B. Devani, C. J. Shishoo, R. R. Kulkarni, V. M.
                Rakholia, V. S. Bhadti, S. Ananthan, M. G. Dave, and V. A.
                Shah, *Indian J. Chem., Sect. B* **25**, 489 (1986).
86IJC(B)709   S. B. Barnela and S. Seshadri, *Indian J. Chem., Sect. B* **25**, 709
                (1986).
86IJC(B)957   H. K. Gakhar, R. Gupta, and J. K. Gill, *Indian J. Chem., Sect. B* **25**,
                957 (1986).
86JAP(K)6150983 I. Ueda, M. Kato, and T. Ogawara, Jpn. Kokai 61/50,983 (1986) [*CA*
                105, 78958 (1986)].
86JAP(K)61115083 E. Takeuchi and T. Sato, Jpn. Kokai 61/115,083 (1986) [*CA* **105**,
                226621 (1986)].
86JCR(S)232   P. D. Kennewell, R. M. Scrowston, I. G. Shenouda, W. R. Tully,
                and R. Westwood, *J. Chem. Res., Synop.,* 232 (1986).
86JCS(P1)1215   R. S. Atkinson and M. J. Grimshire, *J. C. S. Perkin 1,* 1215 (1986).
86JCS(P1)2295   B. C. Uff and B. L. Joshi, *J. C. S. Perkin 1,* 2295 (1986).
86JHC43   P. Molina, A. Arques, I. Cartagena, and M. V. Valcarcel, *J. Hetero-*
                cycl. Chem. **23**, 43 (1986).
86JHC53   A. D. Dunn and K. I. Kinnear, *J. Heterocycl. Chem.* **23**, 53 (1986).
86JHC833   R. Murdoch, W. R. Tully, and R. Westwood, *J. Heterocycl. Chem.*
                23, 833 (1986).
86JHC1359   V. St. Georgiev, G. A. Bennett, L. A. Radov, D. K. Kamp, and L. A.
                Trusso, *J. Heterocycl. Chem.* **23**, 1359 (1986).
86KFZ690   L. N. Vostrova, T. A. Voronina, T. L. Karaseva, S. A. Gernega,
                E. I. Ivanov, A. M. Kirichenko, and M. Yu. Totrova, *Khim.-*
                Farm. Zh., **26**, 690 (1986) [*CA* **106**, 32976 (1987)].
86MI1   S. Botros, M. M. El-Enany, F. A. Amine, and L. N. Soliman, *Bull.*
                Fac. Pharm. (Cairo Univ.) **25**, 41 (1986) [*CA* **109**, 211005 (1988)].
86MI2   R. K. Saksena, A. Ali, P. Kant, and M. Kumar, *Indian Drugs* **24**, 16
                (1986) [*CA* **106**, 149259 (1987)].
86MI3   H. A. Hammouda and S. M. Hussain, *Sulfur Lett.* **4**, 51 (1986) [*CA*
                106, 84538 (1987)].

86MI4	K. C. Liu and M. K. Hu, *Tai-wan Yao Hsueh Tsa Chih* **38**, 85 (1986) [*CA* **108**, 56049 (1988)].
86MI5	K. C. Liu, S. W. Hsu, and M. K. Hu, *Tai-wan Yao Hsueh Tsa Chih* **38**, 242 (1986) [*CA* **108**, 94490 (1988)].
86MI6	M. E. Suh, *Yakhak Hoe Chi* **30**, 203 (1986) [*CA* **107**, 198236 (1987)].
86RRC365	E. I. Georgescu, F. Georgescu, M. D. Gheorghiu, P. Filip, and M. Petrovanu, *Rev. Roum. Chim.* **31**, 365 (1986) [*CA* **106**, 176314 (1987)].
86SC35	F. Ricciardi and M. M. Joullie, *Synth. Commun.* **16**, 35 (1986).
86T4481	D. Ranganathan, F. Farooqui, D. Bhattacharyya, S. Mehrotra, and K. Kesavan, *Tetrahedon* **42**, 4481 (1986).
86URP1235866	V. A. Portnyagina, V. K. Karp, I. S. Barkova, F. P. Trinus, and N. A. Mokhort, U. S. S. R. Pat. Su 1,235,866 (1986) [*CA* **105**, 208916 (1986)].
86USP4576750	S. M. Pitzenberger, U. S. Pat. 4,576,750 (1986) [*CA* **106**, 42855 (1986)].
86USP4588812	G. A. Saeva and V. S. Georgiev, U. S. Pat. 4,588,812 (1986) [*CA* **105**, 226618 (1986)].
86USP4593029	M. C. Venuti and J. J. Bruno, U. S. Pat. 4,593,029 (1986) [*CA* **106**, 18594 (1987)].
86USP4594191	M. G. Bock, B. E. Evans, R. M. Freidinger, and S. M. Pitzenberger, U. S. Pat. 4,594,191 (1981) [*CA* **105**, 133917 (1986)].
87AP166	K. C. Liu and M. K. Hu, *Arch. Pharm.* (*Weinheim, Ger.*) **320**, 166 (1987).
87AP569	K. C. Liu and L. Y. Hsu, *Arch. Pharm.* (*Weinheim, Ger.*) **320**, 569 (1987).
87AP1276	J. W. Chern, K. C. Liu, F. J. Shish, and C. H. Chan, *Arch. Pharm.* (*Weinheim, Ger.*) **320**, 1276 (1987).
87BSB797	J. Garin, C. Guillen, E. Melendez, F. L. Merchan, J. Orduna, and T. Tejero, *Bull. Soc. Chim. Belg.* **96**, 797 (1987) [*CA* **108**, 204587 (1988)].
87CZ373	H. Klein and G. Zinner, *Chem.-Ztg.* **111**, 373 (1987).
87EGP(D)251983	J. Spindler, G. Kempter, and M. Klepel, Ger. Pat. (East)DD 251,983 (1987) [*CA* **109**, 92993 (1988)].
87EUP236251	I. Kanmacher, J. F. Stambach, L. Jung, C. Schott, J. C. Stoclet, and C. Heitz, Eur. Pat. 236,251 (1987) [*CA* **109**, 6429 (1988)].
87H1841	Y. Ahmad, T. Begum, I. H. Qureshi, Atta-ur-Rahman, K. Zaman, C. Xu, and J. Clardy, *Heterocycles* **26**, 1841 (1987).
87H2371	J. Garin, C. Guillen, E. Melendez, F. L. Merchan, J. Orduna, and T. Tejero, *Heterocycles* **26**, 2371 (1987).
87IJC(B)983	P. S. N. Reddy and C. N. Raju, *Indian J. Chem., Sect. B* **26**, 983 (1987).
87JHC107	E. A. Adegoke, B. I. Alo, and O. B. Familoni, *J. Heterocycl. Chem.* **24**, 107 (1987).
87JHC227	E. A. A. Hafez, Z. E. S. Kandeel, and M. H. Elnagdi, *J. Heterocycl. Chem.* **24**, 227 (1987).
87JMC295	G. H. Jones, M. C. Venuti, R. Alvarez, J. J. Bruno, A. H. Berks, and A. Prince, *J. Med. Chem.* **30**, 295 (1987).
87JOC1644	G. Buchi, *J. Org. Chem.* **52**, 1644 (1987).
87JOC1810	E. E. Schweizer, J. E. Hayes, A. Rheingold, and X. Wei, *J. Org. Chem.* **52**, 1810 (1987).

87JOC2469   J. W. Tilley, D. L. Coffen, B. H. Schaer, and J. Lind, *J. Org. Chem.* **52**, 2469 (1987).
87KGS1527   L. M. Yun, K. O. Nazhimov, S. Masharipov, R. A. Samiev, S. A. Makhmudov, S. S. Kasymova, S. N. Vergizov, K. A. V'yunov, and Kh. M. Shakhidoyatov, *Khim. Geterotsikl. Soedin.*, 1527 (1987) [*CA* **109**, 92919 (1988)].
87KGS1673   I. I. Popov, S. L. Boroshko, B. A. Tertov, S. P. Makarov, A. M. Simonov, and B. Ya. Simkin, *Khim. Geterotsikl. Soedin.*, 1673 (1987) [*CA* **109**, 110365 (1988)].
87LA103   P. Molina, A. Arques, and M. V. Vinader, *Liebigs Ann. Chem.*, 103 (1987).
87MI1   K. C. Liu and L. Y. Hsu, *Tai-wan Yao Hsueh Tsa Chih* **39**, 21 (1987) [*CA* **109**, 230935 (1988)].
87MI2   K. C. Liu and L. Y. Hsu, *Tai-wan Yao Hsueh Tsa Chih* **39**, 54 (1987) [*CA* **109**, 230936 (1988)].
87MI3   K. C. Liu and M. K. Hu, *Tai-wan Yao Hsueh Tsa Chih* **39**, 204 (1987) [*CA* **109**, 230937 (1988)].
87MIP1   Laboratorio Farmaceutico Quimico-Lafarquim S. A., Span. Pat. 555,233 (1987) [*CA* **108**, 131855 (1988)].
87SC1449   P. Molina, A. Arques, M. L. Garia, and M. V. Vinader, *Synth. Commun.* **17**, 1449 (1987).
87USP4670434   M. C. Venuti, U. S. Pat. 4,670,434 (1987) [*CA* **107**, 96736 (1987)].
87USP4713383   J. E. Francis and K. O. Gelotte, U. S. Pat. 4,713,383 (1987) [*CA* **109**, 129041 (1988)].
88CCC329   M. F. Abdel-Megeed and A. Teniou, *Collect. Czech. Chem. Commun.* **53**, 329 (1988).
88CZ135   W. Ried and S. Aboul-Fetouh, *Chem.-Ztg.* **112**, 135 (1988).
88EGP(D)253623 K. Kottke, H. Kuehmstedt, I. Graefe, H. Wehlan, and D. Knoke, Ger. Pat (East)DD 253,623 (1988) [*CA* **109**, 170464 (1988)].
88EGP(D)258232 S. Leistner, R. Simon, G. Wagner, L. Vieweg, D. Lohmann, and G. Laban, Ger. Pat. (East)DD 258,232 (1988) [*CA* **110**, 114,856 (1989)].
88EGP(D)258815 J. Spindler, G. Kempter, C. Schroeder, and M. Klepel, Ger. Pat. (East)DD 258,815 (1988) [*CA* **110**, 173248 (1989)].
88GEP(O)3634532 G. Hamprecht, G. Seybold, N. Meyer, and B. Wuerzer, Ger. Pat. Offen. 3,634,532 (1988) [*CA* **109**, 73466 (1988)].
88IJC(B)342   P. S. N. Reddy and P. P. Reddy, *Indian J. Chem., Sect. B.* **27**, 342 (1988).
88IJC(B)578   J. S. Rao, P. Neelakantan, and S. N. Rao, *Indian J. Chem., Sect. B* **27**, 578 (1988).
88JHC1399   H. Bartsch and T. Erker, *J. Heterocycl. Chem.* **25**, 1399 (1988).
88JMC466   J. W. Tilley, B. Burghardt, C. Burghardt, T. F. Mowles, F. J. Leinweber, L. Klevans, R. Young, G. Hirkaler, K. Fahrenholtz, *et al., J. Med. Chem.* **31**, 466 (1988).
88JMC1220   S. Clements-Jewery, G. Danswan, C. R. Gardner, S. S. Matharu, R. Murdoch, W. R. Tully, and R. Westwood, *J. Med. Chem.* **31**, 1220 (1988).
88JOC1873   Y. Nakagawa and R. V. Stevens, *J. Org. Chem.* **53**, 1873 (1988).
88M1405   H. G. Henning and H. Haber, *Monatsh. Chem.* **119**, 1405 (1988).
88MI1   A. O. Abdelhamid, N. M. Abed, and A. M. Farag, *An. Quim., Ser. C* **84**, 22 (1988) [*CA* **109**, 190360 (1988)].

88MI2   K. C. Liu and M. K. Hu, *Chung-hua Yao Hsueh Tsa Chih* **40,** 117
                (1988) [*CA* **111,** 57667 (1988)].
88RRC291   A. K. El-Shafei and A. M. El-Sayed, *Rev. Roum. Chim.* **33,** 291
                (1988) [*CA* **109,** 190359 (1988)].
88RRC981   M. F. Abdel-Megeed and A. Teniou, *Rev. Roum. Chim.* **33,** 981
                (1988) [*CA* **112,** 20971 (1990)].
88S336   C. V. R. Sastry, K. S. Rao, V. S. H. Krishnan, K. Rastogi, and M. L.
                Jain, *Synthesis,* 336 (1988).
89CC835   A. Kamal and P. B. Sattur, *J. C. S. Chem. Commun.,* 835 (1989).
89DOK628   V. M. Kisel, V. A. Kovtunenko, A. V. Turov, A. K. Tyltin, and F. S.
                Babichev, *Dokl. Akad. Nauk SSSR* **306,** 628 (1989) [*CA* **112,**
                555759 (1989)].
89IJC(B)120   M. K. A. Ibrahim, *Indian J. Chem., Sect. B* **28,** 120 (1989).
89IJC(B)200   N. Tiwari, B. Chaturvedi, and Nizamuddin, *Indian J. Chem., Sect.
                B* **28,** 200 (1989).
89IJC(B)274   S. K. P. Sinha ad P. Kumar, *Indian J. Chem., Sect. B* **28,** 274 (1989).
89JHC595   C. Y. Shiau, J. W. Chern, J. H. Tien, and K. C. Liu, *J. Heterocycl.
                Chem.* **26,** 595 (1989).
89JHC713   N. P. Peet, *J. Heterocycl. Chem.* **26,** 713 (1989).
89JHC1495   J. Reisch, C. O. Usifoh, and J. O. Oluwadiya, *J. Heterocycl. Chem.*
                26, 1495 (1989).
89JPR537   C. Bischoff and E. Schroeder, *J. Prakt. Chem.* **331,** 537 (1989).
89KGS272   I. I. Popov, S. L. Boroshko, B. A. Tertov, and E. V. Tyukavina,
                Khim. Geterosikl. Soedin., 272 (1989) [*CA* **111,** 232723 (1989)].
89KGS408   K. V. Fedotov and N. N. Romanov, *Khim. Geterosikl. Soedin.,* 408
                (1989) [*CA* **111,** 153740 (1989)].
89MI1   S. Stankovsky and A. Boulmokh, *Chem. Pap.* **43,** 433 (1989) [*CA*
                112, 118749 (1990).
89MI2   A. M. El-Reedy, S. M. Hussain, A. S. Ali, and F. Abdel-Motty,
                Phosphorus, Sulfur Silicon Relat. Elem. **42,** 231 (1989) [*CA* **112,**
                35791 (1990)].
89MI3   R. Guryn, *Pol. J. Chem.* **63,** 273 (1989) [*CA* **112,** 178873 (1990)].
890PP163   J. Reiter and L. Pongo, *Org. Prep. Proced. Int.* **21,** 163 (1989).
89T4263   P. Molina, M. Alajarin, A. Vidal, M. De la Concepcion Foces-Foces,
                and F. Cano Hernandez, *Tetrahedron* **45,** 4263 (1989).
89USP4871732   N. P. Peet and S. Sunder, U. S. Pat. 4,871,732 (1989) [*CA* **112,** 179036
                (1990)].
90JOC344   M. A. Badawy, S. A. Abdel-Hady, A. H. Mahmoud, and Y. A.
                Ibrahim, *J. Org. Chem.* **55,** 344 (1990).
90UP1   M. A. E. Shaban, M. A. M. Taha, and E. M. Sharshira, unpublished
                results.
90UP2   M. A. E. Shaban, M. A. M. Taha, and H. A. M. Hamouda, unpub-
                lished results.

3(2H)-Isoquinolinones and Their Saturated Derivatives

LÁSZLÓ HAZAI*

Department of Pharmacology, Institute of Experimental Medicine,
Hungarian Academy of Sciences, H-1450 Budapest, Hungary

* Present address: Research Group for Alkaloid Chemistry, Hungarian Academy of Sciences, Technical University, H-1521 Budapest, Gellért tér 4., Hungary.

I. Introduction

Of the lactam derivatives of quinoline and isoquinoline, 2(1*H*)-quinolinones and 1(2*H*)-isoquinolinones (carbostyrils and isocarbostyrils) have long been known. Although over 100 years have passed since Gabriel discussed the tautomerism of 3(2*H*)-isoquinolinone derivatives (**1**) (3-isoquinolones, 3-hydroxyisoquinolines or 3-isoquinolinols), these com-

(**1a**)   (**1b**)

(**2**)   (**3**)

pounds have not been the center of interest for many years. The unsubstituted compound (**1**) was first prepared by Boyer and Wolford (56JOC1297); the synthesis of the 1,4-dihydro derivative, lactam **2**, was achieved by Braun and Reich (25LA225). Compound **3**, saturated in the aromatic ring and synthesized by Basu and Banerjee (35LA243), is also related to the just-mentioned isoquinolinones. (The more saturated derivatives are not within the scope of this review.)

After the previous research, especially the tautomerism studies on **1**, interest in these compounds began to increase only in the last decades. New synthetic methods have been elaborated, and a number of new derivatives were synthesized. Our laboratory contributed to this work by investigating the syntheses and reactions of compounds **1**, **2**, and **3**.

Today almost 300 papers and patents deal with these compounds, probably because organic chemistry can less and less avoid the demand and pressure to produce more and new molecules with potential biological activity. Indeed, among these types of compounds, several molecules also exist that look promising from this point of view (see Section V,B).

Within the limits of this short review, it is impossible to discuss all the relevant references; intention is to give a general survey of the most important results in the chemistry of these heterocyclic lactams.

II. 3(2*H*)-Isoquinolinones

In the literature, different names are used for the title compounds, but not without reason (see Section II,A). In this review, however, in most cases, the name 3(2H)-isoquinolinones is used for the lactam structure to emphasize the connection with derivatives saturated in the hetero ring (see Section III) or in the aromatic ring (see Section IV) of the isoquinolinone skeleton or both.

A. SPECTROSCOPIC PROPERTIES, TAUTOMERISM

The tautomerism of these compounds has been discussed comprehensively [63AHC(1)352; 76MI1]; the latter reference gives a detailed review about the tautomeric equilibria of some derivatives of 1. The conclusions are based largely on UV evidence supported by IR, NMR, X-ray crystallography, and theoretical calculations.

In the course of synthetic work related to laudanosine, which is a 1-benzyl-3(2H)-isoquinolinone, it appeared that four prototropic isomers are possible (72JHC853), demonstrating the complexity of the problem. According to the author, the 3-isoquinolone tautomer (4a) with the o-quinonoid structure predominates over the hydroxy form (4b), and the

(4a)   **(4b)**   **(4c)**   **(4d)**

analogous isochromanone derivative has preference over the **4d** structure. The unsaturated amide (**4c**) cannot be regarded as a proven tautomeric form. Authors from Poland investigated the UV spectra of some 3(2H)-isoquinolinones substituted in position 1 with aryl and aralkyl groups in different solvents (77RC691). It has been found that, for example, the 1-β-naphthyl derivative (**5**) exists exclusively as the lactim tautomer in

(7)

(6)

(5)

diethyl ether, carbon tetrachloride, and sulfuric acid; and in benzene, ethanol, and chloroform, this form predominates.

In the case of the 4-chloro-1-phenyl derivative (6) (88S683), more data support the lactim structure. The UV spectrum of 7 (76JMC395) in water has only the lactam band, whereas in diethyl ether, only the lactim structure is shown. In ethanol, the lactam tautomer predominates, but in chloroform, the lactim tautomer predominates.

A number of 3(2H)-isoquinolinone derivatives (8) have been synthesized by Deák and his research group (Section II,B,3) to make possible a general evaluation of their lactam–lactim tautomerism [84ACH(116)303; 89ACH869].

Ar: (substituted)phenyl, pyridyl;

R¹: H, alkyl, aryl, aralkyl, pyridylalkyl;

R²: H, alkyl;

R³, R⁴: H, OMe;

R², R³: benzo [f], R³, R⁴: benzo [g],

R⁴, R⁵: benzo [h] annellation

On the basis of UV spectroscopic data of derivatives with fixed lactam and fixed lactim structures, it was established that the bands characteristic of the lactam form appear between 385 and 438 nm, while those characteristic of the lactim form are found in the range 312–360 nm, depending on the substituents (in 95% ethanol). Coumpound 9, in which the bands are shifted to the higher wavelengths because of the linearly annellated benzene ring is an exception. Most of these compounds exist predominantly in 95% alcohol as the lactim tautomer. In compound 8a, the ratio of the

(8a) R = veratryl
(8b) R = Me
(8c) R = Ph

(9)

R = CH$_2$CONH$_2$

(10)

lactam and lactim forms is nearly equal. In carbon tetrachloride, 8a contains significantly more lactim tautomer than in 95% ethanol. In ethyl acetate, benzene, and even in the dipolar aprotic dimethylformamide (DMF), only the band characteristic of the lactim form of **8a** could be detected. In acetic acid, **8a** exists exclusively in the lactam form. In the case of **8b** and **8c**, according to the spectra recorded in xylene, the compounds can occur in the lactam form even in this solvent.

In the isoquinolinones annellated with a benzene ring at the [*f*] and [*h*] stites, the lactam tautomer predominates, except for the benzo[*g*]isoquinolinone (**9**), which exists in the lactim form [89JCR(S)340] similar to the 1,4-diphenyl derivatives [72JCS(P1)2722]. Cyclopenta[*f*]-isoquinolinone is present exclusively in the lactam form in both water and 95% alcohol (75JMC399).

IR spectra of the crystalline state and of carbon tetrachloride solutions can also give information about the structure of tautomers. In the crystalline state, the lactim structure is characterized by the νC=N band at 1620 cm^{-1} and the region between 3200–2100 cm^{-1} with partial maxima; in carbon tetrachloride, νOH appears at 3525 cm^{-1}. In compounds annellated with a benzene ring (except for **9**), it is possible that the crystalline state consists of a mixture of the lactam and lactim tautomers (unsaturated lactam band at νC=O 1630–1640 cm^{-1} and νC=N at 1620 cm^{-1}); however, these bands usually are not distinct.

After investigating a number of compounds of type **1,** we can state that this type of tautomerism depends both on the substituents and on the solvent and shows considerable variation. Thus, the compounds can be reliably characterized only by characteristic UV spectra or occasionally by IR spectroscopy, especially in carbon tetrachloride.

B. SYNTHESIS

1. From o-Substituted Arylacetic Acid Derivatives

One of the most widespread methods of synthesizing 3(2*H*)-isoquinolinones is the cyclization of esters of *o*-acyl-phenylacetic acids with appropriate amines. These esters with conc. ammonium hydroxide or ethanolic ammonia (Scheme 1) give **11a** and **11b** (52JCS1763). With aqueous methylamine, the *N*-methyl derivative of **11a** can be prepared (64CB667). This type of cyclization has also been achieved with methanolic ammonia [67JCS(B)590] and with liquid ammonia [69JCS(C)1729]. The reaction succeeds without the presence of alkoxy groups in the aromatic ring (77RC691).

In the case of 1-veratryl derivative (**11c**), isoquinolinones unsubstituted at the ring nitrogen can also be synthesized starting from *o*-acylarylacetic

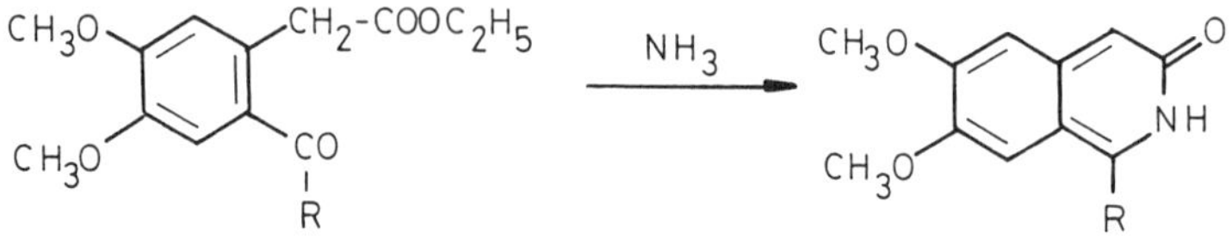

(**11a**) R = Me
(**11b**) R = Ph
(**11c**) R = veratryl

SCHEME 1

acids by heating them with ammonium acetate in acetic acid. Reactions with amines give the N-alkyl derivatives (70JGU249, 70JHC1229; 72JHC853, 72JMC1131; 82JHC1319, 82JHC1469; 84GEP3227741).

In addition, the following o-substituted arylacetic acid derivatives are suitable starting materials for the syntheses of 3(2H)-isoquinolinones: o-formylphenyl-acetamides [71T4653; 80JCS(P1)2013], o-cyanophenyl-acetic esters (68BSF3403), and o-cyanophenylacetic acid chlorides [68AG(E)484].

2. From O-Heterocycles

In preparing 6-alkoxy and 6,7-dialkoxy derivatives, it is convenient to react esters **12** with an acid anhydride in the presence of a Lewis acid or, preferably, perchloric acid to produce a benzopyrylium salt (**14**), presumably via **13** (70KGS200) (Scheme 2). Reaction of **14** with ammonia gives then the isoquinolinols **15** in high yields, where X = alkoxy, R^1 = H, Me, Pr, i-Pr, and R^2 = H or alkyl group (88JMC1363).

Benzopyrones are also useful starting materials for preparing 3(2H)-isoquinolinones. Similarly, **15** and its N-substituted derivatives can be synthesized with ammonia or with amines [70JCS(C)536].

Several authors cited previously (Section II,B,1) have found that, besides the direct cyclization of o-acyl-arylacetic acid derivatives, a suitable method of synthesizing the title compounds involves the reaction of isochromanones with amines. This reaction has been investigated in detail (73JHC317).

3. From 1,4-Dihydro-3(2H)-isoquinolinones

The aromatization of 1,4-dihydro-3(2H)-isoquinolinones resulting in the corresponding 3(2H)-isoquinolinones is a known but not common proce-

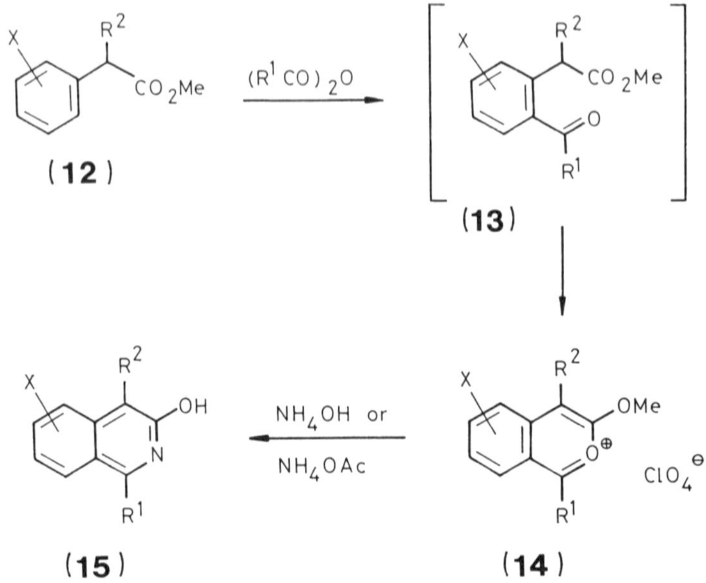

SCHEME 2

dure. These reactions can be achieved usually by heating the starting compounds with sulfur or by refluxing them in diisopropyl-benzene with palladium-on-charcoal (66BSF556; 76JMC395). Some 1,4-disubstituted derivatives give the unsaturated 3(2*H*)-isoquinolinones on heating in DMF or in dimethyl sulfoxide (DMSO) with NaH or with the Avramoff reagent at an elevated temperature (82JHC49).

Methods frequently used for preparing the title compounds are the condensations of the methylene group in position 4 of 1,4-dihydro-3(2*H*)-isoquinolinones and the subsequent rearrangement (Scheme 3).

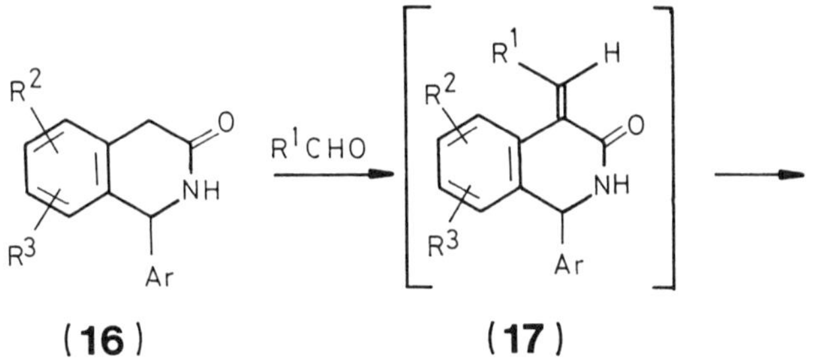

SCHEME 3

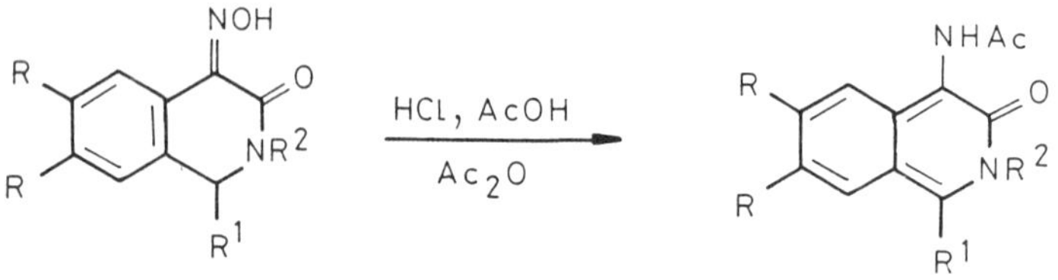

(19)   (20)

SCHEME 4

The scope and limitations of these types of reactions were studied in detail by Hungarian researchers synthesizing a number of derivatives (18) [73ACH(79)113; 77JHC583; 79ACH(102)305; 82JHC49] [R^1 = H, aryl, heteroaryl; R^2,R^3 = H, Me, MeO and benzo-annellation; Ar: (substituted) phenyl, heteroaryl]; the reaction mechanism was also elucidated (Section III,B,1).

3(2H)-Isoquinolinones substituted with an amino group at position 4 (20) (Scheme 4) can be prepared by rearranging 4-oxyimino compounds (19) under conditions similar to those used for Beckmann rearrangement [87JCR(S)95; 88ACH289], where R = H, MeO; R^1 = (substituted) phenyl and R^2 = H, Me. The bifunctional character of these new compounds could be well used after hydrolysis (Section V,A).

4. Miscellaneous Methods

Diazotation of 3-aminoisoquinoline (56JOC1297; 61JOC803) and the alkaline hydrolysis of 3-chloroisoquinoline (70TL1209) give 3(2H)-isoquinolinone (1). Homophthalimides on halogenation can advantageously furnish compounds 1 substituted with a halogen atom in position 1 (52MI1). Reactions between substituted phenylacetamides and chloroacetyl chloride in the presence of a Friedel–Crafts catalyst give 1-halomethyl-3(2H)-isoquinolinones [77JAP(K)77156877]. N-Formyl phenylacetamides are also useful starting materials for synthesizing derivatives of 1 by cyclization in acidic medium (74GEP2330218). Arylacetyl chlorides react with methyl thiocyanate in the presence of tin tetrachloride to yield 1-methylthio-3(2H)-isoquinolinones (76CC695). The reaction was also extended to syntheses of benzo- and pyrrolo derivatives.

A Pomeranz-Fritsch type of cyclization in sulfuric or polyphosphoric acid (Scheme 5) can be used to prepare many of the desired compounds listed here (22) (76GEP2623226; 78H1197; 88JMC1363) (R = Me, Et; R^1 = H, Me, Bu, Ph; R^2 = H, Me, $C_6H_5CH_2$; R^3, R^4, R^5 = H, Cl, Me, MeO). A new modification of intramolecular amidoalkylation, the reac-

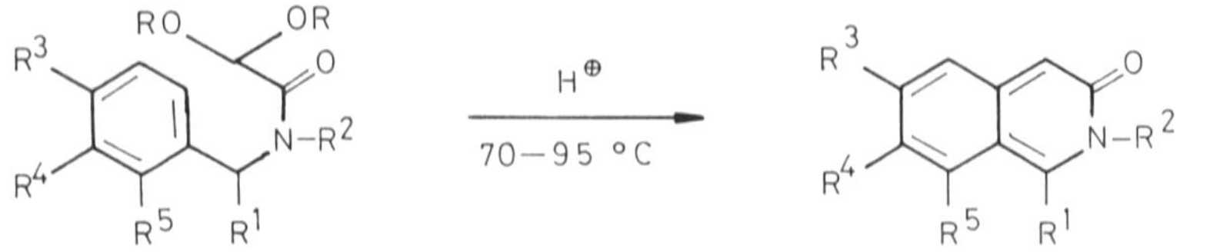

(21)   (22)

SCHEME 5

tions of arylacetic acids and substituted benzamides with POCl$_3$, gives initially N-substituted 3(2H)-isoquinolinones (82S486). The latter compounds can also be obtained by a retro Diels–Alder reaction from the corresponding tricyclic starting derivatives (83H1367) and by Vilsmeier–Haack cyclization of specially substituted anilides (84S349).

C. REACTIONS

1. Alkylation

Previously, in the course of studies on the tautomerism of the title compound **1**, the syntheses of the fixed lactam **23** and the fixed lactim **24** were carried out because of the importance of their UV spectra, which allowed precise assignment of the appropriate wavelength values

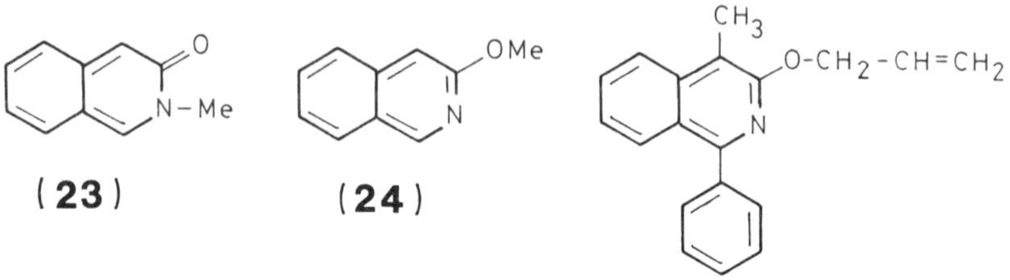

(23)   (24)

(25)

[67JCS(B)590; 69JCS(C)1729; 72JCS(P1)2722]. The usual alkylating agents (methyl iodide, diazomethane) under normal reaction conditions gave mixtures of products alkylated on the nitrogen and oxygen atoms. These mixtures can usually be separated without any difficulty. Reaction with triethyloxonium tetrafluoroborate and subsequent alkylation, e.g., with allyl bromide via the silver salt, gives exclusively the O-alkyl derivatives (e.g., **25**) [67JOC59; 72JMC1131; 77JHC583; 84ACH(116)303]. It is worthy to note that a number of dialkylamino alkyl ethers have been prepared by the reaction of isoquinolinones with alkyl halides in the presence of sodium hydride, and they were separated from the N-alkylated byproducts (84GEP3227741).

2. Reduction, Catalytic Hydrogenation

Several unsaturated 3(2H)-isoquinolinone derivatives have been hydrogenated in ethanol in the presence of PtO_2 catalyst, resulting in the corresponding 1,4-dihydro-3(2H)-isoquinolinones (71T4653) (where R^1 = H, Me; R^2 = R^3 = MeO or R^2, R^3 = methylene group) (Scheme 6). In the case of compound **26** (R^1 = Me), the reduction takes place also with zinc-hydrochloric acid or $NaBH_4$. In the course of the synthesis of laudanosine, the intermediate compound **27** (R^1 = R^2 = R^3 = Me, substituted with a veratryl group in position 1) was prepared under similar reduction conditions (72JHC853). Hydrogenation of the 1-methylthio-3(2H)-isoquinolinone in the presence of Raney-Nickel catalyst gives the unsubstituted parent compound **2** (74FRP2207720).

SCHEME 6

1-Aryl-3(2H)-isoquinolinones (**28**) can be readily hydrogenated in acetic acid in the presence of palladium-on-charcoal (Scheme 7). The end products are mostly the 5,6,7,8-tetrahydroisoquinolinones (**29**), saturated in the homoaromatic ring (77JHC583; 89JHC609), where R = alkyl or aralkyl group. The stereochemistry of the 1,4-dihydro-3(2H)-isoquinolinones (**30**),

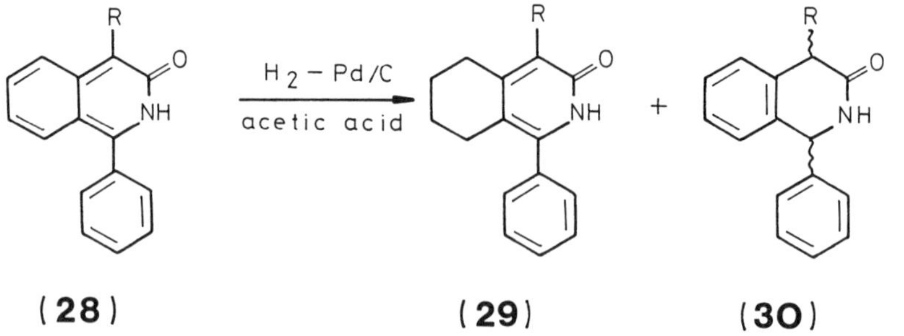

SCHEME 7

isolated as byproducts, was elucidated by means of NMR spectroscopy (88T6861). The derivatives with a fused benzene ring can be hydrogenated analogously. In the case of benzo[*f*]annellation, the 9,10-dihydroiso-

(**31**) R = H

(**32**) R = CH$_2$Ph

(**33**)

quinolinones (**31** and **32**) are obtained, in benzo[*h*]annellation, the 5,6-dihydro compound (**33**) is the main product. In both cases, small amounts of compounds saturated in the hetero ring are also formed (89JHC609).

3. Diels–Alder Reactions

Reactions between 3(2*H*)-isoquinolinones and substituted maleic anhydrides, after hydrolysis of the adducts and oxidative decarboxylation, result in derivatives bridged with an unsaturated carbon chain (**34**)

R^1 = H, alkyl, aralkyl etc.;

R^2, R^3, R^5 = H, Me;

R^4 = H, Me, Ph;

R^6 = H, Me, MeO, F, Cl

(**34**)

(73FRP2111765). The Diels–Alder reactions of unsubstituted and 1-methyl-, 1-veratryl-6,7-dimethoxy-3(2*H*)-isoquinolinones were studied with *N*-phenylmaleimide, maleic anhydride, and tetracyano-ethylene dienophiles [64CB667; 69JCS(C)1729; 70TL1209; 71DIS(B)5268; 72JHC853]; also the stereochemistry of some of the adducts was investigated [69JCS(C)1729].

A number of adducts of substituted 3(2*H*)-isoquinolinones (**35**) have been synthesized by refluxing the starting compounds (heterodienes and

$$X = \quad \diagdown NPh, \quad \diagdown O, \quad \diagdown NCONH_2, \quad \diagdown NH$$

(35)

dienophiles) in xylene for several hours [83ACH(113)237; 84ACH(117)99]. The substituents can be of various kinds; R^1 = (substituted) phenyl, heteroaryl; R^2 = H, alkyl, aralkyl, heteroarylalkyl; R^3 = H, Me; R^4, R^5 = H, MeO; R^3–R^6 = benzo[f]-, [g]-, [h]-annellation. Isoquinolinones substituted on the lactam nitrogen are also suitable as heterodienes [86ACH(121)263]. Contrary to the foregoing reactions, 1,4-additions of the 4-benzyl-1-phenylbenzo[g]isoquinolin-3-ol, existing exclusively in the lactim form, give the 5,10-adducts. The *endo* and *exo* isomers were separated [89JCR(S)340]. Detailed stereochemical investigations of the adducts were done by NMR spectroscopy (85CJC1001).

With the well-known dienophile dimethyl acetylene-dicarboxylate, which contains a C≡C triple bond, the product isolated was that of a Michael addition at the nitrogen atom of the hetero ring [85ACH(120)271].

4. Miscellaneous

The N-methyl derivatives of 3(2H)-isoquinolinones (**36**) are sensitive to oxidation: this reaction takes place via an endoperoxide intermediate (**37**) (70TL1209; 71T4653).

Rearrangement of the oxygen atoms can lead to the formation of products **38** and **39** or, by further oxidation, **40** (82JHC1319; 82JHC1469) (Scheme 8).

3-Allyloxyisoquinoline is rearranged on heating into the 4-ally-3(2H)-isoquinolinone (67JOC59). Other substituted 3(2H)-isoquinolinones (except those with the substituent in position 4) can be nitrated (87USP4714705; 89JMC990) or brominated to 4-nitro- or 4-bromo derivatives.

Some derivatives have also been subjected to photodimerization reactions [69JCS(C)1729; 82JHC1319].

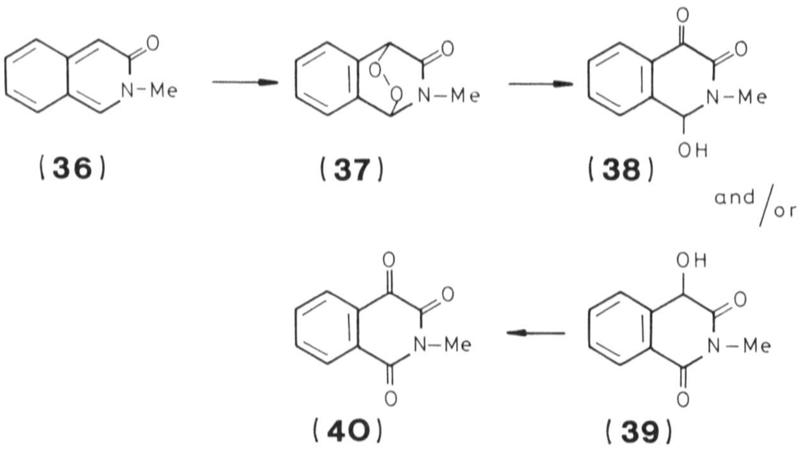

(36)   (37)   (38)

and / or

(40)   (39)

SCHEME 8

III. 1,4-Dihydro-3(2H)-isoquinolinones

1,4-Dihydro-3(2H)-isoquinolinones (2 and its derivatives) (Section I) are structural isomers of the well-known dihydrocarbostyril and dihydro-isocarbostyril nitrogen heterocycles. However in 2, the —CO—NH— lactam group is separated from the aromatic benzene ring by two methylene groups. The effect of this is apparent in the limitation of synthetic methods and even more so in the differences in reactivity.

A. SYNTHESIS

1. Cyclization of N-Substituted Benzylamines

N-Benzylmandelic acid amide (66BSF556) or its O-acyl derivatives (56USP2759936; 57USP2809969) can be converted to the corresponding 4-aryl-1,4-dihydro-3(2H)-isoquinolinones under the conditions of acidic cyclization (H₂SO₄, polyphosphoric acid: PPA). This reaction is also suitable for the syntheses of 4-aryl derivatives condensed with a pyrrole heterocycle (78KGS1279). The same principle was followed in the cyclization of more heavily substituted, for example, cyclic benzylamines (41) (84JMC943), where Ar is a substituted phenyl group (Scheme 9).

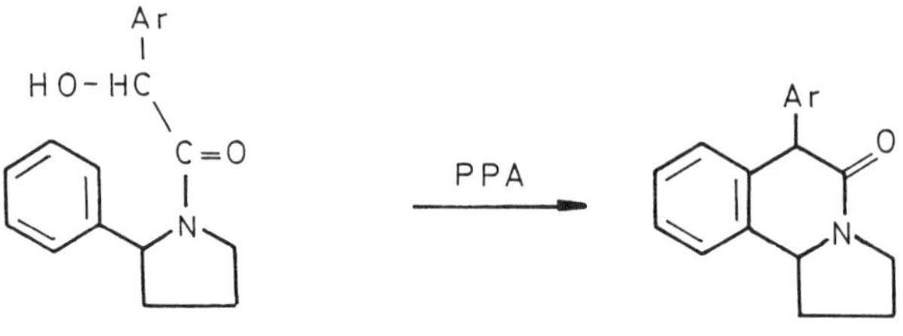

(41)   (42)

SCHEME 9

This type of cyclization extended to properly substituted α-aryl mande-lic acid amides results in 4,4-diaryl-substituted isoquinolinones (52ZOB697; 57ZOB1901; 66KGS734; 68KGS1033; 70KGS954).

N-Chloroacetylbenzylamines (**43**) were subjected to photocyclization by Japanese researchers (Scheme 10). In these examples, the aromatic ring has one or more hydroxy substituents (74TL1181; 77H251, 77T489; 81CPB128). Some chloroacetyl derivatives could be cyclized in the presence of $ZnCl_2$ (87S637).

ortho-Haloaralkyl amines are also used to prepare 1,4-dihydro-3(2H)-isoquinolinones (67JHC149; 81CC1074). N-Trimethylsilyl amide (**45**), un-der the conditions shown in the Scheme 11, gives compound **2** (85JA435). By this method and its modifications, derivatives of **2** with different sub-stituents can be synthesized.

N-Benzyl-α-chloro-α-methylthioacetamides reacting with tin tetrachlo-ride (81S534) and N-benzyldiazoacetamides under acidic cyclization con-ditions (88TL2643) are also appropriate starting materials for synthesizing the title compounds.

Other possibilities are available for achieving the cyclization of N-

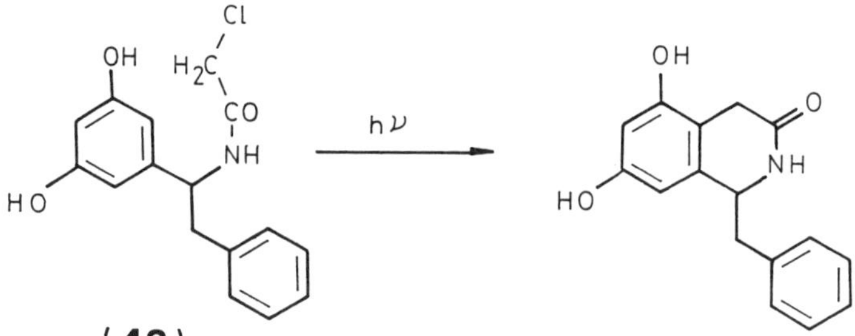

(43)   (44)

SCHEME 10

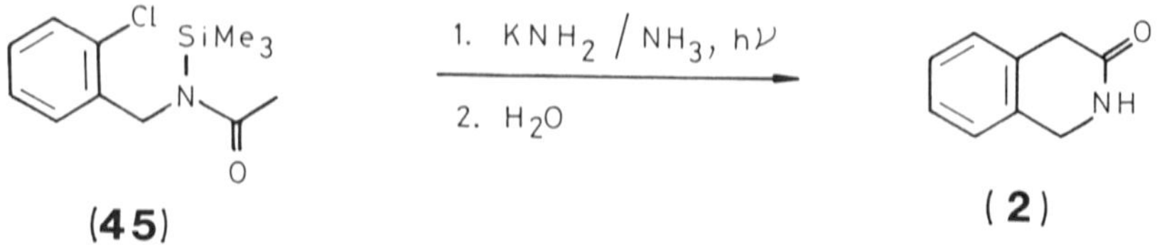

(45)   (2)

SCHEME 11

substituted benzylamines, resulting in isoquinolinones with moderate substitutions. One of these methods was used to synthesize cheryllin (78JA1548).

2. From Arylacetic Acid Derivatives

Most possibilities and most examples in the syntheses of 1,4-dihydro-3(2*H*)-isoquinolinone (2) and its derivatives involve the cyclization of the corresponding arylacetic acid derivatives. The opportunity arises from the lactam structure of compound 2. With suitable reagents, different *o*-aminomethyl-, *o*-hydroxymethyl-, or *o*-chloromethyl-phenylacetic acid derivatives (25LA225; 66JPR12; 69CJC864; 84JHC297) and *o*-hydroxymethyl phenylacetonitriles (85S114) can be cyclized.

A number of 1-aryl derivatives of 1,4-dihydro-3(2*H*)-isoquinolinones with different substituents were synthesized, by Hungarian researchers, via the cyclization of aromatic aldehydes and arylacetonitriles (or amides) through corresponding bisamide intermediates [68ACH(55)125; 69ACH(59)119, 69ACH(60)177; 76ACH(88)87; 86ACH(121)237]. Investigation of the reaction mechanism revealed the details of the cyclization [73ACH(76)299, 73ACH(77)217]; the reaction steps are shown for the example of the parent compound (47) (Scheme 12). In some cases, the intermediate bisamides (46) can be isolated [75ACH(84)477, 75S393]. The method was extended to the synthesis of derivatives substituted in position 1 with a heteroaryl group [79ACH(102)305] and to the preparation of isquinolinones annellated with a benzene ring [81ACH(108)255]. The *N*-methyl derivatives of the lactam can be obtained in a similar manner (67JHC315). Using essentially this procedure, another research group synthesized compounds with various substituents (74GEP2309367).

The synthesis of isoquinolinones by amidoalkylation may also be achieved from other starting materials. *N*-Substituted derivatives were synthesized by the reaction of arylacetyl chlorides and azomethines in the presence of AlCl$_3$ to obtain a number of new compounds, such as 48 (Scheme 13) [75MI1; 78ACH(98)315; 81MI1; 82S216].

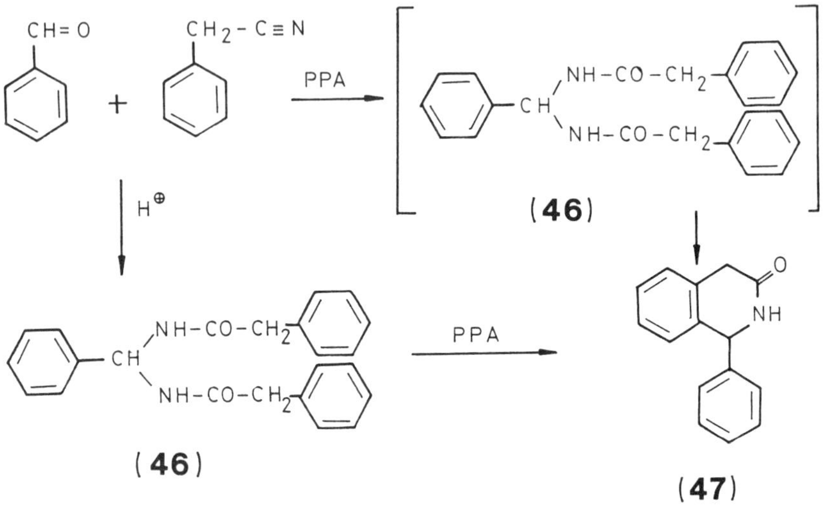

(46)

(47)

SCHEME 12

The method of amidoalkylation was also suitable for the preparation of unsubstituted 1,4-dihydro-3(2H)-isoquinolinone (2) and derivatives substituted in position 1 by other than aryl substituents (80TL569; 87T439). A number of Japanese patents deal with the synthesis of 2. A great variety of cyclization conditions in acidic media were studied in the reaction of

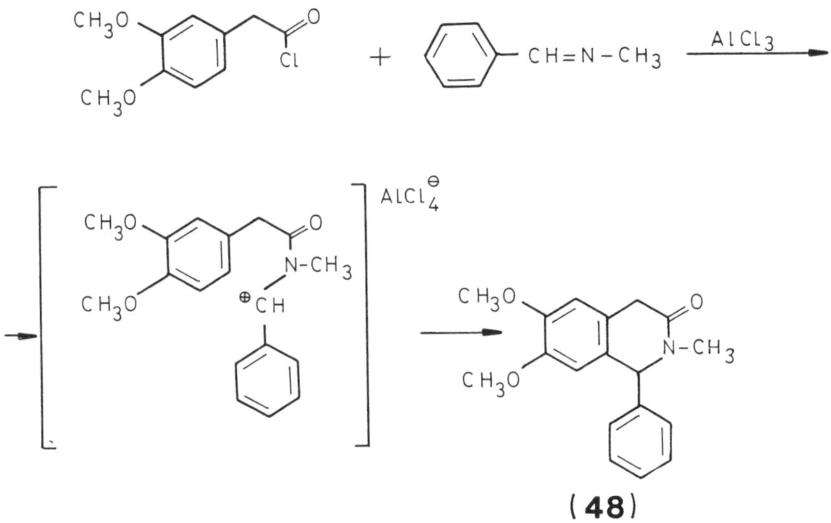

(48)

SCHEME 13

phenylacetamide and formaldehyde (or N-hydroxymethylphenylace-tamide) (e.g., 81H609; 85CPB340; 87H2385), though this reaction had been known previously (70JHC615).

3. Other Special Methods

3-Isochromanone reacts with amines to give the corresponding 1,4-dihydro-3(2H)-isoquinolinones [66JPR12; 67JHC315; 80JCS(P1)2013; 84T4383]. o-Acylphenyl acetic acids can also be used as starting materials for the synthesis of the title compounds (77RC691), as in a reductive amination reaction (71NKZ878; 76BCJ3365).

The Schmidt reaction starting from 2-indanone [70JCS(C)2245; 82T539], and the Beckmann rearrangement of 2-indanone oxime (78JHC321) also gives the isoquinolinones substituted as the starting indanones. Reduction methods are also known for the preparation of 1,4-dihydro-3(2H)-isoquinolinones, e.g., from the unsaturated 3(2H)-isoquinolinones (Section II,C,2). Reduction of homophthalimides initially gives the parent compound 2 or its simple derivatives (69CJC3259; 70JHC615).

B. REACTIONS

1. Reactions on the C-4 Methylene Group

The C-4 methylene group of 1-aryl-1,4-dihydro-3(2H)-isoquinolinones can be regarded, to a certain extent, as an activated methylene group because of the presence of the adjacent lactam carbonyl group and the benzene ring. This is supported by the observation that these compounds react with aromatic aldehydes in the presence of sodium hydride or with dimethylformamide–sodium hydride reagent, upon which they are converted to the unsaturated 3(2H)-isoquinolinone derivatives through the corresponding 4-arylidene or 4-alkylidene intermediates (Section II,B,3). The 4-arylidene intermediates (49) could be isolated in many cases, and the stereochemical structures were elucidated (78LA1103). The role of compound 50 as an intermediate was supported by studies with deuterium labeling (82JHC49). The condensation and rearrangement reaction with pyridinecarboxaldehydes also took place in an acidic medium, in polyphosphoric acid, depending on the anhydride content of PPA [79ACH(102)305]. The mechanism of the rearrangement was elucidated: in the presence of sodium hydride, the hydrogen from position 1 migrated as a hydride anion in an intermolecular reaction to the α-position of the 4-arylidene group. In acidic medium, however, a proton-loss–proton-gain

(49) R = Ar
(50) R = H

(51)

mechanism was proved (82JHC49; 85JHC1491). The latter mechanism was established by means of reactions carried out in perdeuterated poly-phosphoric acid.

The N-methyl derivatives of 1-aryl-1,4-dihydro-3(2H)-isoquinolinones (e.g., **54**) can be lithiated in position 4 with butyllithium. The subsequent reaction, e.g., with p-nitrobenzaldehyde, gives the 4-p-nitrobenzylidene derivative. In this case, the condensation reaction is not followed by H-1 rearrangement [86ACH(121)263].

Compound **51** was synthesized by the interaction of the isoquinolinone unsubstituted in position 4 and dimethylformamide diethyl acetal (81KFZ44, 81KGS511).

A great number of 1,4-dihydro-3(2H)-isoquinolinones react with butyl nitrite in the presence of potassium tert-butoxide to give 4-hydroxyimino derivatives (**19**) (Section II,B,3) [81ACH(106)83; 83ACH(114)355]. The corresponding O-substituted compounds were also synthesized [83ACH(114)69]. These isoquinolinones were converted by further reactions to the corresponding unsaturated 4-amino-3(2H)-isoquinolinones (**20**) (Section II,B,3); these compounds were useful starting materials for synthesizing 1,4-dihydro-3(2H)-isoquinolinones carrying an oxo group [84JCS(P1)619] or amino substituent [86ACH(121)255] in position 4.

The Vilsmeier–Haack reaction on C-4 of 1-aryl-1,4-dihydro-3(2H)-isoquinolinones, followed by an oxidation process (88S680, 88S683), resulted in the key intermediate 3-chloro-4-formylisoquinoline (**53**) (Scheme 14). A number of new isoquinoline derivatives have been prepared from a variety of reactions that use both the halo and aldehyde functions (89H691, 89H707). Compounds **47** and **54** react with dichloromethyleneiminium salt (Scheme 15) to give the corresponding intermediate salts **55** and **56** (86MI1). Further reactions of these salts with aniline or 2-amino-pyridine lead to more condensed heterocycles containing two or three nitrogen atoms (Section V,A). In addition, the reaction of **47** with phosphorus pentachloride gives the 4,4-dichloro derivatives (88S683).

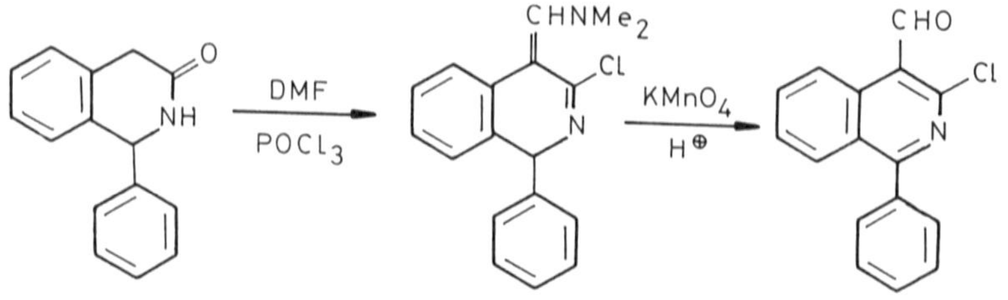

(**47**) (**52**) (**53**)

SCHEME 14

2. *Lactam Carbonyl Substitutions*

A number of new tetrahydroisoquinoline derivatives can be prepared when effecting the C=O → CH_2 transformation by reduction (67JHC315; 70JHC1229; 72JHC853; 77RC691; 78JA1548; 87ACH195). The reduction is usually carried out with $LiAlH_4$ or with borane-tetrahydrofuran. In a special case, the reaction with $LiAlH_4$ gave the 3-hydroxy derivative [84JCR(S)282].

In the course of *O*-alkylation with triethyloxonium tetrafluoroborate, the lactim ethers of 1,4-dihydro-3(2*H*)-isoquinolinones were obtained (75JMC395). These are useful starting materials for reactions with amines resulting in 1,4-dihydro-3-isoquinolylamines (85EUP139296). By means of bifunctional amines, further condensed hetero-ring compounds could be synthesized (73JMC633).

The conversion of a lactam carbonyl to a thiolactam was achieved with

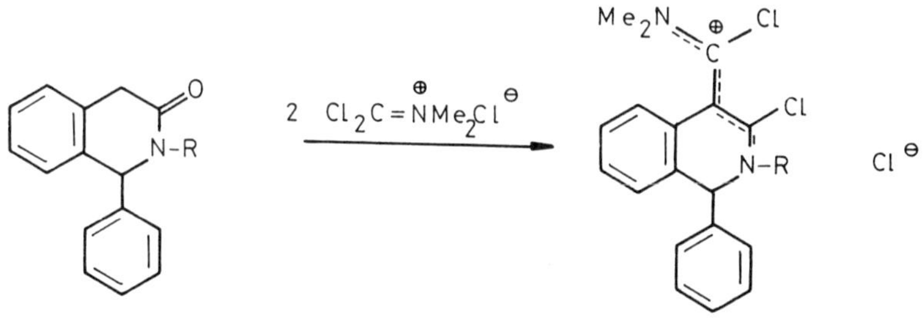

(**47**) R = H (**55**) R = H
(**54**) R = Me (**56**) R = Me

SCHEME 15

phosphorus pentasulfide (88MI1). The thiolactam offers possibilities of further reactions resulting in derivatives similar to those mentioned earlier.

3. Lactam Nitrogen Substitutions, Ring Opening Reactions

In the case of, for example, 1-aryl-1,4-dihydro-3(2*H*)-isoquinolinones, *N*-methylation takes place with dimethyl sulfate without difficulty [86ACH(121)263]. Acylation can also be carried out under the usual reaction conditions; the derivatives **58, 59,** and **60** are thus obtained in

(**57**) R = Me

(**58**) R = Ac

(**59**) R = COOEt

(**60**) R = (substituted) carbamoyl group

satisfactory yields [78ACH(98)247; 79ACH(100)37; 81GEP2948472; 84ACH(116)89; 90ACH607]. In the case of **60** the reactions of the unsubstituted lactam with various isocyanates were used.

The reaction of the *N*-ethoxycarbonyl derivative (**59**) with amines does not give 2-*N*-carbamoylisoquinolinones, but ring opening takes place and, for example, urethane **61** is obtained [78ACH(98)247]. A similar ring opening reaction was found when carbamoyl compounds of type **60** reacted with primary amines to give products such as **62.** In connection with this

(**61**)

(**62**)

latter reaction, kinetic measurements were also made [79ACH(100)37]. Reactions with sterically hindered or secondary amines resulted in the loss of the carbamoyl side chain instead of ring opening.

IV. 5,6,7,8-Tetrahydro-3(2*H*)-isoquinolinones

A. Synthesis

1. *From Cyclohexane Derivatives*

The most widespread methods of synthesizing 5,6,7,8-tetrahydro-3(2*H*)-isoquinolinones involve the use of cyclohexane derivatives as starting materials. Accordingly, the reaction of 2-aminomethylcyclohexanone (**63**) with cyanoacetic esters (35LA243; 47HCA1339; 64CPB1296) or the reaction of 2-acylcyclohexanones (**64**) with malonitrile or cyanoacetamide (53LA110; 68JOC3648; 89JMC351) furnish the products **65** and **66** (Scheme 16). Many varieties of these syntheses are known and patented for compounds with different substituents both in position 1 and on the carbon atoms of the saturated ring. The C-4 nitrile group also offers several possibilities of further reactions (hydrolysis and decarboxylation, preparation of amides, etc.), and these have been used to synthesize a number of derivatives. Instead of a 2-acylcylohexanone (**64**), a 2-hydroxymethylenecyclohexanone, substituted in the methylene carbon atom, can also be used as starting material (74JMC1272). The reaction of β-acylenamines derived from **64** is similarly a suitable method of synthesizing 5,6,7,8-tetrahydro-3(2*H*)-isoquinolinone derivatives, using, for example, cyanothioacetamide to obtain isoquinolinone thiones (82JPR933; 84ZOR2432, 84ZOR2442). Cyanoacetamide containing a cyclohexylidene substituent in the methylene group also reacts with dimethylformamide dimethylacetal to give compounds of type **65** (83KGS1279).

2. *Hydrogenation of 3(2H)-Isoquinolinones*

As discussed previously (Section II,C,2), a great number of 1-aryl-3(2*H*)-isoquinolinones furnish 5,6,7,8-tetrahydro-3(2*H*)-isoquinolinones

(63)

(65) R = H
(66) R = Me

(64)

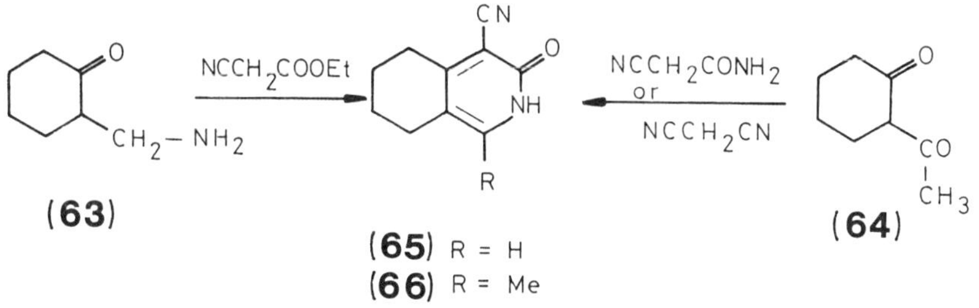

Scheme 16

under the conditions of catalytic hydrogenation. A possible advantage of this method is the absence of a cyano group in the product, a group that is eliminated in many cases (Section IV,A,1). This result, however, can also be regarded as a limitation because the absence of such a functional group reduces the possibility of further transformations.

B. REACTIONS

By reactions of the lactam carbonyl group, the title compounds can be transformed to the corresponding 3-chloro-5,6,7,8-tetrahydroisoquin-olines; suitable reagents are phosphorus pentachloride (68JOC3648), phosphorus oxychloride (47HCA1339; 53LA110), or phenylphosphonic dichloride (74JMC1272; 90ACH601). Bromination takes place similarly (64CPB1296). In one case, when phosphorus oxybromide was used, an aromatization reaction in the saturated ring was observed in addition to bromination in position 3 (58JCS2038). Halo derivatives can be routinely hydrogenolyzed to yield the corresponding 5,6,7,8-tetrahydroisoquin-olines, but this reaction is outside the scope of this review. The exchange of the halogen for an amino group is more useful because new heterorings can be synthesized in this way (Section V,A).

In addition to these reactions, methylation of the lactam carbonyl group via its silver salt (75TL623), and acylation on the oxygen atom can be achieved (77JHC583). The reaction with phenol at an elevated temperature gives the corresponding phenol ether (90ACH601).

V. Applications

A. PREPARING OTHER OR MORE COMPLEX HETEROCYCLES

Some 1-aryl-1,4-dihydro-3(2H)-isoquinolinone derivatives can be con-verted to the corresponding isoindolinones by ring contraction; oxidation with potassium permanganate [86ACH(121)237] of compounds 47 and 54 (Section III,B,1) and treatment with polyphosphoric acid (88ACH289) of several 4-hydroxyimino derivatives (19) (Section II,B,3) give the pre-viously mentioned products.

Several compounds discussed in this review are also suitable for synthesizing more complex heterocycles, as illustrated by the following examples.

Oxazolo- and oxazino-isoquinolines 67 and 68 (88JHC273) can be ob-tained when 4-amino-1-phenyl-3(2H)-isoquinolinone, prepared by the

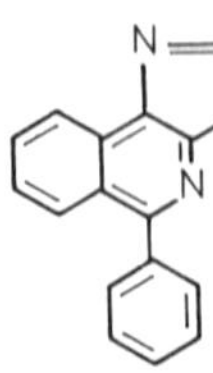

(67)

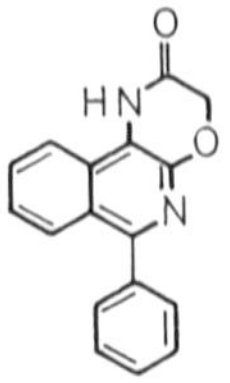

(68)

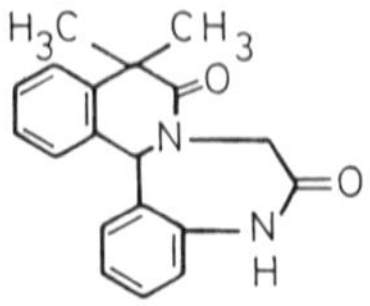

(69)

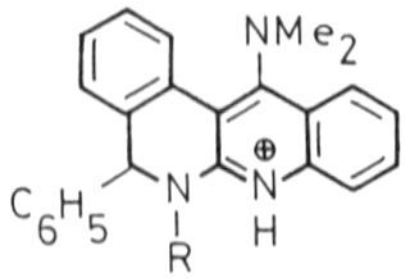

(70)　　R = H, Me

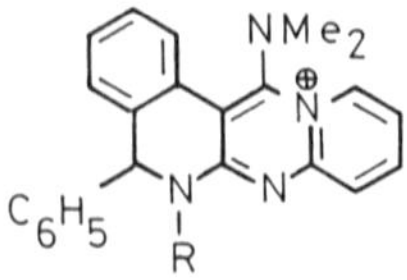

(71)

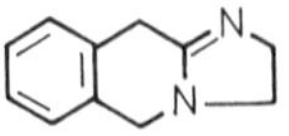

(72)

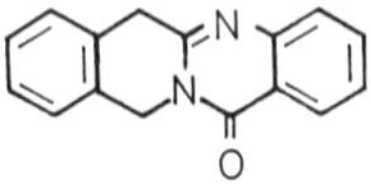

(73)

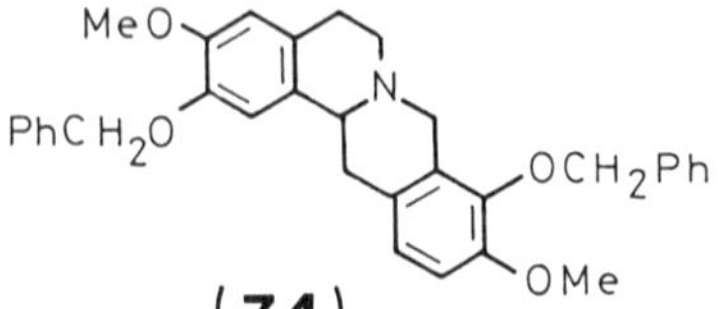

(74)

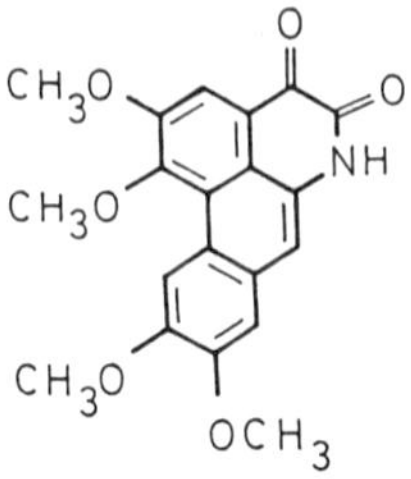

(75)

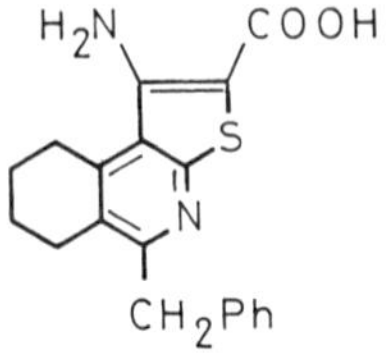

(76)

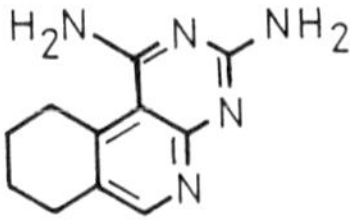

(77)

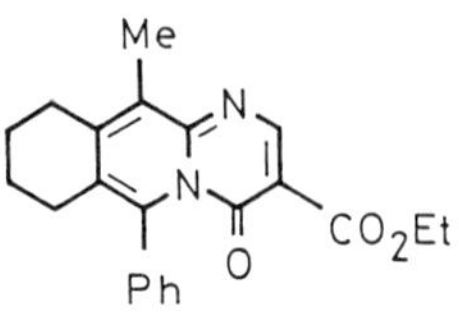

(78)

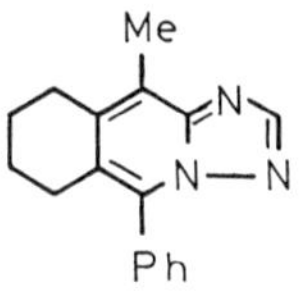

(79)

hydrolysis of **20** (Section II,B,3), is treated with acetic anhydride and chloroacetyl chloride, respectively. Inner alkylation of 4,4-dimethyl-1-(2-chloroacetylaminophenyl)-1,4-dihydro-3($2H$)-isoquinolinone in the presence of sodium hydride gives the isoquino-benzodiazepine derivative **69** (84JHC1045). Compounds **70** and **71** are the end products of the reactions of the 1,3-dichlorotrimethine cyanines **55** and **56** (Section III,B,1) with aniline and 2-aminopyridine (86MI1). 1,4-Dihydro-3($2H$)-isoquinolinone (**2**) (Section I), when reacted with the appropriate bifunctional amines, give, via the lactim ether or thioether and subsequent cyclization of the intermediate enamines, new condensed heterocycles (e.g. compounds **72** and **73**) (73JMC633; 88MI1). Variously substituted 1,4-dihydro-3($2H$)-isoquinolinones are also used to synthesize some alkaloids. For example, **74** was prepared by the cyclization with phosphorus oxychloride of the 7,8-dialkoxyisoquinolinone substituted on the lactam nitrogen with an arylethyl group, followed by a reduction step (79H1327). The starting material of the synthesis leading to **75** is also a properly substituted 1-benzyl-1,4-dihydro-3($2H$)-isoquinolinone derivative, which was transformed to **75** by means of several steps (82JHC1319). 5,6,7,8-Tetrahydro-3($2H$)-isoquinolinones (Section IV) or their 3-chloro and 3-amino derivatives are the key intermediates of the following tricyclic products: **76** (84ZOR2432, 84ZOR2442). **77** (74JMC1272), **78,** and **79** (88MI2).

B. Biological Properties

Several derivatives of the unsaturated and saturated lactams mentioned in this review have been submitted to pharmacological investigations, and various biological activities were found. In this section, an attempt is made to enumerate some derivatives together with their biological properties. Such compounds are shown in the lactam structure.

The structure–activity relationship of 1-alkyl-4-substituted derivatives was investigated (88JMC1363; 89JMC990) and discussed in connection with their cardiotonic (**80**) and renal vasodilating effects (**81**) (depicted in the lactim form by the authors). In a study on papaverine analogues, some 1-benzyl-3($2H$)-isoquinolinones have been synthesized having hypotensive and peripheral vasodilating effects (72JMC1131). Some 1-(substituted)aminomethyl derivatives of 3($2H$)-isoquinolinones had analgesic and antiinflammatory activity [77JAP(K)77156877].

A number of 1-aryl derivatives of 1,4-dihydro-3($2H$)-isoquinolinones proved to be anticonvulsants [77JMC1384; 81ACH(108)255; 87ACH649]. Compound **82** with a basic side chain in position 4 of the 1-aryl group has excellent effect as a new type of antiepileptic. Side chains containing a

(80) R = Me, Et, iPr, etc.

(81) R = $CH_2=CH-CH_2-NH-CO-NH-$

(82)

(83)

piperazine ring were also attached to the lactam nitrogen atom or to position 4 of the aryl group in 1-aryl-1,4-dihydro-3(2H)-isoquinolinones to obtain compounds with antiserotoninergic activity (90ACH607). Similar compounds, such as 4-aryl derivatives of 1,4-dihydro-3(2H)-isoquinolinones with alkylaminoalkyl substitution on the lactam nitrogen or in position 4 are known as central nervous system agents (56USP2759936; 57USP2809969). Hoechst workers have synthesized several 1-aryl-1,4-dihydro-3(2H)-isoquinolinones initially using the substitution possibilities on the C-4 carbon atom to prepare compounds having antiarrhytmic, lipide level reducing, and other biological effects (74GEP2309367; 75GEP2352702). 1,4-Dihydro-3(2H)-isoquinolinones substituted on the lactam nitrogen with a side chain containing ureido and sulfonamido structural units have antidiabetic activity (81GEP2948472).

Some of the tricyclic compounds bridged between positions 1 and 4 (type **34**) (Section II,C,3) were found to have sedative, anticonvulsive, and tranquillant effects (73FRP2111765).

Japanese researchers synthesized several derivatives of 5,6,7,8-tetrahydro-3(2H)-isoquinolinones with cardiotonic activity. Of these compounds, **83** is one of the most potent positive ionotropic agents (89JMC351).

ACKNOWLEDGMENTS

The author would like to thank Professor Gyula Deák, not only for his careful comments, but for his guidance and help during the last twenty years.

References

25LA225	J. Braun and H. Reich, *Justus Liebigs Ann. Chem.* **445,** 225 (1925).
35LA243	U. Basu and B. Banerjee, *Justus Liebigs Ann. Chem.* **516,** 243 (1935).
47HCA1339	E. Schlittler and R. Merian, *Helv. Chim. Acta* **30,** 1339 (1947).
52JCS1763	H. R. Bentley, W. Dawson, and F. S. Spring, *J. Chem. Soc.,* 1763 (1952).
52MI1	W. J. Gensler, *in* "Heterocyclic Compounds" (R. C. Elderfield, ed.), Vol. 4, p. 420. Wiley, New York, 1952.
52ZOB697	P. A. Petyunin, *Zh. Obshch. Khim.* **22,** 697 (1952) [*CA* **47,** 5385 (1953)].
53LA110	H. Henecka, *Justus Liebigs Ann. Chem.* **583,** 110 (1953).
56JOC1297	J. H. Boyer and L. T. Wolford, *J. Org. Chem.* **21,** 1297 (1956).
56USP2759936	M. E. Speeter, U.S. Pat. 2,759,936 (1956) [*CA* **51,** 2883 (1957)].
57USP2809969	M. E. Speeter, U.S. Pat. 2,809,969 (1957) [*CA* **52,** 4697 (1958)].
57ZOB1901	P. A. Petyunin, I. S. Berdinskij, and N. G. Panferova, *Zh. Obshch. Khim.* **27,** 1901 (1957) [*CA* **52,** 4647 (1958)].
58JCS2038	G. A. Swan, *J. Chem. Soc.,* 2038 (1958).
61JOC803	H. E. Baumgarten, W. F. Murdock, and L. E. Dirks., *J. Org. Chem.* **26,** 803 (1961).
63AHC(1)352	A. R. Katritzky and J. M. Lagowski, *Adv. Heterocycl. Chem.* **1,** 352 (1963).
64CB667	H. Plieninger, W. Müller, and K. Weinerth, *Chem. Ber.* **97,** 667 (1964).
64CPB1296	Y. Ban and M. Seo, *Chem. Pharm. Bull.* **12,** 1296 (1964).
66BSF556	J. Gardent and M. Hamon, *Bull. Soc. Chim. Fr.,* 556 (1966).
66JPR12	E. Hoeft and H. Schultze, *J. Prakt. Chem.* **32,** 12 (1966).
66KGS734	P. A. Petyunin and N. G. Panferova, *Khim. Geterotsikl. Soedin.,* 734 (1966) [*CA* **66,** 46304 (1967)].
67JCS(B)590	D. A. Evans, G. F. Smith, and M. A. Wahid, *J. Chem. Soc. B,* 590 (1967).
67JHC149	J. R. Fryer, J. V. Earley, and W. Zally, *J. Heterocycl. Chem.* **4,** 149 (1967).
67JHC315	J. Finkelstein and A. Brossi, *J. Heterocycl. Chem.* **4,** 315 (1967).
67JOC59	H. Win and H. Tieckelmann, *J. Org. Chem.* **32,** 59 (1967).
68ACH(55)125	Z. Csürös, Gy. Deák, and I. Hoffmann, *Acta Chim. Acad. Sci. Hung.* **55,** 125 (1968).
68AG(E)484	G. Simchen, *Angew. Chem., Int. Ed. Engl.* **80,** 484 (1968).
68BSF3403	L. Arsenijevic and V. Arsenijevic, *Bull. Soc. Chim. Fr.,* 3403 (1968).
68JOC3648	F. Freeman, D. K. Farquhar, and R. Walker, *J. Org. Chem.* **33,** 3648 (1968).
68KGS1033	P. A. Petyunin, A. K. Sukhomlinov, and N. G. Panferova, *Khim. Geterotsikl. Soedin.,* 1033 (1968) [*CA* **70,** 68045 (1969)].
69ACH(59)119	Z. Csürös, Gy. Deák, M. Haraszthy-Papp, and I. Hoffmann, *Acta Chim. Acad. Sci. Hung.* **59,** 119 (1969).
69ACH(60)177	Z. Csürös, Gy. Deák, I. Hoffmann, and A. Török-Kalmár, *Acta Chim. Acad. Sci. Hung.* **60,** 177 (1969).

69CJC864   G. Rosen and F. D. Popp, *Can. J. Chem.* **47**, 864 (1969).
69CJC3259   Y. C. Kim, *Can. J. Chem.* **47**, 3259 (1969).
69JCS(C)1729   D. W. Jones, *J. Chem. Soc. C*, 1729 (1969).
70JCS(C)536   J. M. Holland and D. W. Jones, *J. Chem. Soc. C*, 536 (1970).
70JCS(C)2245   V. Askam and R. H. L. Deeks, *J. Chem. Soc. C*, 2245 (1970).
70JGU249   G. N. Dorofeenko and V. G. Korobkova, *J. Gen. Chem. USSR (Engl. Transl.)* **40**, 249 (1970) [*CA* **73**, 15043 (1970)].
70JHC615   D. Ben-Ishai, Z. Inbal, and A. Warshawsky, *J. Heterocycl. Chem.* **7**, 615 (1970).
70JHC1229   I. W. Elliott, *J. Heterocycl. Chem.* **7**, 1229 (1970).
70KGS200   G. N. Dorofeenko, V. G. Korobkova, and S. V. Krivun, *Khim. Geterotsikl. Soedin.*, 200 (1070) [*CA* **76**, 140460 (1972)].
70KGS954   P. A. Petyunin and P. A. Bezuglij, *Khim. Geterotsikl. Soedin.*, 954 (1970) [*CA* **74**, 12975 (1971)].
70TL1209   N. J. Mruk and H. Tieckelmann, *Tetrahedron Lett.*, 1209 (1970).
71DIS(B)5268   N. J. Mruk, *Diss. Abstr. Int. B* **31**, 5268 (1971) [*CA* **75**, 76572 (1971)].
71NKZ878   H. Kotake, A. Kawashiri, S. Miyashita, and H. Kinoshita, *Nippon Kagaku Zasshi* **92**, 878 (1971) [*CA* **76**, 153543 (1972)].
71T4653   N. J. McCorkindale and A. W. McCulloch, *Tetrahedron* **27**, 4653 (1971).
72JCS(P1)2722   D. W. Jones and R. J. Wife, *J. C. S. Perkin 1*, 2722 (1972).
72JHC853   I. W. Elliott, *J. Heterocycl. Chem.* **9**, 853 (1972).
72JMC1131   W. E. Kreighbaum, W. F. Kavanaugh, W. T. Comer, and D. Deitchman, *J. Med. Chem.* **15**, 1131 (1972).
73ACH(76)299   Gy. Deák, K. Gáll-Istók, Zs. Kálmán, and J. Haskó-Breuer, *Acta Chim. Acad. Sci. Hung.* **76**, 299 (1973).
73ACH(77)217   Gy. Deák, P. Frøyen, K. Gáll-Istók, and J. Møller, *Acta Chim. Acad. Sci. Hung.* **77**, 217 (1973).
73ACH(79)113   Gy. Deák and L. Hazai, *Acta Chim. Acad. Sci. Hung.* **79**, 113 (1973).
73FRP2111765   UCB S. A., Fr. Pat. 2,111,765 (1973) [*CA* **78**, 67176 (1973)].
73JHC317   W. E. Kreighbaum, W. F. Kavanaugh, and W. T. Comer, *J. Heterocycl. Chem.* **10**, 317 (1973).
73JMC633   T. Jen, B. Dienel, F. Dowalo, H. Van Hoeven, P. Bender, and B. Loev, *J. Med. Chem.* **16**, 633 (1973).
74FRP2207720   Aspro-Nicholas Ltd., Fr. Pat. 2,207,720 (1974) [*CA* **82**, 72807 (1975)].
74GEP2309367   R. Kunstmann and E. Granzer (Hoechst A.-G.), Ger. Pat. 2,309,367 (1974) [*CA* **82**, 4141 (1975)].
74GEP2330218   J. P. Geerts and R. A. Linz, Ger. Pat. 2,330,218 (1974) [*CA* **80**, 82709 (1974)].
74JMC1272   A. Rosowsky and N. Papathanasopoulos, *J. Med. Chem.* **17**, 1272 (1974).
74TL1181   M. Ikeda, K. Hirao, Y. Okuno, and O. Yonemitsu, *Tetrahedron Lett.*, 1181 (1974).
75ACH(84)477   Gy. Deák, K. Gáll-Istók, and L. Hazai, *Acta Chim. Acad. Sci. Hung.* **84**, 477 (1975).
75GEP2352702   R. Kunstmann and J. Kaiser (Hoechst A.-G.), Ger. Pat. 2,352,702 (1975) [*CA* **83**, 164005 (1975)].

75JMC395   N. G. Kundu, J. A. Wright, K. L. Perlman, W. Hallett, and
                   C. Heidelberger, J. Med. Chem. 18, 395 (1975).
75JMC399   N. G. Kundu, W. Hallett, and C. Heidelberger, J. Med. Chem. 18,
                   399 (1975).
75MI1   N. M. Mollov and P. A. Venkov, Dokl. Bolg. Akad. Nauk 28, 1055
                   (1975).
75S393   Gy. Deák, K. Gáll-Istók, L. Hazai, and L. Sterk, Synthesis, 393
                   (1975).
75TL623   R. J. Chorvat and R. Pappo, Tetrahedron Lett., 623 (1975).
76ACH(88)87   Gy. Deák, K. Gáll-Istók, and L. Sterk, Acta Chim. Acad. Sci.
                   Hung. 88, 87 (1976).
76BCJ3365   H. Kotake, H. Kinoshita, K. Inomata, and H. Masuda, Bull.
                   Chem. Soc. Jpn. 49, 3365 (1976).
76CC695   M. A. Ainscough and A. F. Temple, J. C. S. Chem. Commun.,
                   695 (1976).
76GEP2623226   L. Ghosez, G. Rossey, and F. Didderen (UCB S.A.), Ger. Pat.
                   2,623,226 (1976) [CA 86, 120998 (1977)].
76MI1   J. Elguero, C. Marzin, A. R. Katritzky, and P. Linda, in "The
                   Tautomerism of Heterocycles" (A. R. Katritzky and A. J. Boul-
                   ton, eds.), p. 99. Academic Press, New York, 1976.
77H251   T. Hamada, Y. Okuno, M. Ohmori, T. Nishi, and O. Yonemitsu,
                   Heterocycles, 251 (1977).
77JAP(K)77156877   T. Muro and T. Nakao (Yoshitomi Pharm. Ind., Ltd.), Jpn. Kokai
                   77/156,877 [CA 88, 152448 (1978)].
77JHC583   Gy. Deák, L. Hazai, and G. Tóth, J. Heterocycl. Chem. 14, 583
                   (1977).
77JMC1384   Gy. Deák, M. Dóda, L. György, L. Hazai, and L. Sterk, J. Med.
                   Chem. 20, 1384 (1977).
77RC691   R. Nowicki and A. Fabrycy, Rocz. Chem. 51, 691 (1977).
77T489   M. Ikeda, K. Hirao, Y. Okuno, N. Numao, and O. Yonemitsu,
                   Tetrahedron 33, 489 (1977).
78ACH(98)247   E. Zára-Kaczián and Gy. Deák, Acta Chim. Acad. Sci. Hung. 98,
                   247 (1978).
78ACH(98)315   N. M. Mollov and A. P. Venkov, Acta Chim. Acad. Sci. Hung. 98,
                   315 (1978).
78H1197   H. Fukumi and H. Kurihara, Heterocycles 9, 1197 (1978).
78JA1548   D. J. Hart, P. A. Cain, and D. A. Evans, J. Am. Chem. Soc. 100,
                   1548 (1978).
78JHC321   B. L. Jensen and D. P. Michaud, J. Heterocycl. Chem. 15, 321
                   (1978).
78KGS1279   A. N. Koot, L. G. Yudin, and A. Mumikov, Khim. Geterotsikl.
                   Soedin., 1279 (1978) [CA 89, 215262 (1978)].
78LA1103   G. Tóth, L. Hazai, Gy. Deák, and H. Duddeck, Liebigs Ann.
                   Chem., 1103 (1978).
79ACH(100)37   E. Zára-Kaczián, Gy. Deák, L. Hazai, K. Gáll-Istók, and
                   J. Haskó-Breuer, Acta Chim. Acad. Sci. Hung. 100, 37 (1979).
79ACH(102)305   L. Hazai, Gy. Deák, G. Szabó, and E. Koltai, Acta Chim. Acad.
                   Sci. Hung. 102, 305 (1979).
79H1327   G. D. Pandey and K. P. Tiwari, Heterocycles 12, 1327 (1979).
80JCS(P1)2013   M. P. Carmody, M. Sainsbury, and R. F. Newton, J. C. S. Perkin
                   1, 2013 (1980).

80TL569	D. Ben-Ishai, N. Peled, and I. Sataty, *Tetrahedron Lett.* **21**, 569 (1980).
81ACH(106)83	I. Tikk, Gy. Deák, and G. Tóth, *Acta Chim. Acad. Sci. Hung.* **106**, 83 (1981).
81ACH(108)255	L. Hazai, Gy. Deák, and M. Dóda, *Acta Chim. Acad. Sci. Hung.* **108**, 255 (1981).
81CC1074	S. V. Kessar, P. Singh, R. Chawla, and P. Kumar, *J. C. S. Chem. Commun.*, 1074 (1981).
81CPB128	T. Hamada, Y. Okuno, M. Ohmori, T. Nishi, and O. Yonemitsu, *Chem. Pharm. Bull.* **29**, 128 (1981).
81GEP2948472	Hoechst A.-G., Ger. Pat. 2,948,472 (1981) [*CA* **95**, 132694 (1981)].
81H609	Y. Watanabe, Y. Kamochi, and T. Miyazaki, *Heterocycles*, **16**, 609 (1981).
81KFZ44	V. F. Knyazeva, V. G. Granik, R. G. Glushkov, and G. S. Arutyunyan, *Khim.-Farm. Zh.* **15**, 44 (1981) [*CA* **95**, 115238 (1981)].
81KGS511	V. F. Knyazeva, V. G. Granik, R. G. Glushkov, N. P. Solov'eva, and O. S. Anisimova, *Khim. Geterotsikl. Soedin.*, 511 (1981) [*CA* **95**, 115235 (1981)].
81MI1	A. P. Venkov and N. M. Mollov, *Dokl. Bolg. Akad. Nauk* **34**, 1405 (1981) [*CA* **97**, 92110 (1982)].
81S534	Y. Tamura, J. Uenishi, H. Maeda, and H. Choi, *Synthesis*, 534 (1981).
82JHC49	L. Hazai, Gy. Deák, G. Tóth, J. Volford, and J. Tamás, *J. Heterocycl. Chem.* **19**, 49 (1982).
82JHC1319	L. Castedo, R. J. Estévez, J. M. Saá, and R. Suau, *J. Heterocycl. Chem.* **19**, 1319 (1982).
82JHC1469	L. Castedo, R. J. Estévez, J. M. Saá, and R. Suau, *J. Heterocycl. Chem.* **19**, 1469 (1982).
82JPR933	K. Gewald, H. Schaefer, and P. Bellmann, *J. Prakt. Chem.* **324**, 933 (1982).
82S216	A. P. Venkov and N. M. Mollov, *Synthesis*, 216 (1982).
82S486	A. P. Venkov, L. K. Lukanov, and N. M. Mollov, *Synthesis*, 486 (1982).
82T539	J. C. Gramain, N. Simonet, G. Vermeersch, N. Febvay-Garot, S. Caplain, and A. Lablache-Combier, *Tetrahedron* **38**, 539 (1982).
83ACH(113)237	L. Hazai, Gy. Deák, G. Tóth, A. Schnitta, Á. Szöllősy, and J. Tamás, *Acta Chim. Hung.* **113**, 237 (1983).
83ACH(114)69	I. Tikk, Gy. Deák, and G. Tóth, *Acta Chim. Hung.* **114**, 69 (1983).
83ACH(114)355	I. Tikk, G. Tóth, and Gy. Deák, *Acta Chim. Hung.* **114**, 355 (1983).
83H1367	T. Kappe and D. Pocivalnik, *Heterocycles* **20**, 1367 (1983).
83KGS1279	V. G. Granik, N. I. Smetskaya, N. A. Muhina, I. V. Persyanova, and V. G. Klimenko, *Khim. Geterotsikl. Soedin.*, 1279 (1983).
84ACH(116)89	E. Zára-Kaczián and Gy. Deák, *Acta Chim. Hung.* **116**, 89 (1984).
84ACH(116)303	L. Hazai, Gy. Deák, A. Schnitta, J. Haskó-Breuer, and E. Horváth, *Acta Chim. Hung.* **116**, 303 (1984).
84ACH(117)99	L. Hazai, A. Schnitta, Gy. Deák, G. Tóth, and Á. Szöllősy, *Acta Chim. Hung.* **117**, 99 (1984).
84GEP3227741	E. Konz, H. Kruse, and F. Hock (Hoechst A.-G.), Ger. Pat. 3,227,741 (1984) [*CA* **100**, 174683 (1984)].

84JCR(S)282   E. Zára-Kaczián, Gy. Deák, and G. Tóth, *J. Chem. Res., Synop.* 282 (1984).
84JCS(P1)619   I. Tikk, Gy. Deák, G. Tóth, and J. Tamás, *J. C. S. Perkin 1*, 619 (1984).
84JHC297   P. Sanna and F. Savelli, *J. Heterocycl. Chem.* **21**, 297 (1984).
84JHC1045   K. Gáll-Istók, L. Sterk, G. Tóth, and Gy. Deák, *J. Heterocycl. Chem.* **21**, 1045 (1984).
84JMC943   B. E. Maryanoff, D. F. McComsey, M. J. Costanzo, P. E. Setler, I. J. F. Gardock, R. P. Shank, and C. R. Schneider, *J. Med. Chem.*, **27**, 943 (1984).
84S349   D. J. Le Count, R. J. Pearce, and A. P. Marson, *Synthesis*, 349 (1984).
84T4383   G. G. Black, M. Sainsbury, and A. J. Majeed, *Tetrahedron* **40**, 4383 (1984).
84ZOR2432   Yu. A. Sharanin, A. M. Shestopalov, L. A. Rodinovskaya, V. K. Promonenkov, and V. P. Litvinov, *Zh. Org. Khim.* **20**, 2432 (1984).
84ZOR2442   Yu. A. Sharanin, A. M. Shestopalov, L. A. Rodinovskaya, V. K. Promonenkov, and V. P. Litvinov, *Zh. Org. Khim.* **20**, 2442 (1984).
85ACH(120)271 L. Hazai, A. Schnitta, Gy. Deák, and J. Tamás, *Acta Chim. Hung.* **120**, 271 (1985).
85CJC1001   G. Tóth, A. Almásy, L. Hazai, A. Schnitta, and Gy. Deák, *Can. J. Chem.* **63**, 1001 (1985).
85CPB340   S. Kano, T. Yokomatsu, Y. Yuasa, and S. Shibuya, *Chem. Pharm. Bull.* **33**, 340 (1985).
85EUP139296   Beecham Group PLC, Eur. Pat. Appl. 139,296 (1985) [*CA* **105**, 97347 (1986)].
85JA435   R. R. Goehring, Y. P. Sachdeva, J. S. Pisipati, M. C. Sleevi, and J. F. Wolfe, *J. Am. Chem. Soc.* **107**, 435 (1985).
85JHC1491   L. Hazai, Gy. Deák, J. Tamás, G. Doleschall, and G. Tóth, *J. Heterocycl. Chem.* **22**, 1491 (1985).
85S114   N. S. Narasimhan, R. S. Mali, and B. K. Kulkarni, *Synthesis*, 114 (1985).
86ACH(121)237 L. Hazai and Gy. Deák, *Acta Chim. Hung.* **121**, 237 (1986).
86ACH(121)255 I. Tikk, Gy. Deák, G. Tóth, Á. Szőllősy, and J. Tamás, *Acta Chim. Hung.* **121**, 255 (1986).
86ACH(121)263 L. Hazai, Gy. Deák, Á. Szőllősy, G. Tóth, and I. Bitter, *Acta Chim. Hung.* **121**, 263 (1986).
86MI1   G. Tóth, I. Bitter, G. Bigam, and O. Strausz, *Magn. Reson. Chem.* **24**, 137 (1986).
87ACH195   I. Tikk, Gy. Deák, L. György, P. Sohár, and J. Tamás, *Acta Chim. Hung.* **124**, 195 (1987).
87ACH649   L. Hazai, Gy. Deák, L. György, and M. Dóda, *Acta Chim. Hung.* **124**, 649 (1987).
87H2385   Y. Kamochi and Y. Watanabe, *Heterocycles* **26**, 2385 (1987).
87JCR(S)95   I. Tikk, Gy. Deák, P. Sohár, and J. Tamás, *J. Chem. Res. Synop.*, 95 (1987).
87S637   O. Petrov, V. Ognyanov, and N. M. Mollov, *Synthesis*, 637 (1987).
87T439   D. Ben-Ishai, I. Sataty, N. Peled, and R. Goldshare, *Tetrahedron* **43**, 439 (1987).

87USP4714705   Ortho Pharm. Corp., U.S. Pat. 4,714,705 (1987) [*CA* **108**, 186593 (1988)].
88ACH289   I. Tikk, Gy. Deák, and J. Tamás, *Acta Chim. Hung.* **125**, 289 (1988).
88JHC273   I. Tikk, Gy. Deák, P. Sohár, and J. Tamás, *J. Heterocycl. Chem.* **25**, 273 (1988).
88JMC1363   R. M. Kanojia, J. B. Press, O. W. Lever, Jr., L. Williams, J. J. McNally, A. J. Tobia, R. Falotico, and J. B. Moore, Jr., *J. Med. Chem.* **31**, 1363 (1988).
88MI1   W. Wendelin, H. Keimelmayr, and M. Huber, *Sci. Pharm.* **56**, 195 (1988).
88MI2   K. Gáll-Istók and Gy. Deák, *Stud. Org. Chem. (Amsterdam)* **35**, 279 (1988).
88S680   W. Bartmann, E. Konz, and W. Rüger, *Synthesis*, 680 (1988).
88S683   B. Renger, E. Konz, and W. Rüger, *Synthesis*, 683 (1988).
88T6861   G. Tóth, L. Hazai, Gy. Deák, H. Duddeck, H. Kühne, and M. Hricovini, *Tetrahedron* **44**, 6861 (1988).
88TL2643   G. M. Rishton and M. A. Schwartz, *Tetrahedron Lett.* **29**, 2643 (1988).
89ACH869   L. Hazai, Gy. Deák, J. Hazai-Horváth, and G. Tóth, *Acta Chim. Hung.* **126**, 869 (1989).
89H691   E. Konz and W. Rüger, *Heterocycles* **29**, 691 (1989).
89H707   W. Bartmann, E. Konz, and W. Rüger, *Heterocycles* **29**, 707 (1989).
89JCR(S)340   L. Hazai, G. Tóth, and Gy. Deák, *J. Chem. Res., Synop.*, 340 (1989).
89JHC609   L. Hazai, Gy. Deák, G. Tóth, and J. Tamás, *J. Heterocycl. Chem.* **26**, 609 (1989).
89JMC351   T. Kaiho, K. San-nohe, S. Kajiya, T. Suzuki, K. Otsuka, T. Ito, J. Kamiya, and M. Maruyama, *J. Med. Chem.* **32**, 351 (1989).
89JMC990   R. M. Kanojia, O. W. Lever, Jr., J. B. Press, L. Williams, H. M. Werblood, E. C. Giardino, R. Falotico, and A. J. Tobia, *J. Med. Chem.* **32**, 990 (1989).
90ACH601   L. Hazai, Gy. Deák, and G. Tóth, *Acta Chim. Hung.* **127**, 601 (1990).
90ACH607   E. Zára-Kaczián, L. Hazai, Gy. Deák, L. György, and P. Sohár, *Acta Chim. Hung.* **127**, 607 (1990).

ADVANCES IN HETEROCYCLIC CHEMISTRY, VOL. 52

Directed Metalation of Pi-Deficient Azaaromatics: Strategies of Functionalization of Pyridines, Quinolines, and Diazines

GUY QUEGUINER AND FRANCIS MARSAIS

Institut de Recherches en Chimie Organique Fine,
Insa de Rouen BP 08 76131 Mont Saint Aignan, France

VICTOR SNIECKUS

Guelph-Waterloo Centre for Graduate Work in Chemistry,
University of Waterloo, Waterloo, Ontario N2L 3G1, Canada

JAN EPSZTAJN

Department of Organic Chemistry,
University of Naturowicz A68, 136 Lodz, Poland

I. Introduction

Since the pioneering work by Gilman (39JA109) and Wittig (40CB1197) and the systematic studies by Hauser and his students (64JOC853), the directed metalation reaction has evolved into a powerful method for regioselective efficient functionalization of aromatic compounds [54OR(8)258; 74MI1; 87MI2; 79OR(26)1; 80H(14)1649; 82ACR306; 83S957, 83T2009; 86MI1; 87MI1; 88BSCF67; 90CRV879]. Although the metalation of pi-excessive heteroaromatic systems (furan, thiophene) has also been long recognized and explored (85HC), the application of this ortho functionalization strategy to pi-deficient heteroaromatics (pyridines, quinolines, diazines) has been delayed by evidence that indicated these low lowest unoccupied molecular orbital (LUMO)-level substrates undergo facile nucleophilic attack by RLi and RMgX reagents

[74MI3; 79OR(26)1]. In the 1960s, Chambers [67JCS5045; 69JCS(C)1700] and Abramovitch (67JA1537) first demonstrated ortho metalation processes on highly activated polyfluoropyridines and pyridine *N*-oxides respectively. In the early 1970s, Queguiner [72CR(C)(275)1439], while studying metal–halogen exchange reactions of bromopyridines, provided the first evidence that the bromo substitutent acts as a directed metalation group (DMG), an observation that was extended to other halopyridines and alkoxypyridines by Queguiner [72CR(C)(275)1535], Suschitzky (72CC505), Wakefield [74JOM(69)161], and subsequently by Gribble (80TL4137). Following the early work on the directed ortho metalation (DoM) reaction of aromatic 2-oxazolino (85T837) and tertiary carboxamide (82ACR306), Meyers (78TL227), Epsztajn (80TL4739), Snieckus (80JA1457), and Katritzky (81S127), respectively, demonstrated the application of these DMGs in the pyridine series. During the 1980s, the use of the DoM strategy for substituted pyridines has been reinforced by work in several laboratories. Following the seminal work of Gilman (51JA32), a similar development in the field of quinoline metalation occurred, in alkoxyquinolines and in fluoroquinolines from work by Narasimhan (71T1351) and by Quéguiner [79JOM(171)273], respectively.

Studies to date indicate that a variety of DMGs can be used to activate ortho metalation of pi-deficient heteroaromatic systems, especially pyridine [88AHC(44)199], thus serving as the basis for new and useful synthetic methodologies. The purpose of this review is to update the previous summary (83T2009) concerning the DoM reaction of pyridines, quinolines, pyrimidines, pyrazines, and pyridazines. The review is divided according to the various DMGs (halo, NHCOR and $NHCO_2R$, OR and $OCONR_2$, 2-oxazolino, CONHR, $CONR_2$, and masked RCHO and RCOR, SO_2NR_2), with major emphasis given to the metalation of halo pyridines, an area in which selectivity and mechanistic aspects have been extensively studied. Lithiation chemistry of bare pyridines and pyridine *N*-oxides is then presented, followed by a final section summarizing work on the use of the DoM strategy for the synthesis of natural products and biologically active molecules.

II. Halogen-Based DMGs

A. INTRODUCTION

In 1957, Gilman and Soddy (57JOC1715) first demonstrated the ortho metalation (*n*-BuLi/tetrahydrofuran (THF)/-60°C) of fluorobenzene. Among the halogens, fluoro is the most useful for DoM reactions, while the others tend to favor metal–halogen exchange (Br, I), benzyne formation (Br, Cl), and coupling (Br, Cl) processes [79OR(26)1]. Early reports by numerous workers, in particular Kauffmann [71AG(E)20], showed that halopyridines undergo deprotonation in reactions with strong bases. However, deprotonated intermediates were not detected, and only products resulting from pyridyne intermediates were observed. The first evidence of the formation of an ortho-halo lithio pyridine was provided by Chambers, who showed the conversion of the highly activated polyfluoropyridine 1 into products 3, implicating the intermediacy of 2 (Scheme 1) [65JCS5045; 69JCS(C)1700]. In 1972, comprehensive studies [72CR(C)(275)1439, 72CR(C)(275)1535] on numerous monohalopyridines indicated that F, Cl, and Br serve as good DMGs and, on the basis of variation of reaction parameters, that some useful regio- and chemo-selectivities can be observed.

The exploration of DoM chemistry in haloheteroaromatics is valuable for two reasons: (a) halo derivatives can be readily prepared from acessible amino, hydroxy, and, at times, unsubstituted systems, and (b) halogens can be subsequently induced to undergo a variety of funcionalizations via addition–elimination, metal–halogen exchange, S_{RN1}, and cross coupling reactions. Some of these transformations are more facile than in the aromatic series. Thus, DoM tactics may lead to diverse substituted heteroaromatics, which can be difficult to obtain by more classical means.

B. Chemoselectivity

Reactions of powerful alkyllithiums with halo pyridines, quinolines, and diazines may lead to nucleophilic substitution (by addition–elimination or hetaryne mechanisms), ring opening, halogen-scrambling, and coupling reactions, which compete with the desired DoM process.

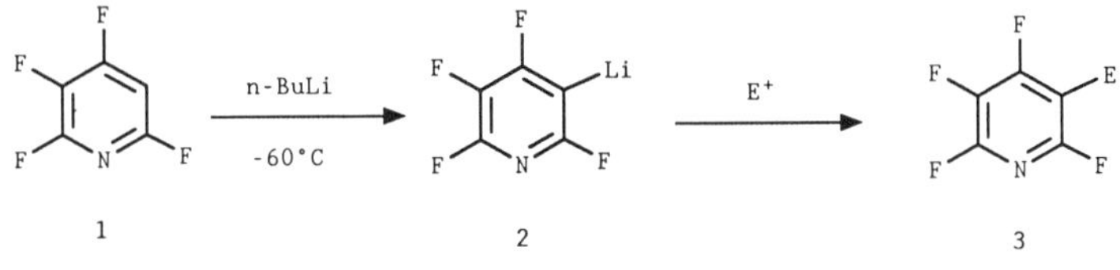

$$E = CHO \ (40\%), \ CO_2H \ (62\%).$$

SCHEME 1

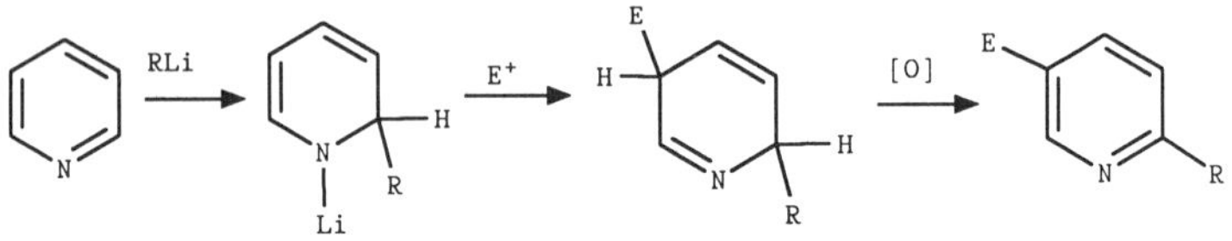

R = n-Bu, Ph .   E = alkyl, Ph$_2$COH, CO$_2$H, R'CO .

SCHEME 2

1. Competitive Addition

a. *Pyridine*. Treatment of pyridine (**4**) with organolithiums leads to dihydro species **5** (Scheme 2) [70JCS(D)478; 74JOC3565]. If, in a subsequent step, electrophiles are added, 2,5-disubstituted products **7** are obtained from spontaneous dehydrogenation of the intermediates **6**. Although 2,5-dihydropyridines were plausible intermediates, intervening oxidation or dismutation reactions prevented their isolation. Nevertheless, they were characterized by IR and ^{1}H-NMR spectroscopies (78JOC3227).

Treatment of 2-fluoro or 2-chloropyridine (**8** or **9**) with *n*-BuLi or its TMEDA complex leads, by nucleophilic C-6 addition, to species **10** which, if protonated or alkylated, provides as the only isolable products the surprisingly stable 2,5-dihydropyridines **11** (Scheme 3) (81JOC4494).

When the metalation of 2-fluoropyridine (**8**) was carried out at lower temperatures ($-60°$C), aside from the product of addition **13**, evidence for the DoM product **12** was obtained by a TMSCl quench experiment (Scheme 4). Complete chemospecificity was achieved only with the more selective LDA base at $-75°$C. [81JOC4494, 81JOM(215)139]. Whereas the "soft" character of *n*-BuLi favors nucleophilic reactivity, the "hard" LDA leads to preferential protophilic attack. The stronger kinetic basicity

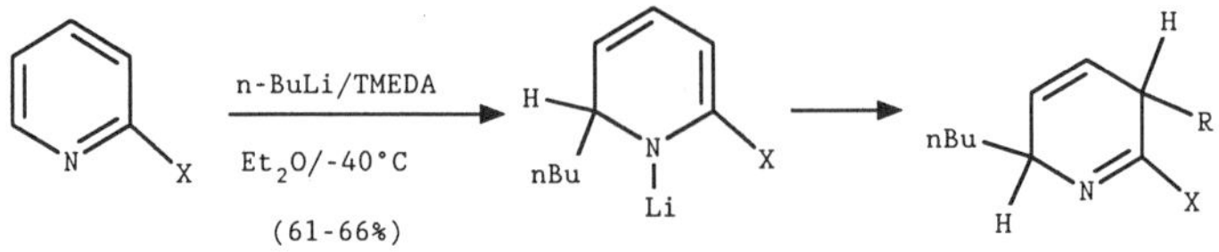

X = F, Cl.   R = H, D, Me, Et, i-Pr.

SCHEME 3

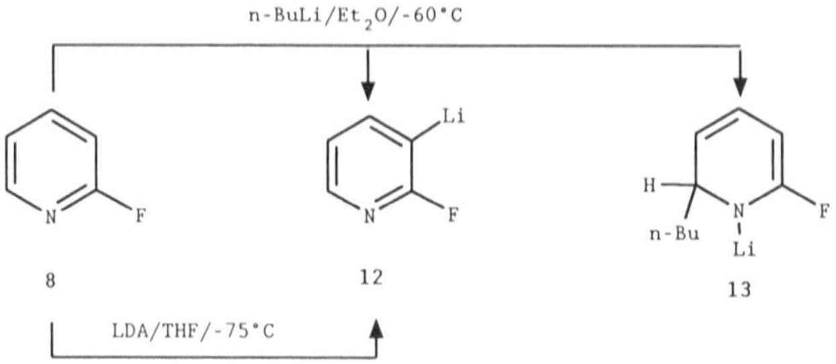
n-BuLi/Et$_2$O/-60°C
Li
N
F
N
F
H
n-Bu
N
Li
F
8
12
13
LDA/THF/-75°C

SCHEME 4

of lithium dialkylamides towards alkyllithiums is in good agreement with experimental results [60AG(E)91]. While the latter are thermodynamically more basic, their reactivity is lower due to their existence as oligomers.

b. *Quinoline*. Treatment of 2-fluoroquinoline (**14**) with *n*-BuLi/TMEDA in Et_2O leads to the formation of 2-*n*-butylquinoline (**15**) (Scheme 5) [79JOM(171)273]. In the more basic THF solvent, some DoM reaction was observed, as evidenced, after TMSCl quench, by the isolation of 2-fluoro-3-trimethysilylquinoline (**16**) (16%) in addition to **15** (20%). At lower temperatures (−110°C), slightly improved ortho metalation **16** (29%) to addition **15** (8%) product ratio was observed.

The application of the *n*-BuLi/TMEDA conditions on 3-fluoroquinoline (**17**) leads, after proton or TMSCl quench, to products of 1,2-addition **18** (Scheme 6).

Similar results are observed for the same reactions of 5-, 6-, 7-, and 8-fluoroquinolines, except that in the case of the 7-fluoro derivative **19,** product of the DoM process **21** was isolated in low yield together with the addition product **20** (Scheme 7) [79JOM(171)273].

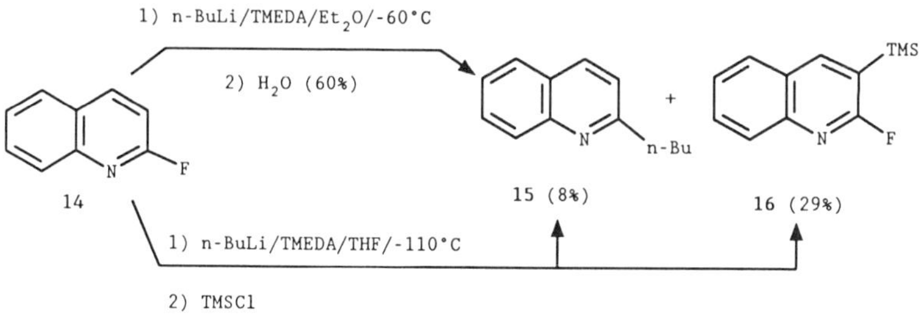
1) n-BuLi/TMEDA/Et$_2$O/-60°C
2) H$_2$O (60%)
N
F
14
N
n-Bu
15 (8%)
+
TMS
N
F
16 (29%)
1) n-BuLi/TMEDA/THF/-110°C
2) TMSCl

SCHEME 5

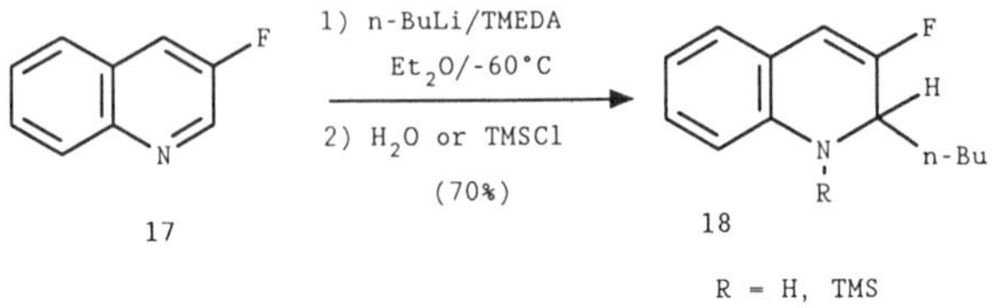

SCHEME 6

As in the case of fluoropyridines, problems of addition are overcome by LDA metalation. Thus 2-, 3-, 5-, 6-, and 7-fluoroquinolines lead, after TMSCl quench, to modest to good yields of products **22–26** of chemoselective DoM reaction (Scheme 8) [79JOM(171)273].

c. *Pyrimidine.* As expected, the pyrimidine nucleus is more sensitive than pyridine to nucleophilic addition. Thus, organolithiums add smoothly to 2- and 5-bromo pyrimidines **27** to give 3,4-dihydro intermediates **28,** which may be characterized by oxidation to compounds **29** (Scheme 9) (65ACS1741).

On the other hand, treatment of **30** with LDA, followed by oxidation, results in the formation of dimer **32** (Scheme 10) (79JOC2081). That this reaction proceeds via **31** by initial ortho metalation was established by performing the reaction between −65 and −10°C in the presence of benzaldehyde, which led to the formation of the carbinol 33.

2. *Competitive Heteraryne Formation*

Kauffmann showed that treatment of halopyridines **34** with an excess of lithium dialkylamides at room temperature leads, via the intermediate 3,4-pyridyne (**35**), to an isomeric mixture of the amide addition products **36** and **37** (Scheme 11) [71AG(E)20]. Such reactions are inconsequential if they are performed at low temperatures where the rate of lithium halide elimination is slow.

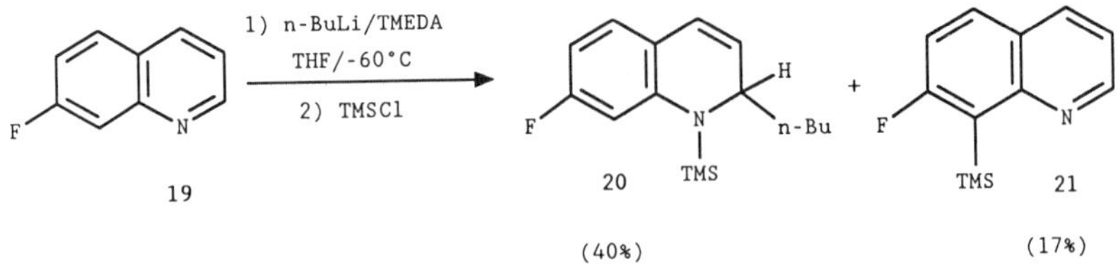

SCHEME 7

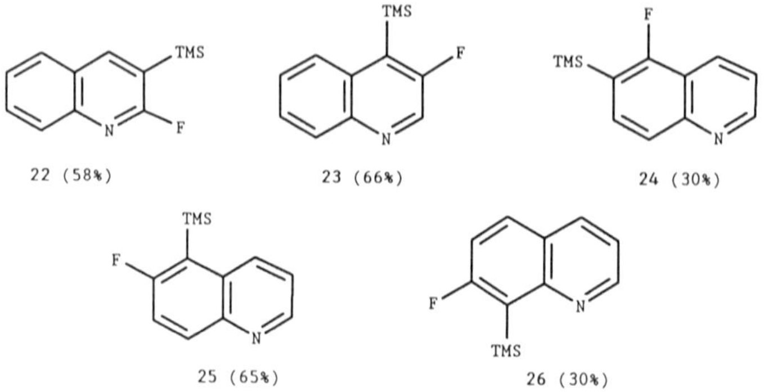
TMS
F
N
22 (58%)
23 (66%)
24 (30%)
25 (65%)
26 (30%)

SCHEME 8

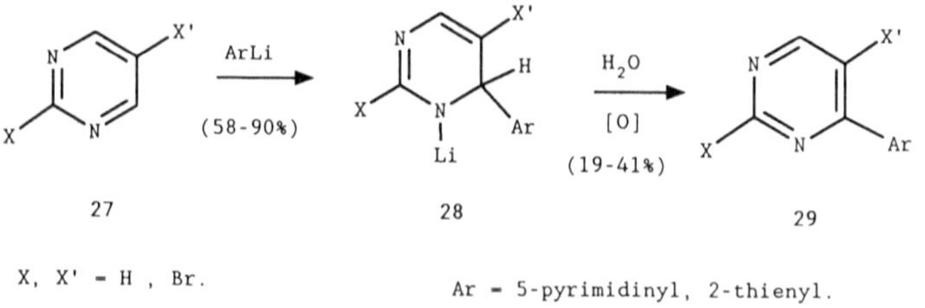
X'
X
N
ArLi
(58-90%)
H
Ar
Li
H2O
[O]
(19-41%)
27
28
29
X, X' = H , Br.
Ar = 5-pyrimidinyl, 2-thienyl.

SCHEME 9

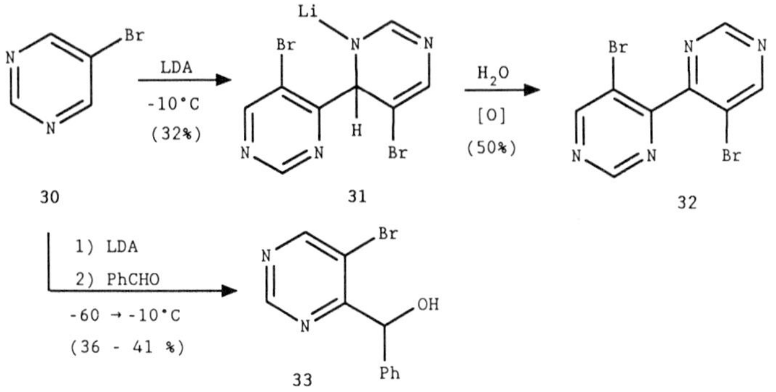
Br
N
LDA
-10°C
(32%)
Li
H
H2O
[O]
(50%)
30
31
32
1) LDA
2) PhCHO
-60 → -10°C
(36 - 41 %)
OH
Ph
33

SCHEME 10

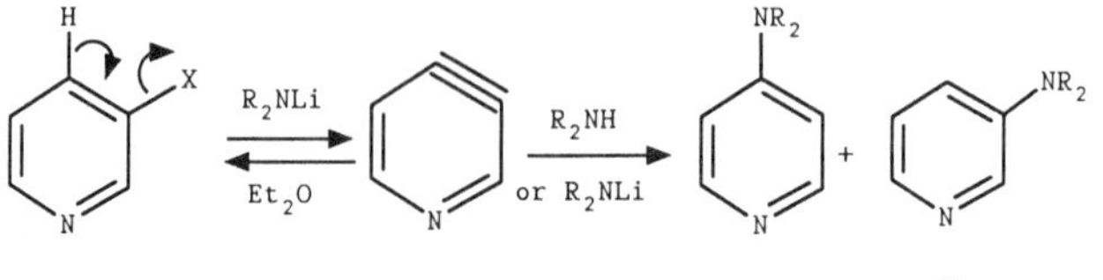

34   35   36   37

X = F, Cl, Br

SCHEME 11

Thus, 3-fluoro-4-lithiopyridine, prepared from **38** by *n*-BuLi/TMEDA metalation, is stable to $-20°C$ and only at room temperature undergoes elimination to 3,4-pyridine, which can be trapped with furan to give cycloaddition product **39** (Scheme 12) [72CR(C)(275)1535].

Similarly, 4-lithiated 3-bromo and 3-chloro pyridines generated from substrates **40,** are stable between -60 and $-40°C$, and lithium halide elimination to 2-fluoro-3,4-pyridyne occurs only upon warming to room temperature, as evidenced by the formation of adduct **41** (Scheme 13) [72CR(C)(275)1439, 72CR(C)(275)1535].

On the other hand, LDA metalation of 3-bromopyridine (**42**) at $-70°C$ yields, after hydrolysis, a mixture of 3- and 4-substituted products (**43** and **44**) in addition to starting material **42** (Scheme 14) (82T3035). A potential explanation for these results involves the formation of 3,4-pyridyne, which undergoes nonregioselective attack by amine or lithio amide to give **43** and **44**. An alternative rationalization is the isomerization of **42** into the 4-isomer **45** under the metalation conditions (see Section II,B,4), followed by the conversion of either isomer into the radical anions **46** which, via the caged radical pairs **47,** is converted into **43** and **44** (Radical Anion-Radical Pair = RARP pathway).

Ortho-lithiated halobenzenes, prepared by lithium-bromide exchange, undergo rapid benzyne formation at -30, -40, and $-50°C$ for bromo, chloro, and fluoro derivatives [57JA(79)2625]. They normally require

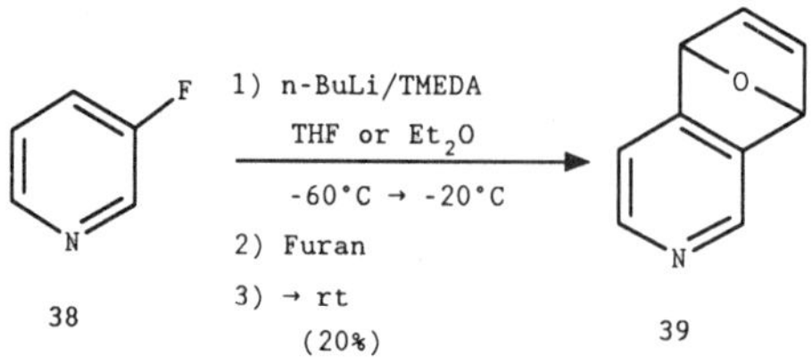

1) n-BuLi/TMEDA
   THF or Et$_2$O
   $-60°C \rightarrow -20°C$
2) Furan
3) $\rightarrow$ rt
   (20%)

38   39

SCHEME 12

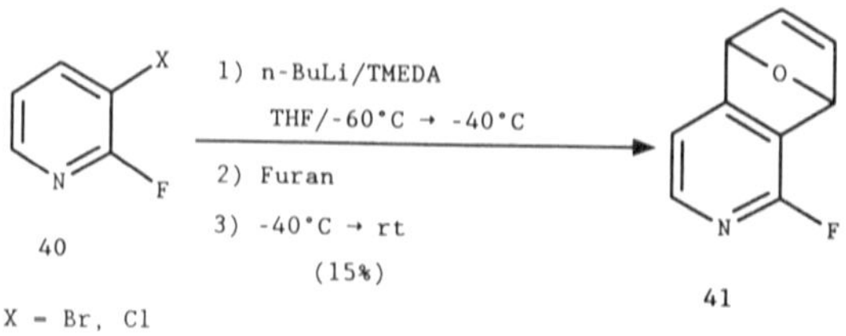

SCHEME 13

−100, −90, and −60°C, respectively, to achieve synthetically useful ortho-functionalization [56JA(78)2217, 57JOC1715]. Thus, based on the previous results, it appears that ortho lithiated halopyridines are less prone to undergo lithium halide elimination and thus serve as a synthetically useful species for preparing substituted pyridines.

3. Competitive Ring Opening Reactions

Metalation of 2-bromopyridine (48) with 1 equiv. of LDA leads to the formation of a relatively stable (−60°C) 3-lithiated species 51, which can be trapped with TMSCl to give 49 (Scheme 15) [82JCR(S)278]. However,

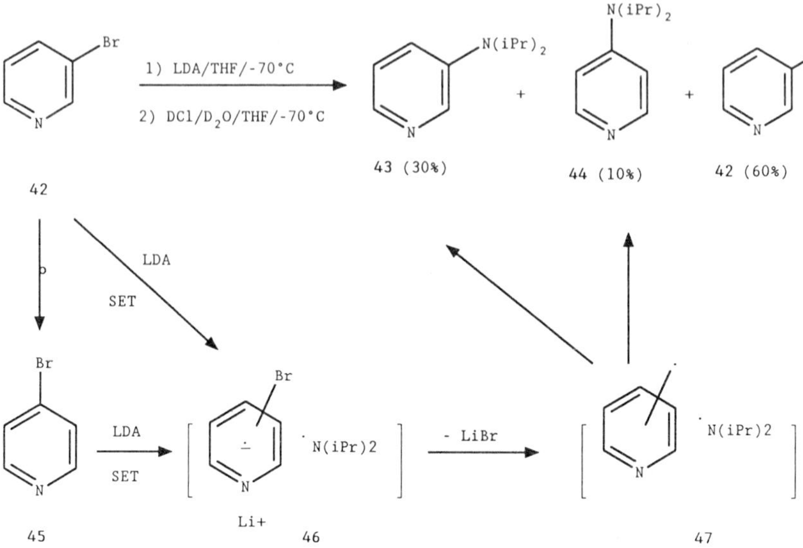

SCHEME 14

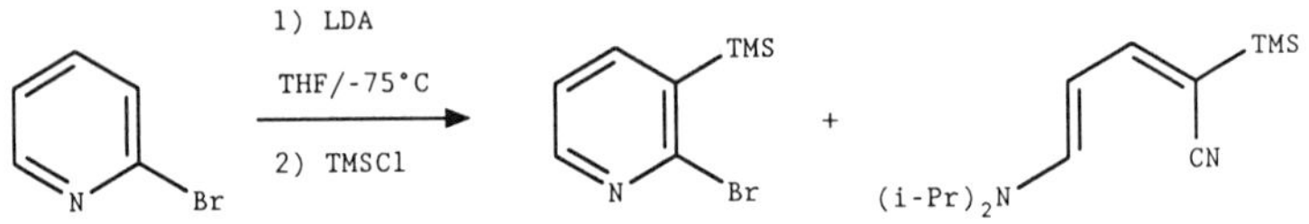

SCHEME 15

LDA equiv.	49	Yields	50
1	55%		0%
4	40%		30%

when an excess of LDA is used in an attempt to shift the equilibrium towards lithiated species, a competitive ring opening reaction leading to **50** is observed.

This result can be explained (Scheme 16) by initial C-6 attack of LDA on species **51** to give **52,** which undergoes a cycloreversion process to the aza-triene **53.** The latter undergoes loss of LiBr to the species **54,** which is quenched by TMSCl to furnish the E,E-cyano diene **50.**

Analogous ring opening reactions have been observed in related reactons: 5-substituted-2-bromopyridines with piperidyllithium in piperidine at −60°C (71TL1875); 5-substituted-2-bromo and 2-chloropyridines with PhLi in piperidine at 30°C (73TL1887); 1,3-dialkyl-2-aminopyridinium iodides with *n*-BuLi at −78°C (82T1169); and 6-bromo-2-lithiopyridine with trialkylboranes at −40 to 0°C (74JA5601).

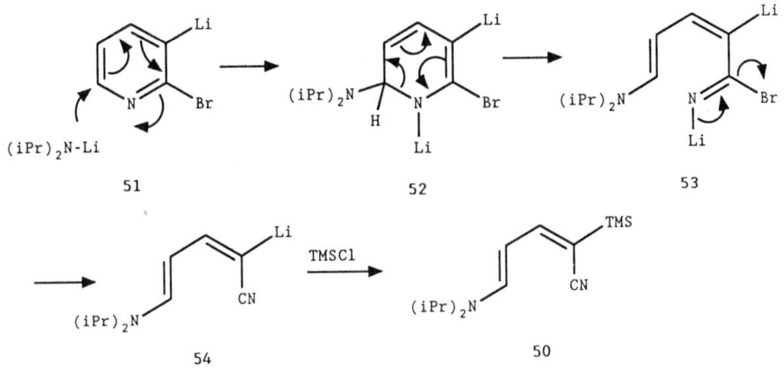

SCHEME 16

4. Competitive Halogen Scrambling

In 1972, Quéguiner and co-workers established that treatment of 3-bromopyridine in excess with n-BuLi at low temperatures results in the formation of a halogen migration product, the 4-bromo isomer [72CR(C)(275)1439].

a. *Halogen Scrambling and Metal–Halogen Exchange.* These early experiments indicated that treatment of 2-halo (F, Cl, Br) -3-bromopyridines with n-BuLi yielded, after addition of electrophiles, mixtures of 2-halopyridines, 4-substituted 3-bromo-2-halopyridines, and 3-substituted 4-bromo-2-halopyridines in various proportions, depending on reaction conditions (solvent, temperature, time, equiv. of RLi). For example, while exposure of **55** to 1 equiv. of n-BuLi for short periods of time followed by acetone quench led in high yield to **56**, warming the reaction mixture to −40°C for 5 min. and similar quench led to a mixture of **11, 57,** and **58** (Scheme 17) [72CR(C)(275)1439]. If 0.5 equiv. of n-BuLi was used, in addition to **59** (60%), the selective formation of the bromine migration product **58** (40%) was observed (85T3433).

Similar behavior of lithium–bromide exchange (−60°C) and homo-transmetalation (metalation of a substrate by one of Li derivatives) (−40°C) was observed in reactions of 4-bromo-3-fluoro- and 3-bromo-2-fluoropyridine with n-BuLi (86T2253). Under similar conditions, 2,3-dibromopyridine undergoes partial bromine migration, yielding, after

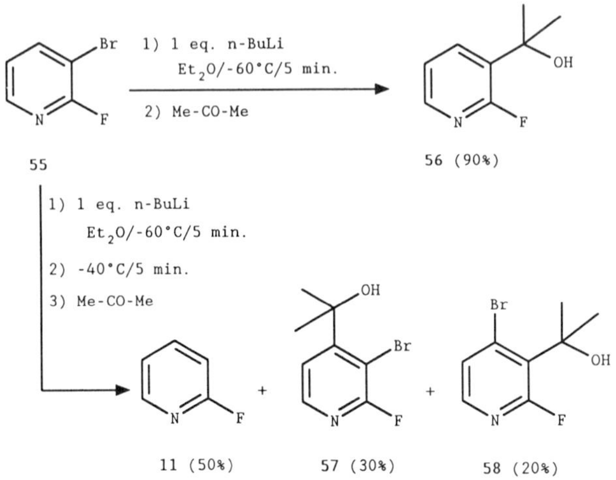

SCHEME 17

treatment with *n*-BuBr, 2-bromopyridine (60%), 4-butyl-2,3-dibromopyridine (15%), and 3-butyl-2,4-dibromopyridine (25%) (85T3433).

b. *Mechanism.* A rationalization of the bromide scrambling reaction is given in Scheme 18 (86T2253; 82T3035; 85T3433). Initiation involves bromo–lithium exchange between **59** and *n*-BuLi (0.5 equiv.) to give 3-lithio species **60** (step i), which undergoes equilibrium homotransmetalation with starting material **59** to species **61** and **62** (step ii). Species **62**

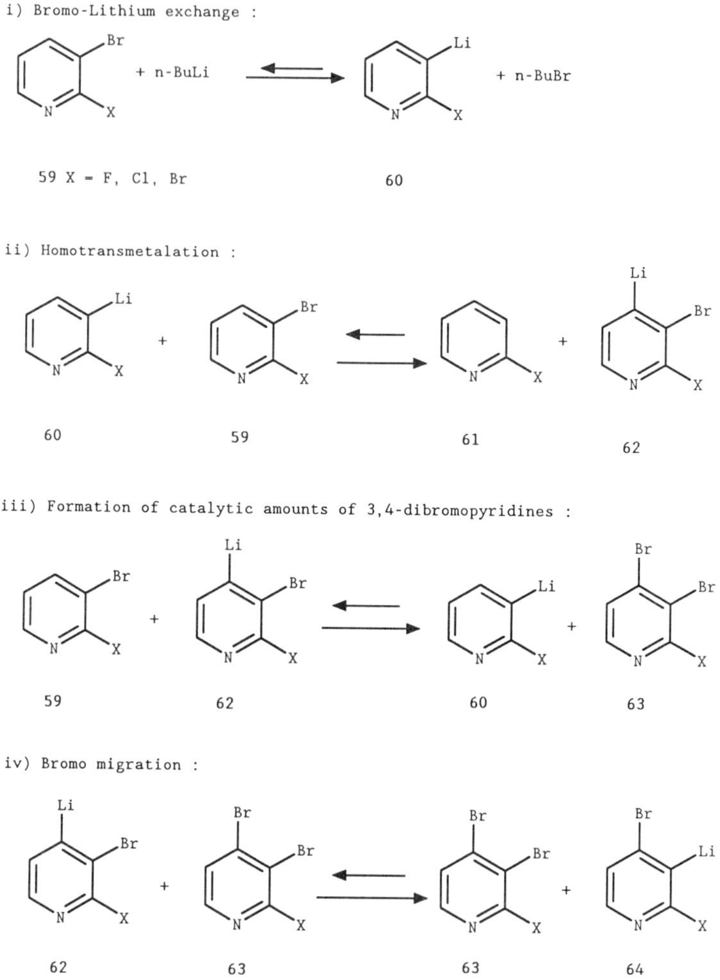

i) Bromo-Lithium exchange :

ii) Homotransmetalation :

iii) Formation of catalytic amounts of 3,4-dibromopyridines :

iv) Bromo migration :

SCHEME 18

then enters into an equilibrium with substrate **59** to afford **60** and the 3,4-dibromopyridine **63** (step iii). A further equilibration of **62** with **63** gives stabilized 3-lithio intermediate **64** (the driving force being the stability of the 3-lithio species **64** due to the electron-withdrawing effects of the 2- and 4- halogens) and regenerates **63** (step iv), thus making the overall process catalytic in this 3,4-dibromopyridine derivative.

Similar pathways have been proposed for the isomerization of 2,3- and 3,4-dibromoquinolines (KNH$_2$/NH$_3$/THF/−75°C) (73RTC304) and 1,2,4-tribromobenzene (amide, alkoxide) (68JA810; 72ACR139).

c. *Halogen Scrambling and DoM*.   Under somewhat different conditions, halogen scrambling of 3-bromopyridine (**42**) may also be observed (Scheme 19) (82T3035). Thus treatment of **42** with 1 equiv. of LDA at −60°C followed by quenching with MeI and warming to −50°C for 5 min. leads to the formation of DoM product **65,** migrated product **66,** and recovered starting material **42.** Since traces of bromine were shown to catalyze the reaction, a reasonable mechanism (**67** + **68→68** + **69**) for this process implicates the catalytic effect of 3,4-dibromopyridine (**68**). The same phenomenon was observed in reactions of 3-bromo-2-chloro and 3-bromo-2-fluoro pyridine [90JOM(382)319; 91JOC(s1)].

The rearrangement of 5-bromo-3-methoxy-2-phenylimidazolo[1,2-*a*] pyridine (**70**) into the 8-isomer **73** also undoubtedly involves a sequential metalation, transmetalation, and metal–halogen exchange pathway involving intermediates **71–75** and the noncatalytic generation of the 5,8-dibromo derivative **75** (Scheme 20) (83S987).

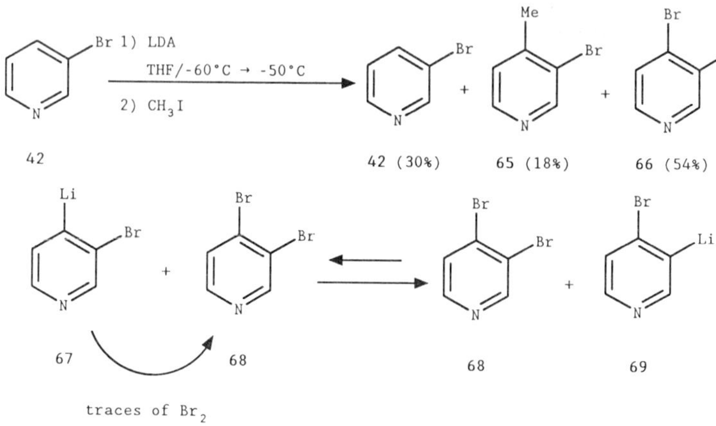

SCHEME 19

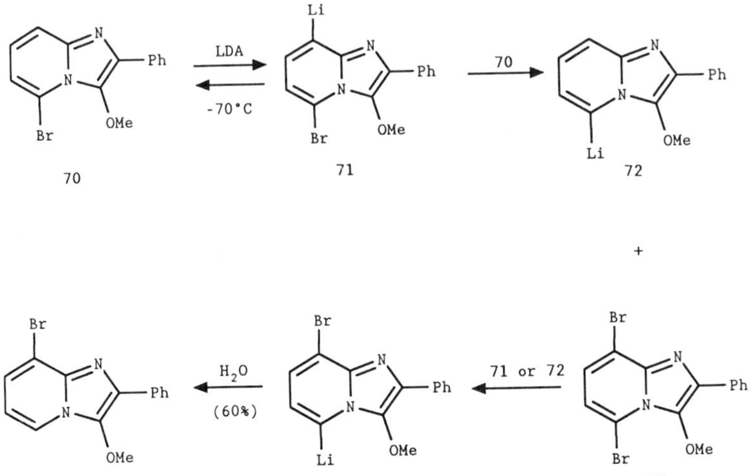

70 71 72 73 74 75

SCHEME 20

5. *Coupling Reactions*

Pyridine, quinoline, and isoquinoline (74TL2373) as well as 5,5′-bipyrimidine and 3,3′-bipyridine [75AG(E)713] undergo dimerization under the action of LDA at −70°C. In all cases, homocoupled products ortho to heterocyclic nitrogen are produced in good yields when the reaction is carried out in Et_2O in the presence of hexamethylphosphoramide (HMPA). Illustrative of the reaction mechanism proposed, pyridine (**4**) undergoes 2-lithiation and reaction with itself to give the addition product **76,** which upon aerial oxidation during work up leads to 2,2′-bipyridyl (**77**) (Scheme 21). Attempts to intercept the carbanionic intermediate were not successful.

Similar results were observed with other halopyridines. Thus, 3-bromopyridine gives 3,3′dibromo-4,4′-bipyridine (LDA/Et_2O/HMPA/−

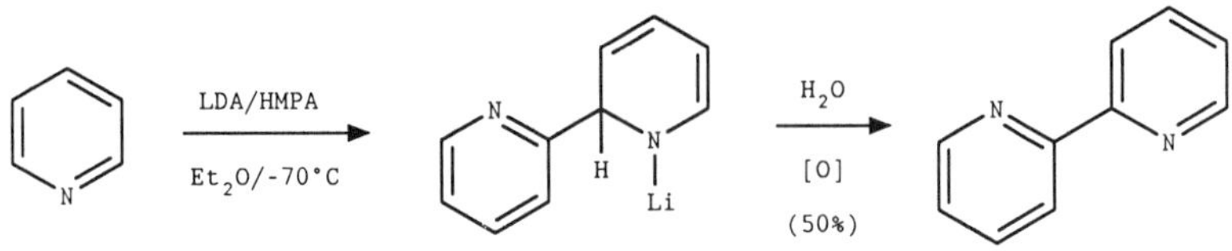

4 76 77

SCHEME 21

100°C) [79AG(E)1], and 3-fluoropyridine affords 3,3'-difluoro-4,4'-bipyridine (25%) and other dimeric derivatives (KNH2/liq. NH3) (64TL3207).

A dihydroypridine intermediate can be isolated in the LDA/HMPA reaction of 3-chloropyridine (**78**) (Scheme 22) [81JOM(216)139; 86PC2]. Thus, metalation at −70°C, followed by addition of HMPA (2 equiv.) to the suspension of the 4-lithio intermediate **81**, gives a mixture of starting materials **78** and **79**. D_2O quench leads to the dihydro dimer **80**, whose structure was confirmed by ^{1}H- and ^{13}C-NMR spectroscopy. Complete formation of **80** was favored by warming the reaction mixture to room temperature. The formation of **80** was also observed upon reaction of **78** with LDA/2 equiv. HMPA. The coupling reaction can be most simply rationalized by a SET route between equilibrating **81** and **78**, leading to the formation of the solvent caged RARP species **82** which collapses to **79**.

Other reactions of electron-poor aza-heterocycles that are suspected to involve a SET mechanism include 1-lithiodithiane with pyridine (73CL1307) and 1,8-naphthyridine (78ZC382), and LDA with pyridine (82JOC599).

C. REGIOSELECTIVITY

DoM reactions of DMG-bearing pyridines can be complicated by the potential competitive DMG effects of the pyridine ring nitrogen atom.

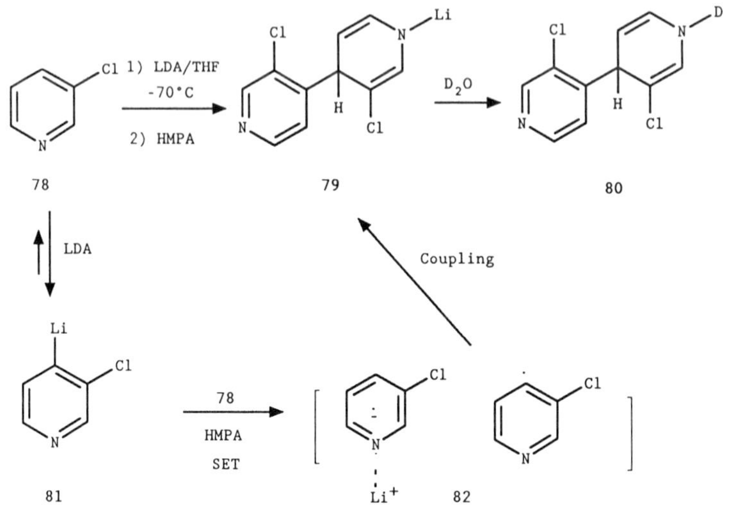

SCHEME 22

Thus, 2- and 4-DMG substituted pyridines may lead to 3-, 6- and 2-, 3-metalation results, respectively, while 3-DMG pyridines may provide 2-, 4-, and 6- metalation products. In general, the ring nitrogen always exhibits a weak DMG in 2- and 4-DMG pyridines, the major regioselectivity effect being manifested in 3-DMG pyridines of which the 3-fluoro and 3-chloro pyridines have been mostly investigated.

1. *DoM Reaction of 3-Fluoropyridine with n-BuLi*

Systematic studies on 3-fluoropyridine, the first mono halopyridine to be shown to undergo the DoM process [72CR(C)(275)1535], showed that metalation regioselectivity was dependent on reaction conditions (solvent, temperature, time, metalating agent).

a. *Solvent Effects.* Whereas metalation of 3-fluoropyridine (**38**) with *n*-BuLi gave poor yields and low regioselectivity, treatment with the *n*-BuLi/TMEDA complex under carefully specified conditions followed by quenching with 3-pentanone or TMSCl gave 2- or 4-substituted products **85** and **86** via the intermediates **83** and **84,** respectively (Scheme 23) [72CR(C)(275)1535]. The results indicate that 2-lithio species *83* predominates in Et_2O solution, while the corresponding 4-lithio intermediate **84** is obtained in THF.

The regioselective DoM effects can be rationalized in terms of kinetic and thermodynamic control of the reaction (83T2009). The relative thermodynamic acidity $(NaNH_2/NH_3/-25°C)$ of pyridine hydrogens

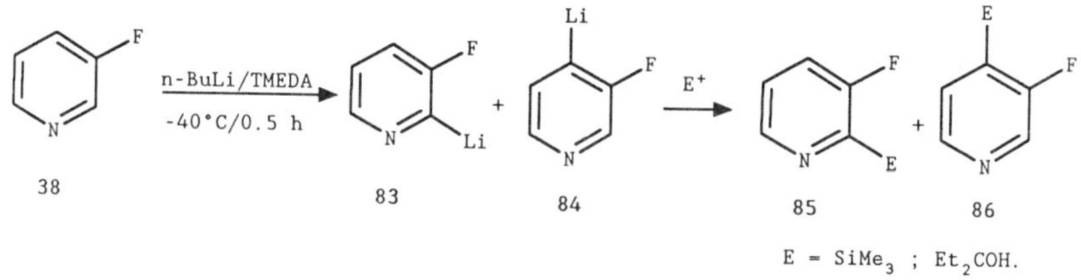

E = $SiMe_3$; Et_2COH.

	85	Yields %	86	
Solvent	E = Et_2COH	TMS	Et_2COH	TMS
Et_2O	65	68	10	3
THF	0	0	50	75

SCHEME 23

(H_4 : H_3 : H_2 = 700 : 72 : 1) has been rationalized on the basis of two effects: (a) an electrostatic repulsive interaction between the N unshared electron pair and the C—Li bond of the developing 2-lithiated species and (b) the enlarged internal ring bond angle NC_2C_3 compared to $C_3C_4C_5$, which causes reduction in the s-character of the C_2—H bond (67CC55; 69JA5501, 69T4331). This geometry was shown to prevail in 3-fluoropyridine [76JSP(59)216]. H/D exchange rate data on 3-chloro-pyridine using MeONa/MeOD shows H_4 : H_2 = 53 : 1 (69T4331).

On the basis of these considerations, proton abstraction in 3-fluoropyridine using n-BuLi/TMEDA complex in THF is expected to occur at the most acidic C4 site, as observed. On the other hand, in Et_2O, which is a more weakly basic solvent than THF, coordination of base with ring nitrogen (87) is favored and leads by a proximity effect to C_2-proton abstraction and formation of 83 (Scheme 24). Addition of THF to a solution of 83 in Et_2O at $-60°C$ did not result in the formation of the isomeric species 84, thus suggesting that C4-lithiation in THF is a kinetically con-trolled process that does not proceed via the 2-lithiated species 83.

b. *Theoretical Considerations.* Complete neglect of differential over-lap (CNDO/2) calculations have been performed on free 3-fluoropyridine and its complex with methyllithium to gain insight into coordination effects between the ring nitrogen and the lithium atom (Fig. 1) (83T2009). In free 3-fluoropyridine, a lower election density at H_4 than at H_2 is calculated. Approach of MeLi to the ring nitrogen along the unshared electron pair axis leads to simutaneous modification of the electron densities at both H_2 and H_4, together with the total energy of the resulting adduct. The greater stability of the adduct was found for a N—Li distance of 2.1 Å, at which point the electron densities at H_2 and H_4 were inverted compared to those of the free molecule.

c. *Temperature Effects.* In order to determine regioselectivity preference in the DoM reaction of 3-fluoropyridine (38), temperature ef-fects in lithiation were investigated (Scheme 25) (83T2009). Thus, metal-

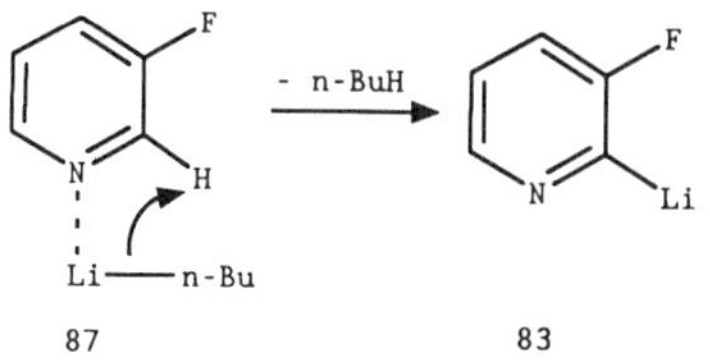

87   83

SCHEME 24

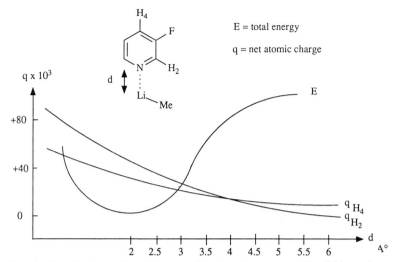

FIG. 1. Coordination effects between the ring nitrogen and the lithium atom on 3-fluoropyridine.

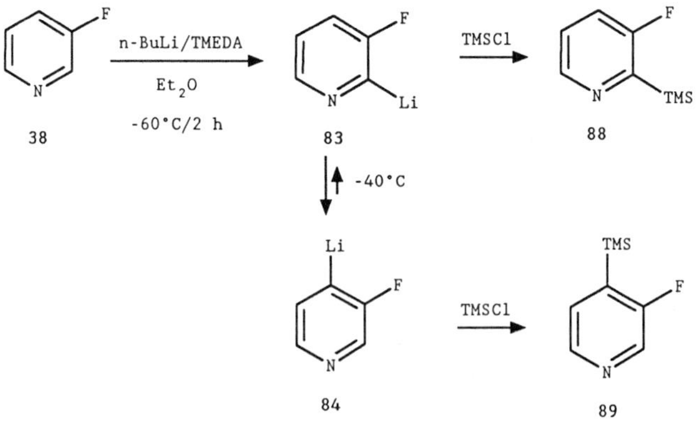

Reaction time, h	88 Yields % 89	
0	60	5
1	50	16
2	40	30
4	20	50
6	5	70

SCHEME 25

ation in Et_2O at $-60°C$ for 2 hr and warming of the resulting lithio species **83** to $-40°C$ was followed by variation in time before quenching with TMSCl to give products **88** and **89**. The results show that equilibration of **83** with **84** occurs as a function of time favoring the latter (thermodynamic) over the former (kinetic) species at longer reaction times without a significant change combined yield.

d. *Isomerization Mechanism.* Two reasonable mechanisms for the isomerization of 3-fluoro-2-lithiopyridine (**83**) to the corresponding 4-lithio isomer **84** are (a) homotransmetalation in which the isomerization involves small amounts of unreacted 3-fluoropyridine (**38**) (Scheme 26) [84TL(40)2107] or (b) an equilibrium of **83** with starting material **38** and the dilithio species **90** in catalytic amounts (76JOC1564; 79JOC4612). Attempts to increase the isomerization rate in Et_2O at $-60°C$ with additional 3-fluoropyridine (**38**) were unsuccessful, thus discrediting the homotransmetalation mechanism. On the other hand, when the metalation was performed in the same solvent with a stoichiometric ratio of n-BuLi/ TMEDA complex, a small amount of 2,4-bis(trimethylsilyl)-3-fluoropyridine was isolated thus suggesting the mechanism which involves the dilithio species **90**.

e. *Metalation with n-BuLi/DABCO.* In spite of the variation of reaction parameters, complete C2-regioselectivity in the n-BuLi/TMEDA metalation of 3-fluoropyridine was never observed. A change of conditions to n-BuLi/DABCO $(Et_2O/-75°C/1$ hr) followed by silylation afforded 2-silylated derivative **88** (80%) with less than 0.1% of the corresponding 4-product **89** (83T2009). The improved regioselectivity may be related to

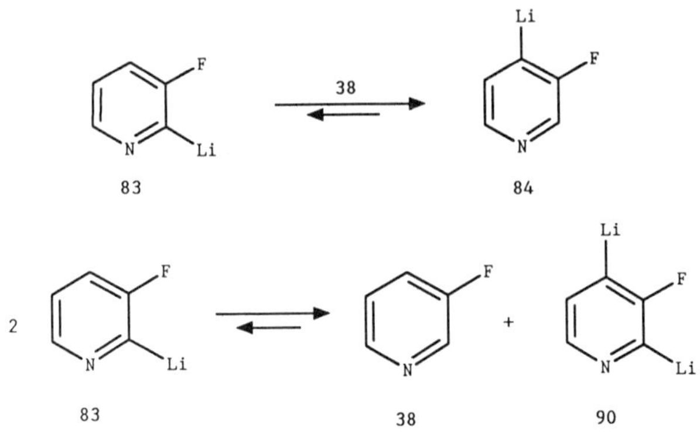

SCHEME 26

the poor solubility of the 2-lithio species **83** in the solution containing DABCO relative to that containing TMEDA. Thus, the isomerization of **83** to **84** is slower because of the low concentration of the former species in the medium containing DABCO. Similar regioselectivity was observed for the lithiation of 3-chloropyridine [81JOM(216)139; 86TH1].

2. *Metalation with Lithium Dialkylamides*

The poor nucleophilic character of lithium dialkyl amides promted their early use in pyridine DoM reactions [79JOM(171)273, 79PC1; 80TL4137; 85TH1].

a. *Choice of Lithium Dialkylamide.* In a survey of lithium dialkylamides, prepared under standard conditions (*n*-BuLi/THF/0°C), for the metalation of 3-chloropyridine (**78**), LDA was shown to be the most efficient in formation of the 4-lithio species **81** as evidenced by TMSCl quench experiments leading to product **91** (Scheme 27) [81JOM(216)139].

b. *Regioselectivity.* Although not studied extensively, regioselectivity in DoM reactions of 3-chloro and 3-fluoropyridines may vary as a function of the introduced electrophile. Thus, metalation of **78** with LDA (1 equiv.) in THF at −60°C [81JOM(216)139] and −78°C (80TL4137) followed by TMSCl quench leads to the 4-tetramethylsilane (TMS) product **91** in 96% and 98% yields, respectively. Similarly, metalation of **78** with LDA (prepared at −78°C) followed by iodine quench affords a mixture of products **92–94** in which the 4-iodo derivative **92** predominates (Scheme 28). However, minor products corresponding to **93** and **94** were not obtained using other electrophiles.

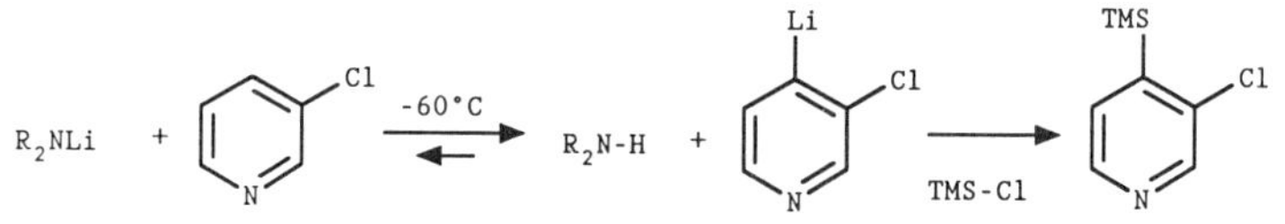

R_2NLi	PipLi	$(C_6H_{11}N)_2Li$	i-Pr(C_6H_{11})NLi	LiTMP	LDA
Yield %	0	18	35	60	96

SCHEME 27

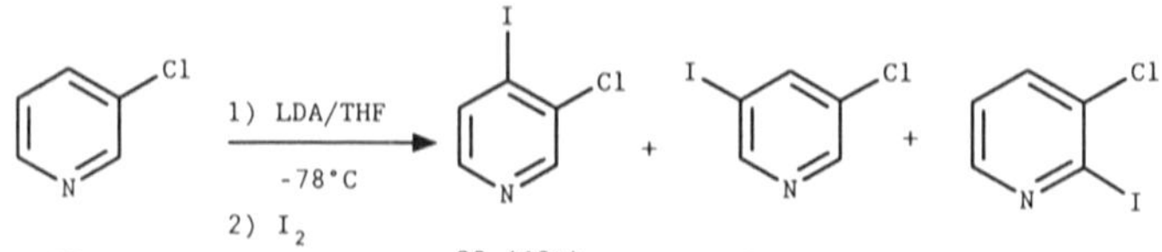
1) LDA/THF
-78°C
2) I_2

78 → 92 (62%) + 93 (2%) + 94 (2%)

SCHEME 28

3-Fluoropyridine also gave 4-substituted products in high yield (75–87%) using LDA [(1 equiv.)/THF/0.5–4hr] metalation conditions (80TL4137; 83T2009). Thus, metalation of 3-chloro and 3-fluoropyridine with LDA appears to occur rapidly at the most acidic C4 site. Compared to deprotonation with alkyllithium reagents, which is slow and irreversible, LDA metalation appears to be a fast equilibrium process.

c. *Amount of Base.* The DoM reaction of 3-fluoro or 3-chloropyridine **95** with LDA is very sensitive to excess of base (Scheme 29) [81JOM(216)139; 83T2009]. As gleaned from the Scheme 29, reaction of **95** with greater than 1 equiv. of LDA followed by silylation leads to the formation of 2,4-disubstituted product **98** in variable amounts in addition

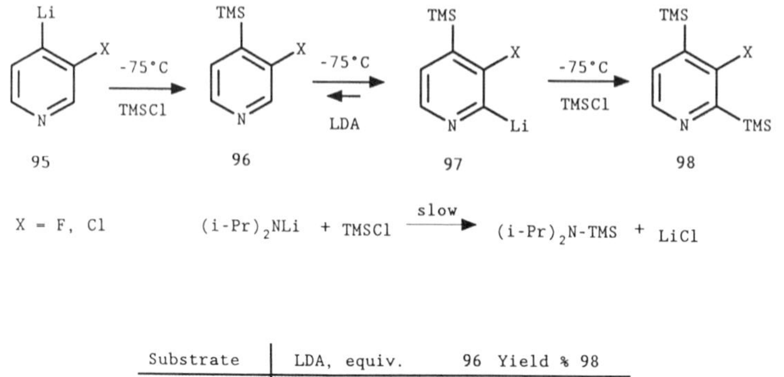
-75°C
TMSCl
-75°C
LDA
-75°C
TMSCl
$(i\text{-Pr})_2NLi$ + TMSCl
slow
$(i\text{-Pr})_2N\text{-TMS}$ + LiCl

95 96 97 98

X = F, Cl

Substrate	LDA, equiv.	96 Yield %	98
	1	75	0
95, X = F	1.1	78	7
	2	16	70
	1	96	0
95, X = Cl	1.1	97	0
	1.5	95	5
	2	65	35

SCHEME 29

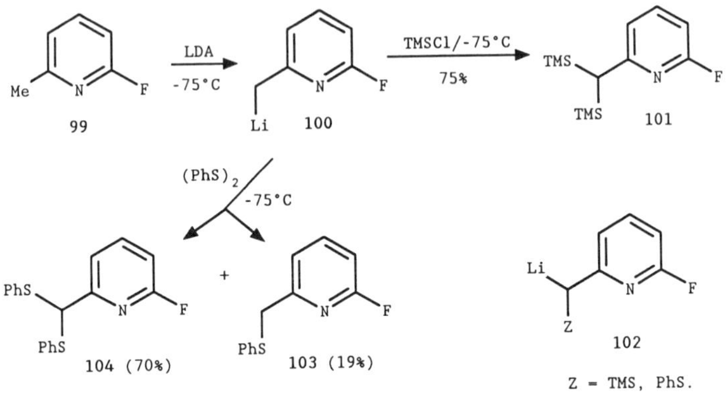

SCHEME 30

to the expected, usually major, 4-substituted derivative **96**. This is un-doubtedly due to the occurrence of a second lithiation of the initial product (**96→97**) by LDA, which reacts slowly with TMSCl electrophile at low temperatures [84TL(25)495; 88JOC1367].

  d. *Competition between Aromatic and Activated Methyl Sites.* The well-recognized high acidity of 2- and 4-alkyl hydrogens in pyridines sug-gests that competition between aromatic and side-chain deprotonation under the conditions of the DoM reaction may be expected. Thus, lithia-tion of 2-fluoro-6-methylpyridine (**99**) with LDA affords the more sta-bilized 6-picolyl anion **100**. This anion reacts with diphenyldisulfide to give **103**, which is transformed into the bis-phenylthio derivative **104** via an α-phenylthio anion intermediate **102** (Scheme 30) (87UP1). Similarly, the use of excess of LDA and TMSCl led to the formation of the bis-TMS product **101**.

  Similar results were observed in the DoM reaction of 3-chloropyridine (**78**) (Scheme 31) [81JOM(216)139]. Thus, treatment with LDA and methyl

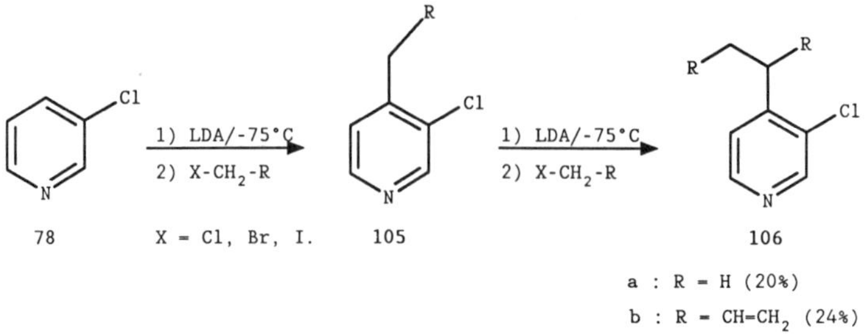

a : R = H (20%)
b : R = CH=CH₂ (24%)

SCHEME 31

iodide gave a mixture of 3-chloro-4-methylpyridine (105a) (40%) and 3-chloro-4-ethylpyridine (106a) (20%). Furthermore, alkylation with allyl chloride afforded 3-chloro-4-(1,5-hexadien-3-yl)pyridine (106a) (24%) as the sole isolable product.

3. DoM of Haloquinolines

The propensity of nucleophilic attack of fluoro and chloro quinolines by RLi reagents dictates the use of LDA for DoM processes that normally occur at the most acidic sites. Thus, LDA metalation of 3-chloro and 3-fluoro quinolines leads, after TMSCl quench, to 4-substituted products [79JOM(171)273]. Furthermore, 4-chloro [89JHC1589] and 5-fluoro [79JOM(171)273] quinolines, under similar conditions, lead to 3- and 6-TMS derivatives, respectively, without the formation of peri-substituted products, as observed in the corresponding naphthalene systems. 6- and 7-Fluoroquinolines likewise furnish 5- and 8-TMS haloquinolines, respectively [79JOM(171)273] (Scheme 8). Finally, 8-fluoroquinoline resists LDA metalation, a result which may be explained by the presence of a strong cation chelating site in the vicinity of the fluorine and nitrogen atoms.

4. DoM of Halopyrimidines

Application of the equilibrium-shift procedure (addition of LDA to a mixture of a halo derivative and a carbonyl electrophile) to 5-bromopyrimidine (30) and benzaldehyde leads to formation of the 4-substituted product 33 (Scheme 32) (79JOC2081). Similarly, the reaction

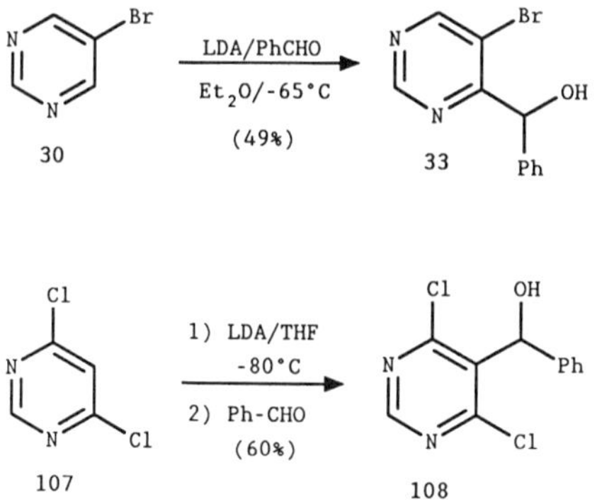

SCHEME 32

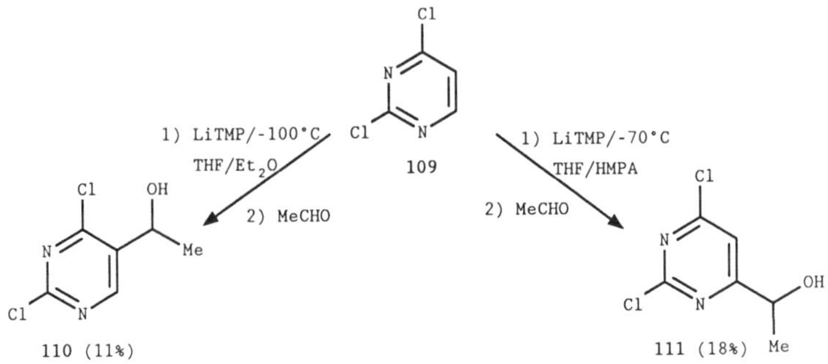

SCHEME 33

between 4,6-dichloropyrimidine (**107**) and benzaldehyde leads to product **108** in good yield (86S886).

Regioselectivity in the LiTMP metalation of 2,4-dichloropyrimidine (**109**) was shown to be dependent on temperature and solvent (Scheme 33) (91JHC). In THF/Et$_2$O at $-100°$C, metalation followed by acetaldehyde quench gave the 5-substituted (kinetic) product **110**, whereas in THF/HMPA mixture at $-70°$C, treatment with the same electrophile afforded the corresponding 6-substituted (thermodynamic) derivative **111**. Minimum neglect of differential overlap (MNDO) calculations support the observed results.

5. DoM of Halopyrazines

In the sole case studied, LiTMP deprotonation of 2-chloropyrazine (**112**) followed by electrophile quench gave the 3-substituted derivatives **113** in good yields (Scheme 34) (88S881).

6. DoM of Pyridazines

The LiTMP-mediated DoM reaction of 2,6-dichloropyridazine (**114**), using several electrophiles, leads to 3-substituted products **115** in variable yields (Scheme 35) (90JHC1377).

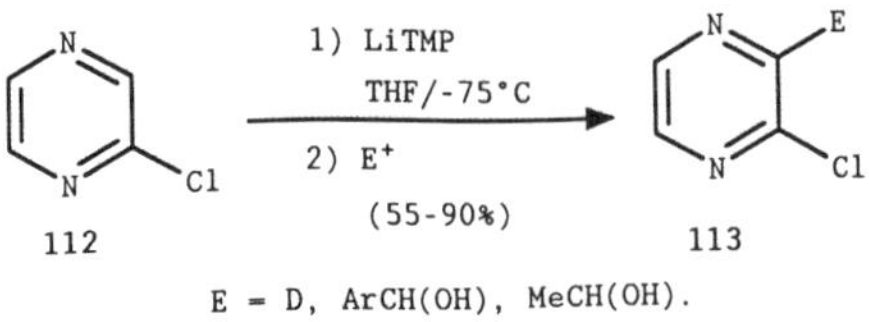

E = D, ArCH(OH), MeCH(OH).

SCHEME 34

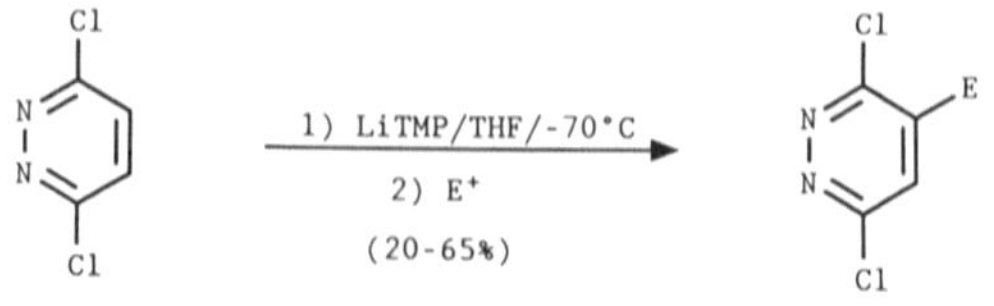

114   E - TMS, I, CHO, RR'C(OH).   115

SCHEME 35

D. SYNTHETIC APPLICATIONS

Halogen functionality in heteroaromatics undergoes a variety of trans-formations by nucleophilic substitution: metal–halogen exchange, radical substitution, and cross coupling reactions. The connecting link between these processes and the DoM reaction is therefore a powerful tool in heterocyclic chemistry.

1. The DoM-Nucleophilic Substitution Link

2- and 4-Halogen substituents are readily attacked by amino, oxygen, and sulfur nucleophiles.

a. *Synthesis of ortho-Carbon Substituted Haloheteroaromatics.* Lith-iation of 2-fluoropyridine (11) followed by condensation with aliphatic, aromatic, and heteroaromatic aldehydes leads to the formation of 3-substituted carbinols 116 which, upon oxidation and displacement with amines, afford, via 117 intermediates, the corresponding 2-amino-3-ketones 118 in short order and good overall yields (Scheme 36) [81JOM(215)139]. 4-Fluoro, 4-chloro, (86S886; 88JHC81) and 2-chloro-pyridine [90JCS(P1)2409] also undergo these reactions to afford analogous products. These methods compare very favorably with more classical

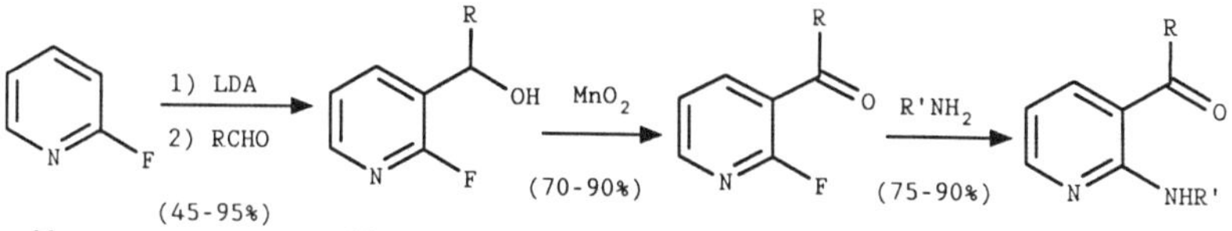

R = Me, Ar, Het. R' = H, Me.

SCHEME 36

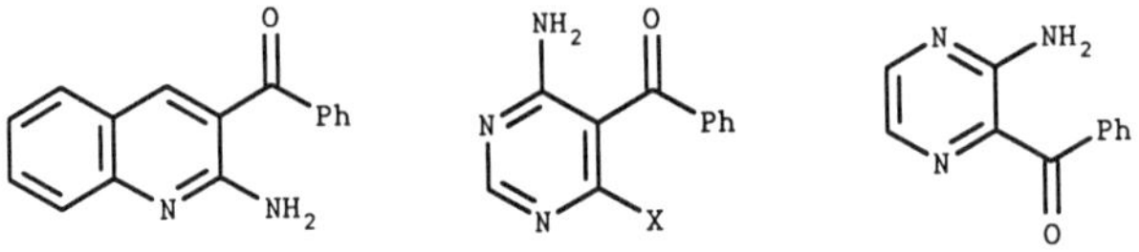

119 (20%)   120 (55%)   121 (45%)

X =Cl, NH$_2$.

SCHEME 37

routes which are longer and proceed in lower yield, for example, the preparation of (2-amino-3-pyridyl)phenylmethanone (67USP3314941).

Using a similar strategy, *ortho*-aminoketones of quinoline **119** (89JHC1589), pyrimidine **120** (86S886), and pyrazine **121** (88S881) were obtained (Scheme 37).

The utility of some of these products for side-chain modification has also been demonstrated. For example, 4-chloro-3-formylpyridine undergoes the Hantzsch reaction to give a 4-(3-chloropyridine)-1,4-dihydropyridine derivative (88%). (4-Chloro-3-pyridyl)diphenylmethanol affords aza-analogues (65–70%) of the antimycotic agent chlortrimazol by reaction with thionyl chloride and imidazole derivatives (88JHC81).

Amminolysis of 3-halopyridines, generally a difficult reaction, can be effected via *N*-oxide derivatives. Thus, metalation-aryl aldehyde condensation on 3-fluoropyridine (**38**) affords carbinols **122** which, upon standard sequential oxidation reactions, affords the *N*-oxide ketone **123** (Scheme 38) (84TH1). Treatment with dimethyl amine afforded the 3-amino derivative **124,** thus completing this high overall yield sequence.

Heterocyclic *ortho*-aminoketones may also be obtained from DoM reactions using amino DMGs (see Section III). *ortho*-Haloketones are also useful starting materials for oxygenated heterocycles. Thus, treatment of the 2-fluoro or 2-chloropyridin-3-ylphenylmethanones **132** with MeONA

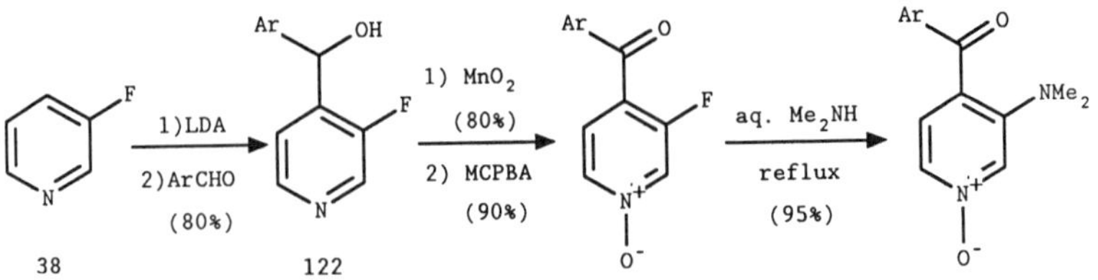

SCHEME 38

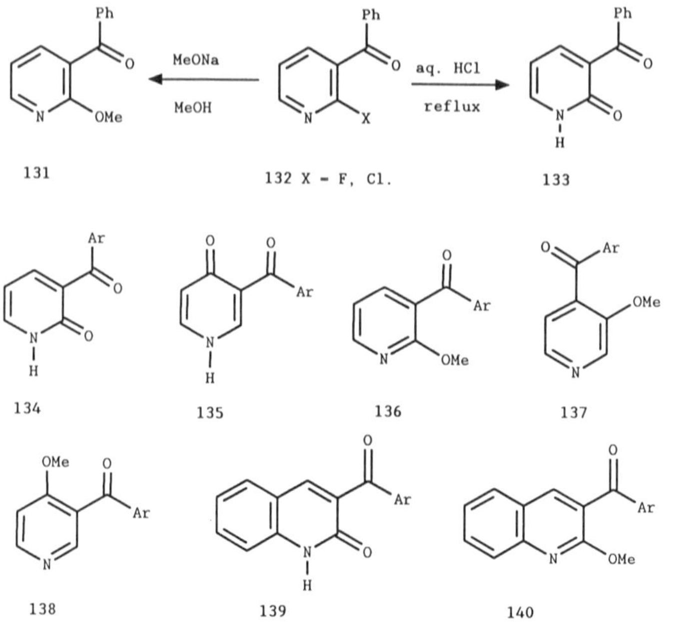

SCHEME 39

or aq. HCl leads to the 2-methoxypyridines **131** and the 2-pyridones **133** (Scheme 39) [90JCS(P1)2409]. Identical reactions of 2-, 3-, and 4-halopyridines and 2-chloroquinoline lead to a variety of oxygenated derivatives (**134–140**) [84TH1; 86S886; 88JHC81; 90JCS(P1)2409].

2-Quinolone-3-carboxylic acid (**143**) is readily obtained from **141** via the chloro acid **142** by a metalation, carboxylation, hydrolysis sequence (Scheme 40) (89JHC1589).

b. *Synthesis of Condensed Heteroaromatics.* Heteroaromatic ortho-halo ketones are demonstrably useful intermediates for the synthesis of a variety of polyheterocyclics.

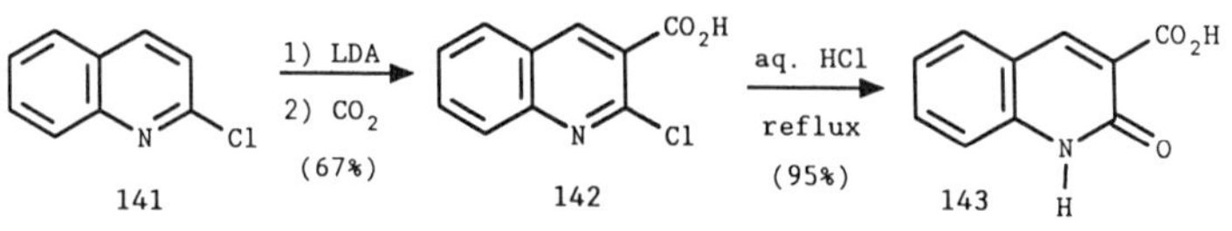

SCHEME 40

i. *1,6-Naphthyridines and pyrimido[4,5-b]quinolinones.* Nucleophilic displacement by aniline on chloropyridine **125** and dichloropyrimidine **128** leads, respectively, to the amines **126** and **129** in high yields (Scheme 41) (87AP704). These, upon treatment with PPA, furnish the 1,6-naphthyridine **127** and pyrimido[4,5-*b*]quinoline-4-one **130** in equally high yields. A compound of the type **125** is also useful for the preparation of pyrido-1,4-diazepine derivatives (see Section VIII).

ii. *Heteroring condensed pyrones.* Short routes to heterocyclic ring condensed pyrones **147** [88JHC81; 90JCS(P1)2409] and **151** (89JHC1589) have been achieved starting with halopyridine **8** and haloquinoline **141** systems (Scheme 42). Thus, metalation–formylation of **8** gave **144** which, upon hydrolysis or methoxide displacement, afforded the 2-pyridinone **145** and 2-methoxypyridine **148,** respectively. Knoevenagel chain extension afforded the acrylic acids **146** and **149**, respectively, which both were converted into the 8-azacoumarin **147** by treatment with pyridinium hydrochloride at high temperatures. A similar sequence on 4-chloropyridine led to the synthesis of the 6-azacoumarin (88JHC81), while its application to 2-chloroquinoline (**141**) afforded, via the aldehyde **150,** the benzo analogue **151** (89JHC1589).

iii. *Aza-xanthones.* The synthesis of aza-xanthone **154** was achieved by taking advantage of intramolecular nucleophilic substitution reactions of ortho-halo(*ortho*-methoxybenzoyl) pyridine (Scheme 43) [84TH1; 90JCS(P1)2409]. Thus, as before, metalation–condensation on **8** followed by oxidation led to the 3-(*ortho*-alkoxyaroyl) pyridine **153**. Treatment with pyridinium hydrochloride afforded the aza-xanthone **154**.

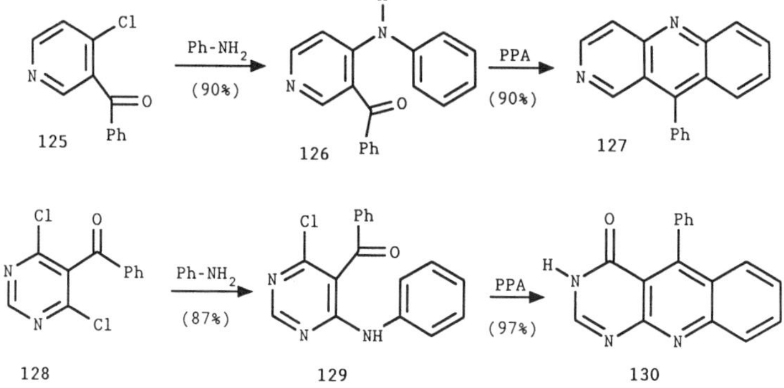

SCHEME 41

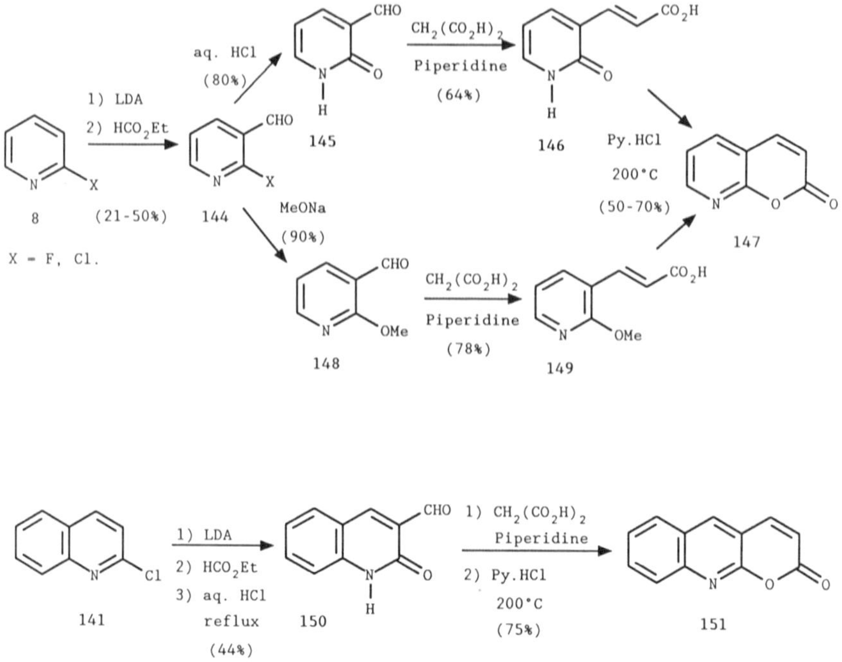

SCHEME 42

SCHEME 43

Analogue routes from 4-fluoropyridine (88JHC81), 2-chloroquinoline (89JHC1589), 2-chloropyrazine (88S881), and 3,6-dichloropyridazine (90JHC1377) lead, respectively, to condensed aza-xanthones **155–158** (Scheme 44).

iv. *Aza-acridones*. The tactic of intramolecular nucleophilic substitution has also been applied in the synthesis of aza-acridones **161** (Scheme 45)

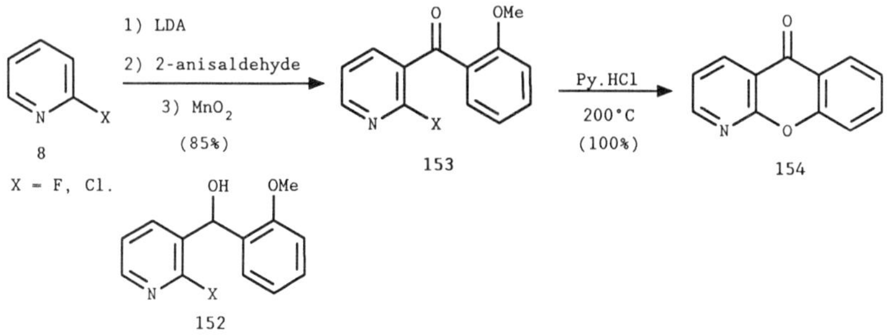

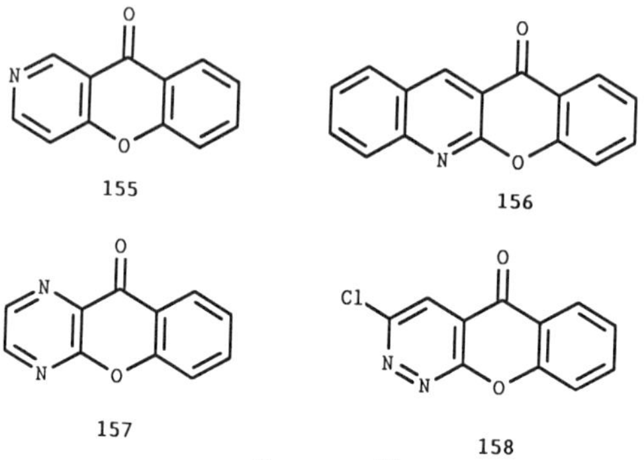

155

156

157

158

SCHEME 44

[90JCS(P1)2409]. Thus, LDA metalation of 2-fluoro or 2-chloro-pyridine **8** followed by condensation with 2-nitrobenzaldehyde leads to **159** which, upon oxidation, provides the ketone **160**. Catalytic reduction results in spontaneous cyclization to afford aza-acridone **161** in quantitative yield.

v. *Aza-indoles.* DoM-mediated introduction of side chain with electrophilic character into 2-fluoropyridine (**11**) allows, through the vehicle of nucleophilic amine for fluoride exchange, the formation of aza-indole **164** (Scheme 46) (88TH1). Thus, LDA metalation followed by quenching with 1-methoxy-2-propanone yields the alcohol **162** which, upon treatment with methyl amine, affords the 2-amino derivative **163**. Exposure to HCl results in cyclization and dehydration to give **164** in excellent yield.

2. *DoM-Oxidation via Boronic Acids*

In a method not generally explored to date, sequential treatment of 2-chloroquinoline (**141**) with LDA and trimethyl borate followed by acid

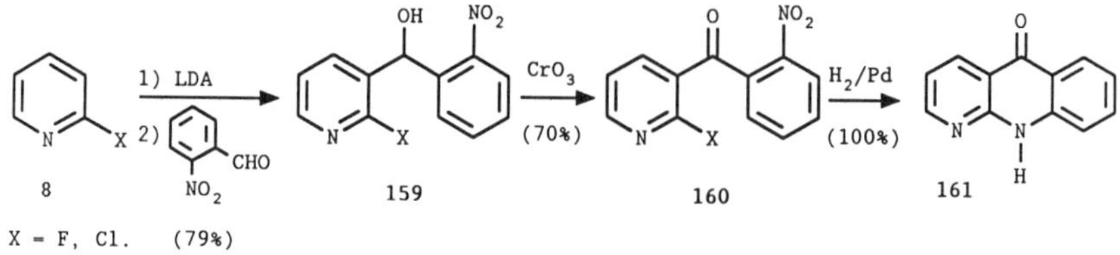

SCHEME 45

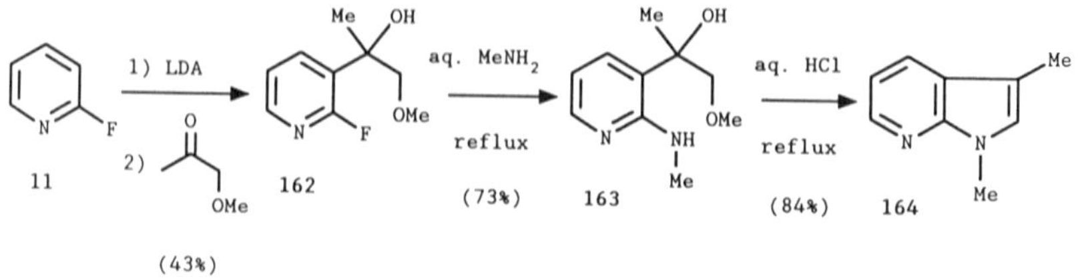

SCHEME 46

hydrolysis leads to the 3-boronic acid **165** which, by hydrogen peroxide oxidation, is converted into the 3-hydroxy derivative **166** (Scheme 47) (89JHC1589). Acid-induced hydrolysis furnishes the 2-quinolone **167**. This method, while broadly developed in aromatic chemistry, has not been extensively applied to heteroaromatic substrates.

3. DoM-Elimination Reactions

The generation of pyridyne intermediates, an area which has been extensively investigated (67MI1), usually involves ortho-disubstituted pyridine precursors. A favorite route involves metal–halogen exchange of ortho-dihalogenated pyridines, which lead by *in situ* lithium halide elimination into pyridyne species that are trapped by appropriate dienes. For example, Li/Hg treatment of the isomeric chloro and bromo pyridines **168a** and **168b** in the presence of furan affords, via 2,3- and 3,4-pyridynes, adducts **169a** and **169b**, which can be further aromatized into quinoline (**170a**) and isoquinoline (**170b**) (Scheme 48) (74MI1).

The scope of this strategy is greatly restricted by the tedious preparation of ortho-dihalo precursors. This problem is circumvented by the applica-

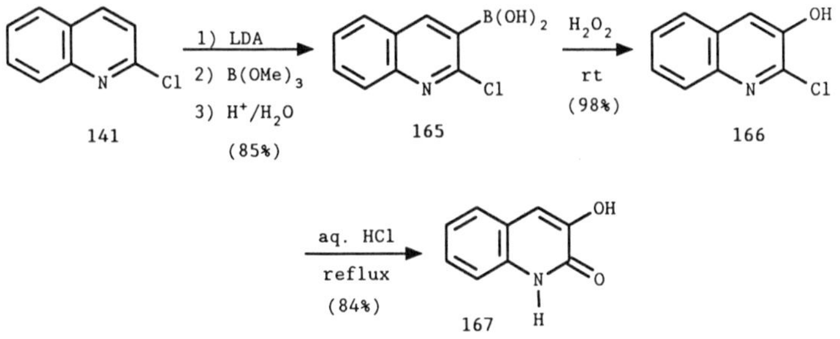

SCHEME 47

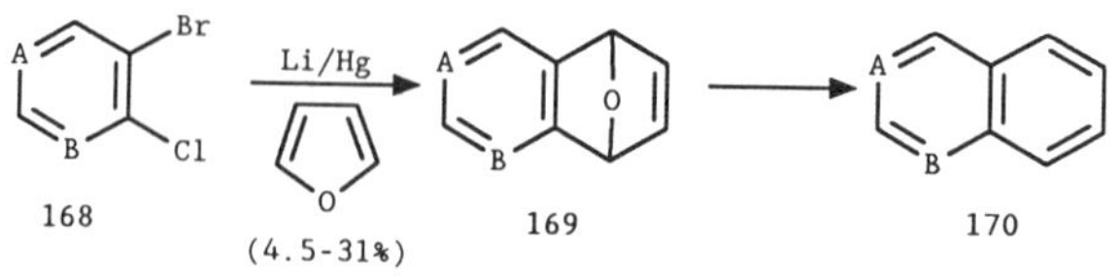

168

(4.5-31%)

169   170

a : A - CH ; B - N.
b : A - N ; B - CH.

SCHEME 48

tion of DoM reactions to halopyridines. Thus, *n*-BuLi metalation of 2-unsubstituted or 2-substituted 3-fluoro or 3-bromopyridines **171,** followed by addition of furan and *N*-methylpyrrole, leads when warmed to room temperature, to the bicyclic products **172** in modest yields via a 3,4-pyridyne species (Scheme 49). In this manner, 3,4-pyridyne has been generated from 3-bromopyridine [72CR(C)(275)1439] and 3-fluoropyridine [72CR(C)(275)1535]; 2-fluoro-3,4-pyridyne from 3-bromo-2-fluoropyridine [72CR(C)(274)719]; and 2-chloro-3,4-pyridyne from 2-chloro-3-fluoropyridine [72CR(C)(275)1535].

Similar reaction conditions have been applied to 4-fluoro and 4-chloropyridine to give the cycloaddition product **39** (Scheme 50) (88JHC81).

The interfering addition of lithium dialkylamides to the pyridyne intermediate can be circumvented by using the bulky amide LiTMP. Using this base, the reaction of 3-bromopyridine (**42**) with 2,7-bis(trimethylsilyl)isobenzofuran (**174**) in refluxing Et$_2$O affords adduct **175** in modest but useful yields (Scheme 51) (85JOC2746). The bis-TMS subsituents in **174** inhibit competitive deprotonation at the acidic 2,7-sites of isobenzofuran. Desilylation followed by deoxygenation affords benz[*g*]isoquinoline (**176**). Analogous reactions have been achieved between **42** and 1-ethoxy-3-(trimethylsily)isobenzofuran (87JOC792) and 1,2,7-trimethylisoindole (79JOC1519).

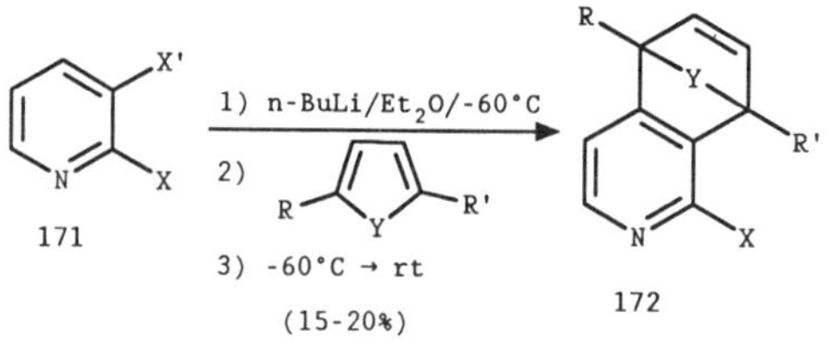

171

1) n-BuLi/Et$_2$O/-60°C
2)
3) -60°C → rt

(15-20%)

172

X = H, F, Cl; X' = F, Br. Y = O, NMe; R, R' = H, Me.

SCHEME 49

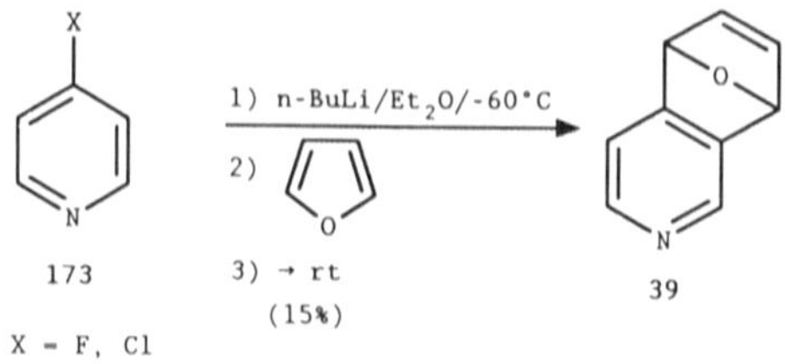

X = F, Cl

SCHEME 50

Direct formation of aza-anthraquinones **181** has been achieved using *in situ* generated lithio cyanophthalide **177** (a 1,4-dipole equivalent) and 3,4-pyridyne **178** (Scheme 52) [88H(27)2643]. Thus, addition of 3-bromopyridine derivative to a solution of LDA and **177** at −40°C leads, when warmed to room temperature, to aza-anthraquinones **181** in good yields via intermediates **179** and **180**. This type of reaction has also been applied to 4-bromoquinoline to give benzo[*d*]-2-azaanthraquinone in 60% yield [88H(27)2643].

The scope of this methodology is restricted to the generation of 3,4-pyridyne. To date, analogous formation of 2,3-pyridyne from 2-halopyridines has not been reported.

4. *DoM Metal–Halogen Exchange Reactions*

2-Bromo-3-substituted and 3-bromo-4-substituted pyridines, obtained by DoM processes, can serve in metal–halogen exchange reactions, thus providing versatile methodology for pyridine functionalization by two different electrophilic reactions. For example, LDA metalation of **48** followed by MeI quench gives **182** which, when subjected to *n*-BuLi exchange conditions and treatment with benzaldehyde, affords the 2,3-dicarbon functionalized pyridine **183** (Scheme 53) [82JCR(S)278; 84TH1]. Presumably, an attempt to carry out metal–halogen exchange on a bromo

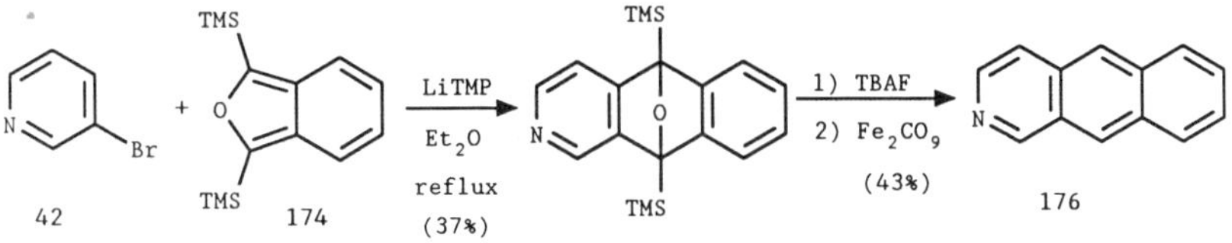

SCHEME 51

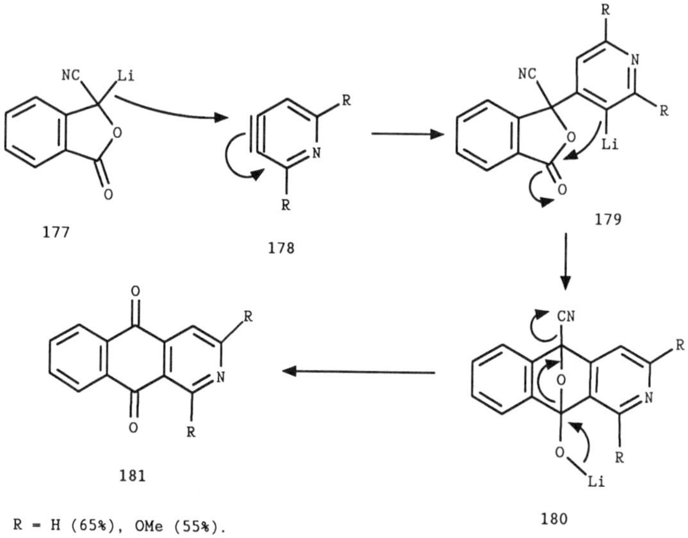

SCHEME 52

2- or 4-methylpyridine would suffer from competitive side-chain deprotonation, although such experiments appear not to have been reported.

DoM–metal–halogen exchange and halogen dance processes may also be promising routes to pyridine functionalization. Thus, low temperature LDA metalation of 2-fluoropyridine (**11**) followed by sequential metalation, addition of trace amounts of bromine, warming to −40°C, and hydrolysis affords 4-bromo-2-fluoropyridine (**186**) via the isomeric intermediates **184** and **185** (Scheme 54) [90JOM(382)319; 91JOC(s1)]. *t*-BuLi-induced metal–halogen exchange and acetaldehyde quench gave the alcohol **187,** which was converted by oxidation into the 4-acetyl derivative **188.**

5. *DoM-S$_{RN1}$ Reactions*

Although the S$_{RN1}$ reaction is of considerable utility in heterocyclic synthesis (80JOC1546, 80TL1943), its scope is restricted by the poor

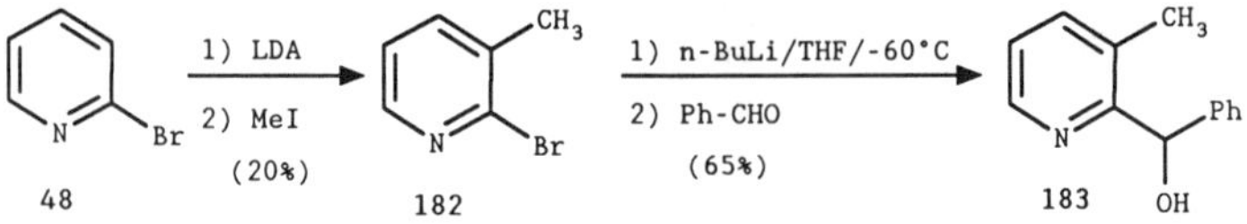

SCHEME 53

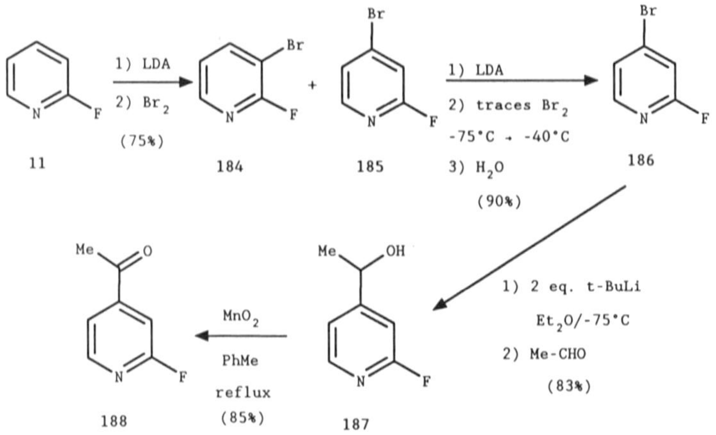

SCHEME 54

availability of specific halo heteroaromatics. The regiospecific formation
of substituted halo heterocycles by the DoM process allows expanded
synthetic utility of the S_{RN1} process.

In a clear demonstration of such advantage, 2-fluoropyridine (**11**) has
been converted into the 3-substituted derivative **190** (Scheme 55)
(88JOC2740). Thus, metalation of **11** followed by iodination gave the
3-iodopyridine **189** which, upon the application of standard S_{RN1} reaction
conditions, furnished **190** in almost quantitative yield. This reaction adds
an umpolung dimension to substition chemistry of halopyridines formed
by DoM processes.

The lack of reactivity of 3-halo substituents under non-radical nucleophilic substitution conditions allows differential functionalization of pyridines by 3-umpolung and 2-nucleophilic substitution processes. Thus,
treatment of 2-fluoro-3-iodopyridine (**189**) with oxygen or amine nucleophiles affords products **191** which, upon subjection of S_{RN1} reactions with
carbon, phosphorus, and sulfur systems, give 2,3-difunctionalized pyridines **192** (Scheme 56) (88JOC2740).

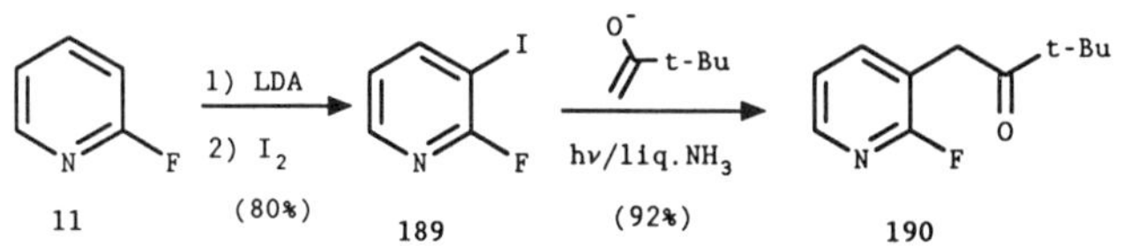

SCHEME 55

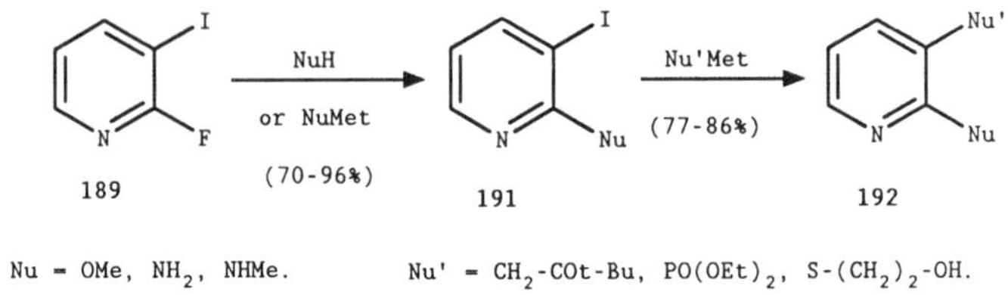

Nu = OMe, NH$_2$, NHMe.   Nu' = CH$_2$-COt-Bu, PO(OEt)$_2$, S-(CH$_2$)$_2$-OH.

SCHEME 56

Certain 2-amino-3-ketones **193** also serve, via the agency of hydrochloric acid, as intermediates for the synthesis of 7-azaindole derivatives **194** (Scheme 57) (88JOC2740).

6. DoM Cross-Coupling Reactions

Transition metal catalyzed cross coupling procedures have superceded classical Ullmann reactions for the construction of unsymmetrical biaryls [82ACR395; 86AG(E)508]. The Suzuki process involving cross coupling between aryl boronic acids and aryl bromides constitutes a most promising new methodology (81SC513), whose connection to the DoM reactions has been systematically developed (85TL5997; 90TL1665; 90JCS(P1)2611).

In an application of the Suzuki process, 2-chloroquinoline (**141**) has been converted into the condensed heterocycle **197** (Scheme 58) (89JHC1589). Thus, metalation, trimethyl borate quench, and hydrolysis affords the stable 3-boronic acid **195** which, upon subjection to cross coupling with *ortho*-iodo aniline in the presence of Pd(0) catalyst and base, affords the 3-arylquinoline **196**. Acid catalysis converts this material into the indolo[2,3-*b*]quinoline (**197**) in 35% overall yield.

III. Nitrogen-Based DMGs

The poor DMG properties of amino, arylamino, and dialkylamino groups in aromatic DoM reactions [66JOC2047; 67JOM(10)171;

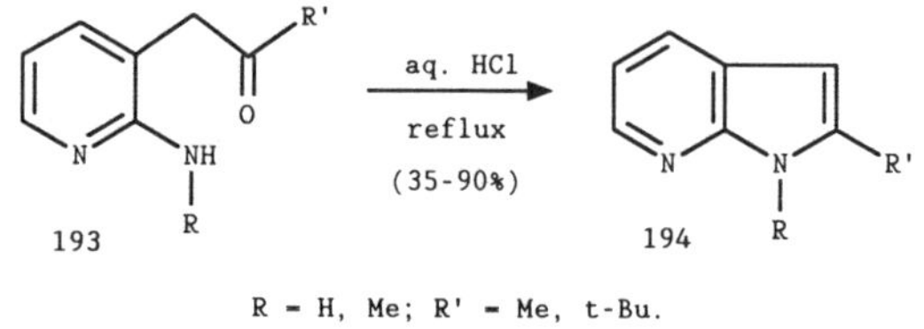

R = H, Me; R' = Me, t-Bu.

SCHEME 57

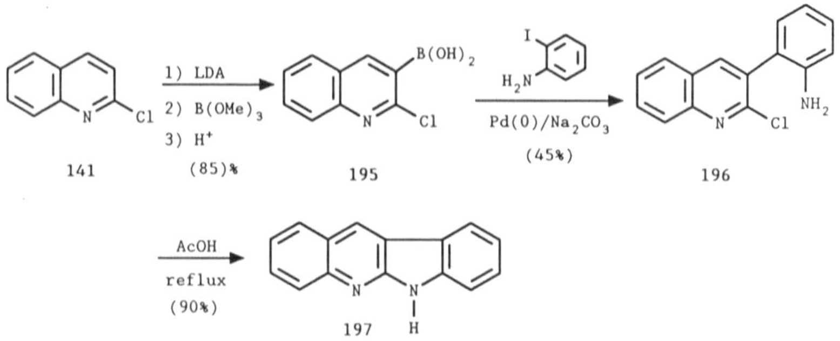

SCHEME 58

70JOC1288] has been attributed to the unavailability of the N-lone pair for coordination with metalating agents due to resonance effects [79OR(26)1]. Nevertheless, these groups exert a weak acidifying effect on ortho hydrogens compared to other noncoordinating functions, such as halogen. In a seminal study, Führer and Gschwend demonstrated the deprotonation of N-pivaloyl aniline **198** to the highly coordinated dianion **199**, using 2 equiv. of n-BuLi (Scheme 59) (79JOC1133). Treatment with a variety of electrophiles allowed the preparation of ortho functionalized derivatives in good yields. Subsequent work by Muchowski (80JOC4798) showed that N-t-Boc (N-CO_2t-Bu) also serves as a useful DMG (84CC1304), while the urea group (NHCONR_2) was found to be weaker [88JOM(354)273]. In a recent development, the DoM reaction of diarylamine carbamic acids have been demonstrated (86T4027).

A. AMINOPYRIDINES

The initial study of aminopyridine metalation was carried out on 3-methoxy-5-pivaloylaminopyridine (**200**), readily prepared from the corresponding pyridine (Scheme 60) (81JOC3564). In this case, the combined activating effects of the two DMGs lead to metalation at the in-between site. Using optimum conditions (n-BuLi/THF/$-25°C$) followed by quench with a variety of electrophiles at $-70°C$ led to products **201**. Addition

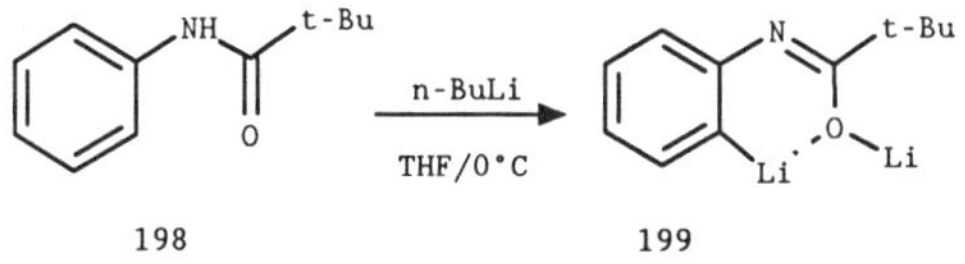

SCHEME 59

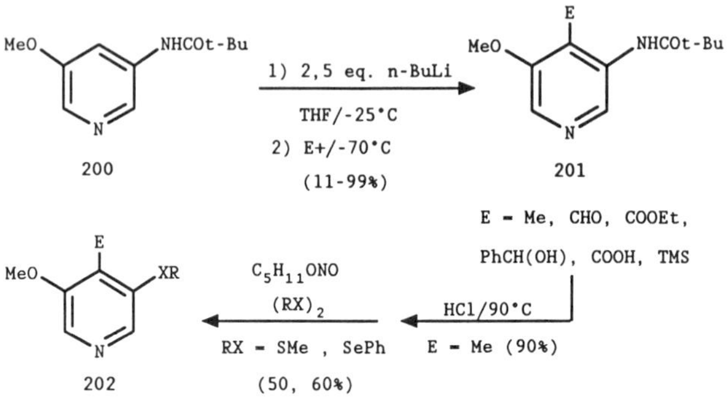

SCHEME 60

reactions were not observed, even at 0°C, which may be explained by the high LUMO energy of pyridines with electron-donating substituents, which makes them comparable to benzenes. In support of this contention, 3-ethoxypyridine can be lithiated in refluxing Et$_2$O without competitive addition (82S235). The new products **201**, upon hydrolysis, are converted into the amino derivatives which, by Sandmeyer reaction, are transformed into the 3-thio and 3-selenopyridines **202**.

That the combined DMG effects of OMe and *N*-pivaloyl are not necessary was demonstrated by the metalation of the simple 2-, 3-, and 4-*N*-pivaloyl pyridines **203** to give, after electrophile quench, products **204** (Scheme 61) (82S499; 83JOC3401; 89JHC105). Using *n*-BuLi (2.5 equiv./ TMEDA/Et$_2$O/ − 10°C) or *n*-BuLi (2.5 equiv./THF/0°C) conditions, a variety of substituted amino pyridines **205–207** were thereby prepared. In all cases, regiospecific metalation at the most acidic site was observed. For example, metalation of the 3-*N*-pivaloyl isomer **203** at −40°C occurred at

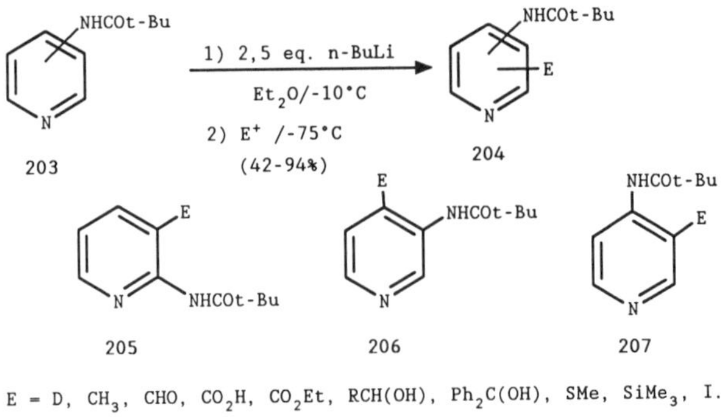

SCHEME 61

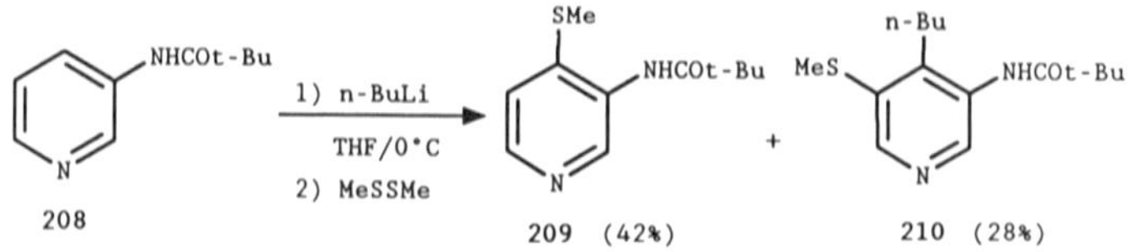

SCHEME 62

the 4-position, a result which can be compared to the reaction of 3-alkoxypyridines, which show 2-metalation. This may be attributed to the preferential coordination of the RLi reagent with the lithiated pivaloylamino group, rather than the pyridine ring nitrogen inducing 4-deprotonation. On the other hand, in 3-alkoxypyridines, the low coordination ability of the alkoxy moiety allows complexation of the lithiating agent to the ring nitrogen, thus favoring 2-deprotonation (82S235).

However, some competition between DoM reaction and addition was observed for 3-pivaloyl pyridine **208** when dimethyl disulfide was used as the electrophile (Scheme 62) (83JOC3401). In addition to the expected product **209**, compound **210**, a result of *n*-BuLi addition followed by dimethyl disulfide attack on the resulting 1,4-dihydropyridine and oxidation during work up, was isolated in 28% yield. Such addition could be decreased, for this and other electrophiles, by operating under the following conditions: slow addition of *n*-BuLi (2 equiv.) to a cold (−70°C) suspension of the substrate in Et$_2$O/TMEDA, then warming to −10°C and cooling to −70°C (89JHC105) before addition of the electrophile.

2-*N*-*t*-Boc pyridine (**212**), readily obtained from the corresponding amino derivative **211**, undergoes smooth metalation with *t*-BuLi at −78°C to give, after quenching with several electrophiles, products **213** (Scheme 63) (84CC1304).

In the case of 2-*N*-*t*-Boc pyridine (**214**), a competitive experiment with the corresponding *N*-pivaloyl system (**215**) showed, on the basis of the

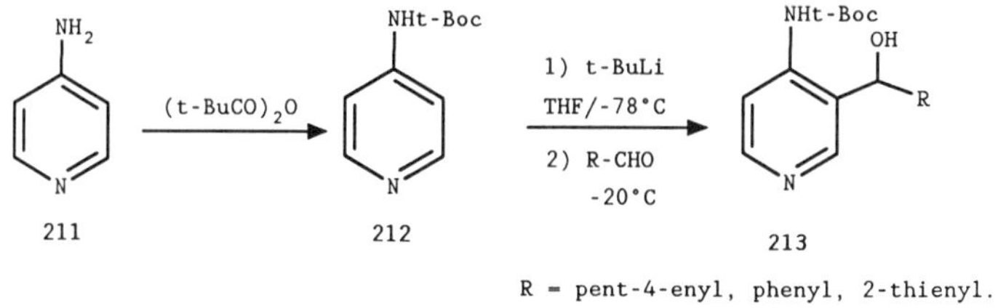

R = pent-4-enyl, phenyl, 2-thienyl.

SCHEME 63

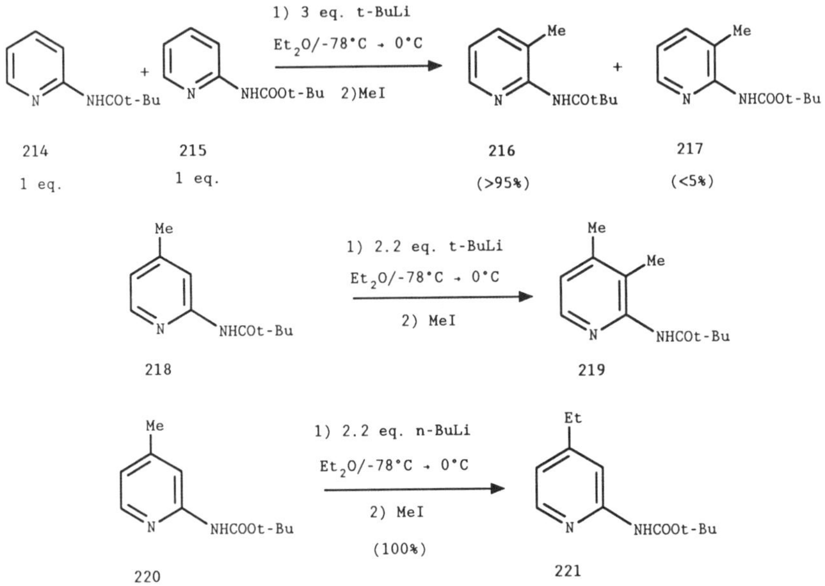

SCHEME 64

isolated products **216** and **217**, that the former is by far the more powerful DMG (Scheme 64) (85PC1). This was also demonstrated in a competition of ring vs. lateral metalation. Thus, *N*-pivaloyl 4-methylpyridine (**218**), upon treatment with 2.2 equiv. of *t*-BuLi in Et$_2$O followed by MeI quench, gave selectively the ring methylated product **219**. The corresponding *N*-*t*-Boc derivative **220** provided exclusively, under the same condition, the 4-methyl alkylated material **221**.

Additional competition experiments were performed on methyl, chloro, and fluoro derivatives of 2-*N*-pivaloyl pyridine (83JOC3401). Metalation and dimethyl disulfide quench of 2-*N*-pivaloylamino picolines **222** gave results highly dependent on conditions (Scheme 65) (83JOC3401). Use of *t*-BuLi/Et$_2$O/−78°C conditions on the 2-amino-4-picoline and 2-amino-6-picoline provided good yields of ring substituted products **223** without interference of lateral metalation. On the other hand, metalation of these substrates with *n*-BuLi/THF/−0°C led to mixtures of ring **223** and lateral **224** products. As may be predicted, no lateral metalation of the 2-amino-5-picoline was observed, and a high yield of ring thiomethylated product was isolated.

Metalation of the 5- and 6-chloropivaloylamino pyridines **225** was shown to occur cleanly giving, after dimethyl disulfide quench, high yields of

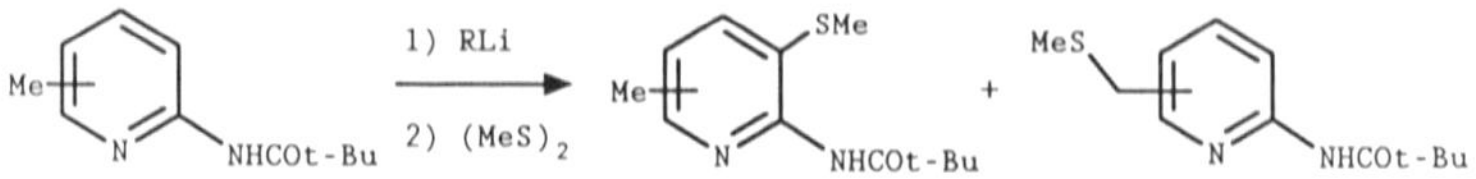

R	Conditions	223 Yield, %	224
4-Me	n-BuLi/THF/0°C	40	51
4-Me	t-BuLi/Et$_2$O/-78°C	74	0
5-Me	n-BuLi/THF/0°C	94	0
6-Me	n-BuLi/THF/0°C	59	27
6-Me	t-BuLi/Et$_2$O/-78°C	74	0

SCHEME 65

products **226** (Scheme 66) (83JOC3401). Rather careful conditions had to be used for the 5-chloro derivative **225** to avoid nucleophilic addition.

Competition between fluoro and N-pivaloyl DMGs was demonstrated in the metalation of **227** as evidenced by dimethyl disulfide quench experiments (Scheme 67) (83JOC3401). While exclusive ortho-to-N-pivaloyl deprotonation was shown to occur using t-BuLi/Et$_2$O/-78°C conditions, almost equal metalation to the two DMGs was observed under n-BuLi/THF/0°C.

Several thiazolo[4,5-b]pyridines, which may be considered as 2-protected-3-aminopyridines, undergo 4-deprotonation and provide, by reaction with a variety of electrophiles, good yields of functionalized products **231** (Scheme 68) (89TL183). In the case of the 6-chloro derivative **230**, exclusive formation of the 4-substituted product **231** was observed.

B. AMINOQUINOLINES

On the basis of considerable documentation regarding the reactions of quinolines with organometallics, the aminoquinolines were expected to

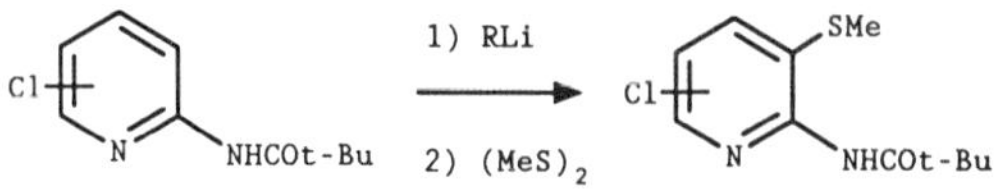

Cl	Conditions	Yield, %
5-Cl	t-BuLi/THF/-78°C	86
6-Cl	n-BuLi/THF/-20°C	78

SCHEME 66

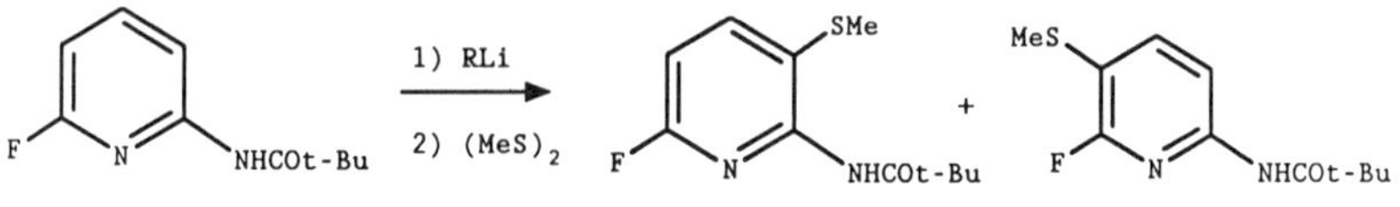

Conditions	228	Yield, %	229
n-BuLi/THF/0°C	45		30
t-BuLi/Et₂O/-78°C	70		0

SCHEME 67

undergo more facile nucleophilic substitution than the corresponding pyridines.

Under *n*-BuLi (3 equiv./Et₂O/0°C) conditions, 2-*N*-pivaloylaminoquinoline (**232**) undergoes smooth 3-lithiation as demonstrated by iodine **233** and D₂O quench experiments (Scheme 69) (86S670). On the other hand, using the *n*-BuLi/TMEDA complex resulted in equal metalation and addition, as evidenced by the formation of **233** and **234** in almost equal amounts.

Using the optimum conditions, a variety of 3-substituted 2-aminoquinolines **235** were prepared from **232** in variable yields (Scheme 70).

Metalation and protonation of 3-*N*-pivaloylaminoquinoline (**236**) gave predominantly the 2-addition product **237** which, for characterization purposes, was oxidized to **238** (Scheme 71) [88JOM(354)273]. Variation of condition bases (*n*-BuLi, *n*-BuLi/TMEDA, *t*-BuLi) gave similar results.

However, the equilibrium-shift process (simultaneous addition of LDA and TMSCl) to the 3-*N,N*-dimethyl urea **239** afforded the 2-TMS derivative **240** in high yield (Scheme 72) [88JOM(354)273]. This was the first use of this DMG group.

4-*N*-pivaloylaminoquinoline (**241**) did not undergo deprotonation under *n*-BuLi/TMEDA/-50°C conditions. Raising the temperature to -35°C

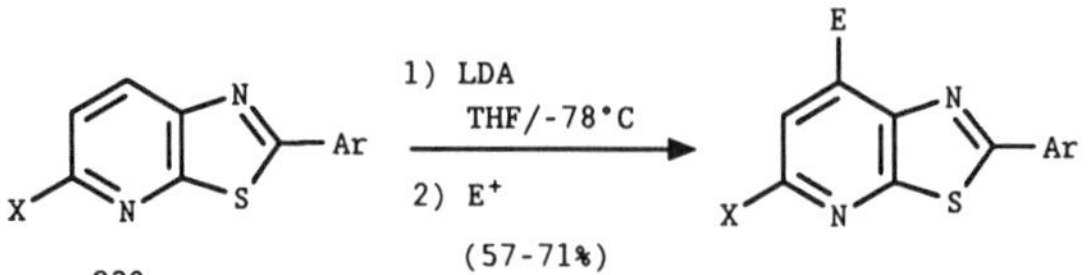

X = H, Cl; Ar = Ph.   E = MeCH(OH), PhCH(OH), S-Me, TMS.

SCHEME 68

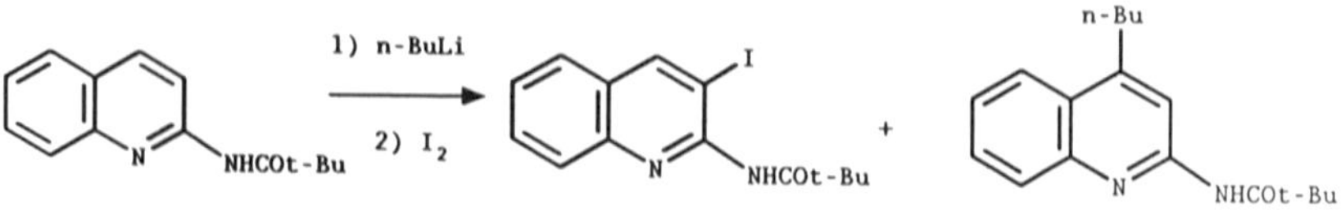

		1) n-BuLi			
232	Conditions	233	Yield, %	234

Conditions	233	234
n-BuLi/Et$_2$O/0°C	90	0
n-BuLi/TMEDA/Et$_2$O/0°C	45	55

SCHEME 69

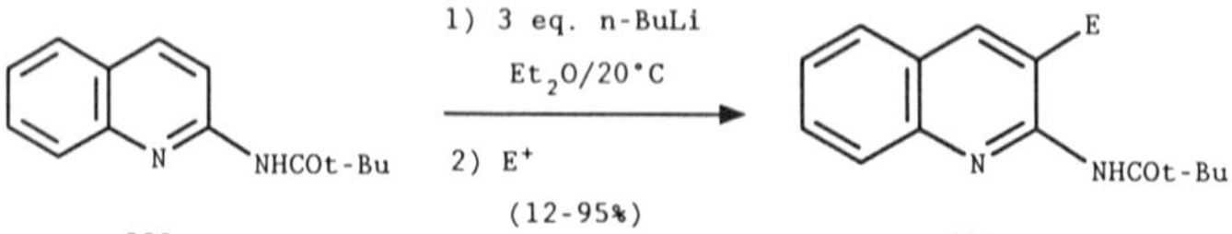

1) 3 eq. n-BuLi
   Et$_2$O/20°C
2) E$^+$
(12-95%)

232   235

E = D, I, TMS, SMe, COOEt, CEt$_2$(OH).

SCHEME 70

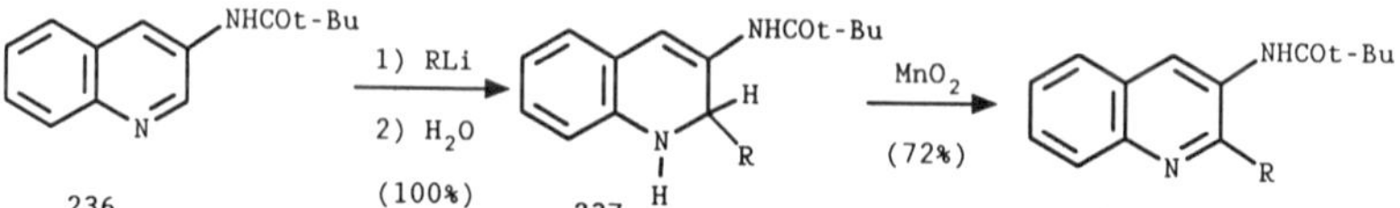

1) RLi
2) H$_2$O
(100%)

MnO$_2$
(72%)

236   237   238

RLi = n-BuLi/TMEDA/THF/20°C or t-BuLi/THF/-70°C.

SCHEME 71

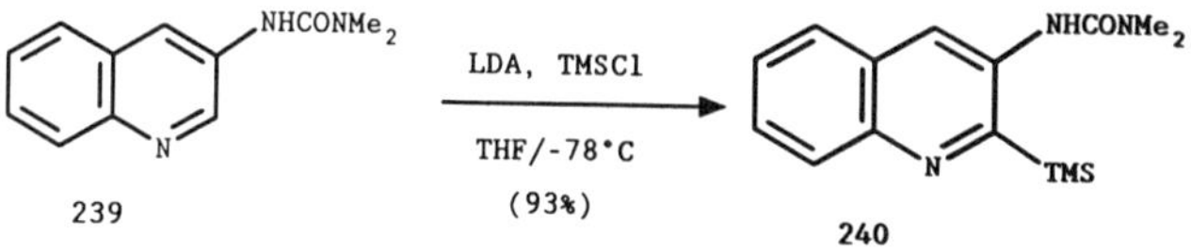

LDA, TMSCl

THF/-78°C
(93%)

239   240

SCHEME 72

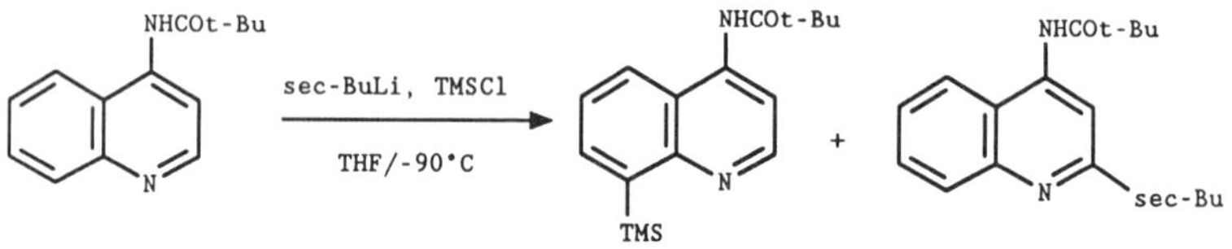

SCHEME 73

resulted in 2-addition only [88JOM(354)273]. However, using the more basic *sec*-BuLi at −90°C, followed by TMSCl quench, gave, aside from the addition product **243**, substantial amounts of the 8-TMS derivative **242** (Scheme 73). This interesting regioselectivity may be related to the complexation ability of the ring nitrogen. Unfortunately, carbonyl electrophiles failed to undergo reaction with metalated **241**.

C. SYNTHETIC APPLICATIONS

The DoM reaction of amino substituted heterocycles leads to versatile ortho-amino aldehydes, ketones, and carboxylic acids synthons that may be available only by lengthy and tedious classical routes.

1. *ortho-Aminopyridine Carboxaldehydes*

Starting with appropriate *N*-pivaloylaminopyridines **244a-b,** short and efficient synthesis of 2- and 4-aminopyridine-3-carboxaldehydes **246a-b** has been achieved in 50% and 52% overall yields, respectively, via the intermediates **245a-b** (Scheme 74) (83JOC3401).

Similarly, 5-methoxy-3-pivaloylaminopyridine-4-carboxaldehyde was obtained from the corresponding 3-amino-5-methoxypyridine (82-CPB1257).

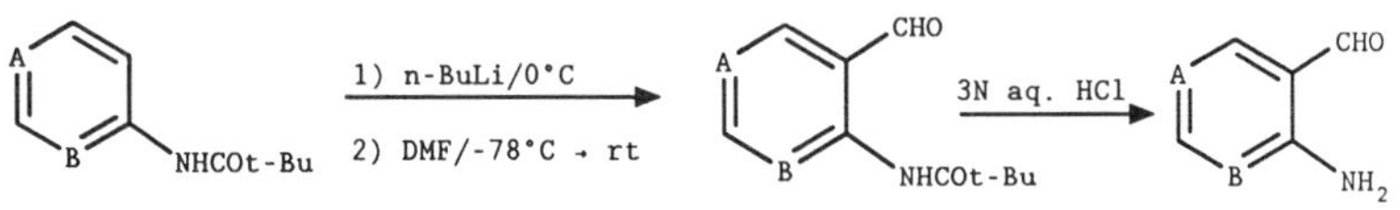

244 a: A = CH, B = N   245 a (54%)   246 a (93%)

244 b: A = N, B = CH   245 b (60%)   246 b (86%)

SCHEME 74

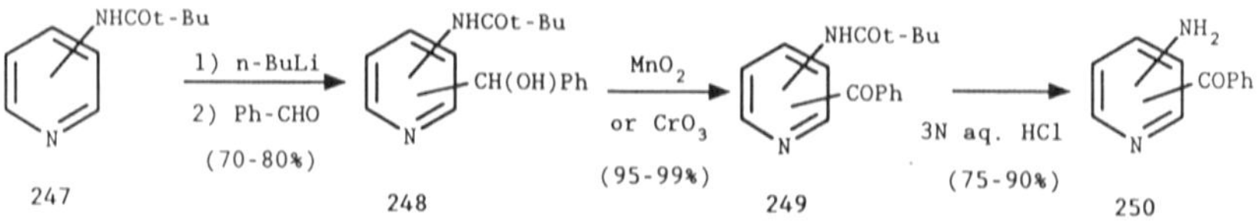

SCHEME 75

2. ortho-Aminopyridine Ketones

A convenient four-step synthesis of 2-amino-3-pyridinyl, 3-amino-4-pyridinyl, and 4-amino-3-pyridinyl phenylmethanones **250** begins with the three isomeric N-pivaloyl pyridines **247** and involves the intermediate carbinols **248** (Scheme 75) (89JHC105).

The synthetic utility of these derivatives towards the preparation of condensed heterocycles has been demonstrated. Thus, treatment of **252** with cyclohexanone under acidic conditions (66JOC3852) (Friedlander reaction) and molten urea affords the naphthyridine **253** and pyridopyrimidinone **251**, respectively (Scheme 76) (89JHC105). Application of these reactions on the isomeric N-pivaloylamino pyridine ketones affords analogue heterocycles **254–257**.

4-Acetyl-3-N-pivaloylaminopyridine (**259**) was similarly prepared from **208**, although a large excess of acetaldehyde (10 equiv.) was required because of its competitive degradation in the reaction medium (Scheme 77) (89JHC105). The resulting carbinol **258** was oxidized by MnO_2 or CrO_3 into the 4-acetyl derivative **259**, a useful intermediate for the preparation of ellipticine alkaloid analogues (see Section VIII).

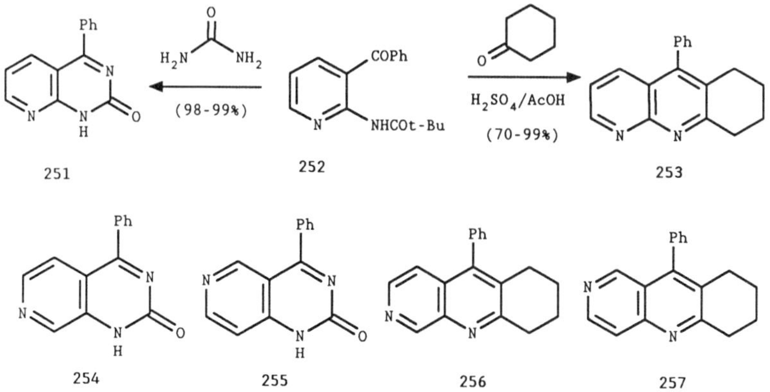

SCHEME 76

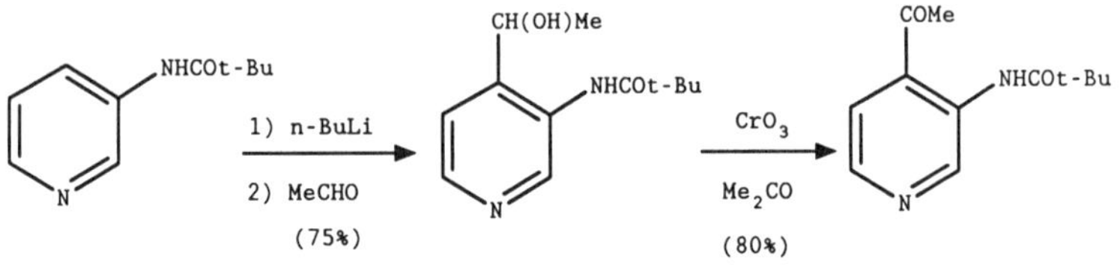

208 258 259

SCHEME 77

3. *ortho-Aminopyridine Carboxylic Acids*

Metalation of 2-*N*-pivaloylaminopyridine (**214**) followed by carbonation leads to the nicotinic acid **260** in excellent yields (Scheme 78) (83JOC3401; 89JHC105). The two isomeric *N*-pivaloylaminopyridines afford the corresponding acids **261** and **262.**

4. *Naphthyridines*

The amino carbinol pyridines **264,** prepared from **263** by DoM reaction and aliphatic or aromatic carboxaldehyde quench, are useful intermediates for condensed naphthyridines **266–268.** The thermolytic reaction proceeds via intramolecular Diels–Alder reactions of the aza-*ortho*-quinodimethane species **265** (Scheme 79) (84CC1304).

Metalation of either 4-*N*-pivaloyl or 4-*N*-*t*-Boc aminopyridines **269** followed by the bis electrophile, 3-chloro-1-iodopropane, furnishes the tetrahydro-1,6-naphthyridine **270** (Scheme 80) (88TL5725). Application of this reaction to the corresponding 2-aminopyridines gives access to the corresponding 1,8-naphthyridine.

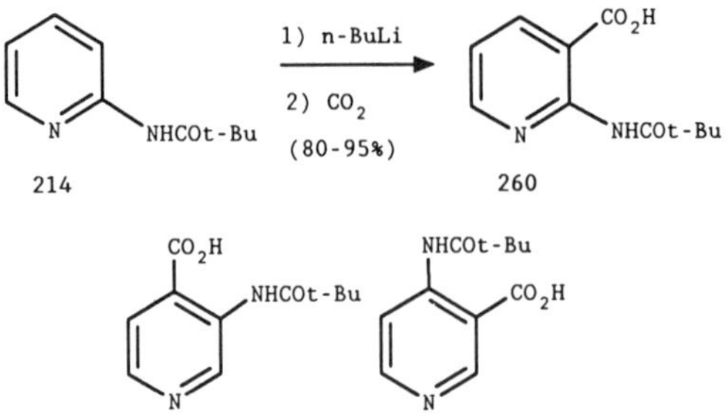

214 260

261 262

SCHEME 78

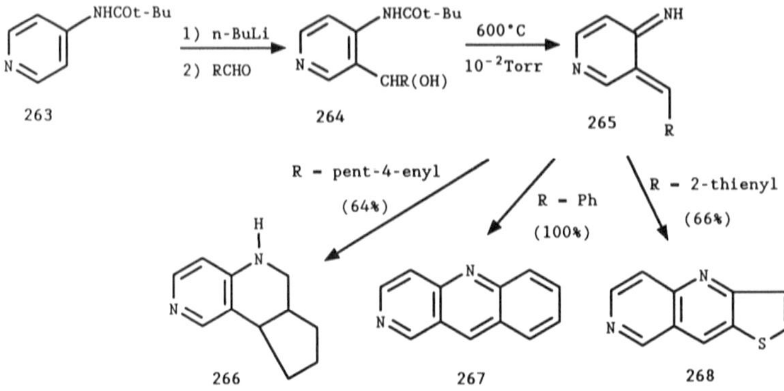

SCHEME 79

5. DoM-Madelung Synthesis of Azaindoles

The ready accessibility of *ortho*-aminopicolines **271a-b** from simple pre-cursors **244a-b** (Scheme 81) (83JOC3401) by the DoM process allows a significant improvement of the classical Madelung synthesis of azaindoles (12CB1128; 65JOC2531). Selective lateral metalation of **271a-b** with *n*-BuLi followed by warming to room temperature leads to **272a-b** in high overall yields.

6. DoM-S_{RN1} Synthesis of Azaindoles

The Madelung and Fischer syntheses (65JA3530, 65JOC2531) are not generally applicable for the construction of azaindoles. Advantage may be taken of the recently developed S_{RN1} reaction (80JOC1546, 80TL1943) in connection with DoM reactions of halopyridines (see Section II,D,5) to form azaindoles. Alternatively, amino DMGs can be used for regiospecific halogen introduction, and the resulting compounds can be used for S_{RN1} processes to achieve the same goal.

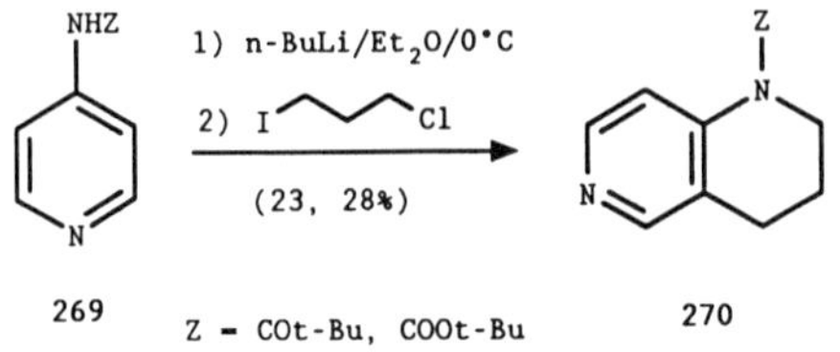

Z = COt-Bu, COOt-Bu

SCHEME 80

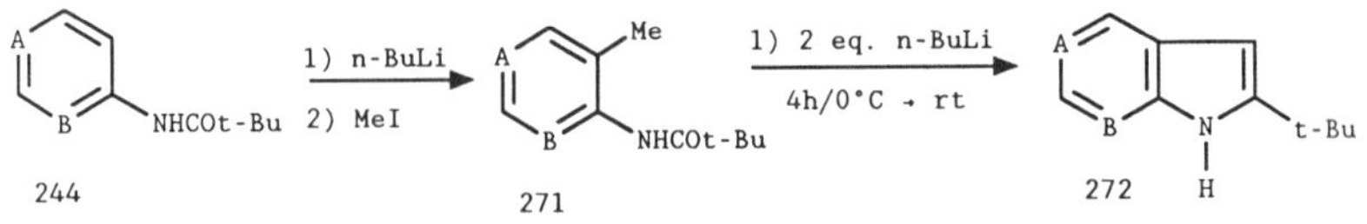

244 a: A = CH; B = N   271 a (72%)   272 a (83%)

244 b: A = N; B = CH   271 b (74%)   272 b (75%)

SCHEME 81

The efficacy of such sequential ortho halogenation-S_{RN1} reactions is demonstrated in the synthesis of 2-substituted 7-azaindoles **275** (Scheme 82) (88JOC2749). Thus, exposure of **273,** obtained by metalation–iodination of 2-*N*-pivaloylaminopyridine, to acetone and *t*-butyl methy ketone enolates under typical S_{RN1} conditions gives ketone **274** in almost quantitative yields. Subjection to 3N aqueous HCl treatment at reflux affords the azaindole **275.** Compounds **276** and **277** were likewise obtained in high yields starting with the appropriate *N*-pivaloylaminopyridines. Unfortunately, the synthesis of the 2-unsubstituted azaindole using acetaldehyde enolate was not achieved due to competitive dehalogenation of **273** under the reaction conditions. Attempts to use acetaldehyde equivalents, such as oximes and hydrazones, also failed to give the desired result.

7. *Thiazolopyridines*

Convenient syntheses of the rare thiazolo[5,4-*c*]pyridine **279** and 4-aminopyridine-3-thiol **(281)** systems have been achieved (Scheme 83) [88CI(L)302]. Thus, metalation of **263** followed by quenching with diisopropylthiocarbonyl disulfide affords, in high yield, compound **278.** Base-induced cyclization affords the thiazolopyridine **279,** while treatment

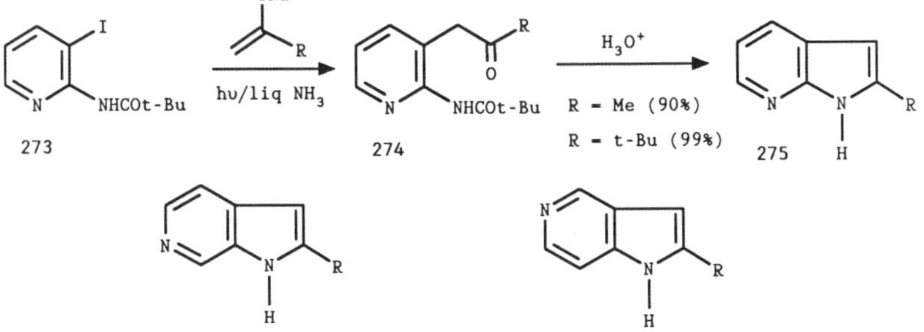

276 (98-99%)   277 (98%)

SCHEME 82

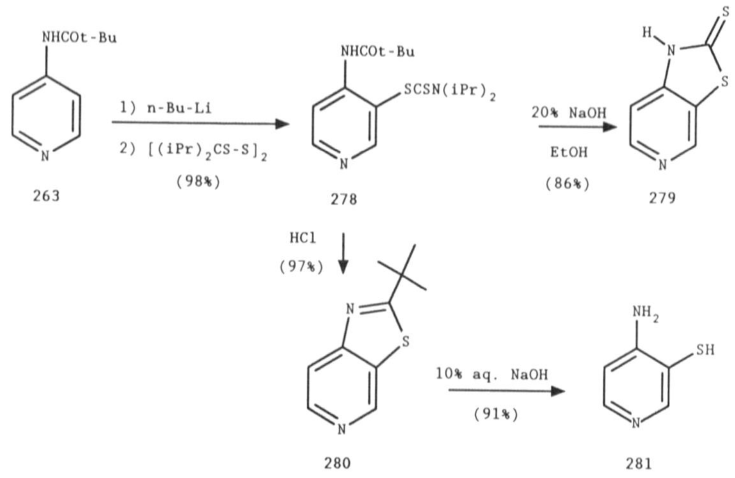

SCHEME 83

with acid leads to **280** which, upon base treatment, provides the aminopyridine thiol **281**.

A similar strategy has been successfully used to synthesize 2-substituted thiazolo[5,4-*c*]pyridines (90JHC563).

8. *Polysubstituted 4-Phenylpyridines*

The 3-amino-4-phenylpyridine **284,** a model unit found in the structure of the antibiotic streptonigrin, has been constructed via a DoM-cross coupling sequence (Scheme 84) [90JCS(P1)2611]. Thus, metalation of the

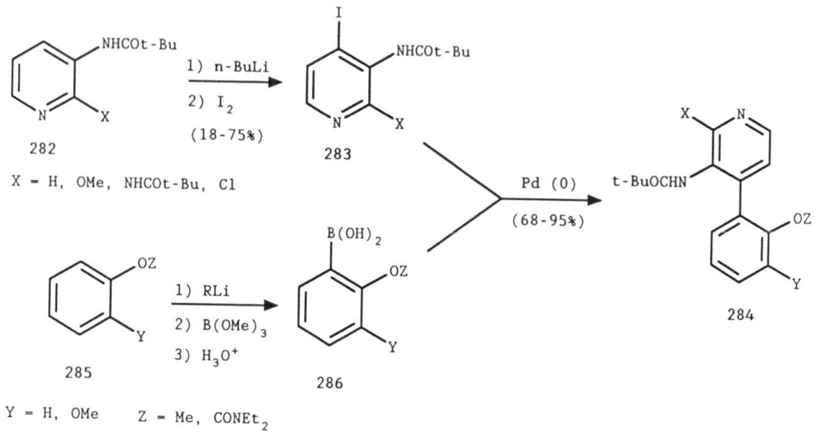

SCHEME 84

diverse 2-*N*-pivaloyl pyridines **282** followed by iodination furnishes the iodopyridine **283**, while metalation of **285** and subsequent trimethyl borate quench and hydrolysis gives the boronic acid **286**. Combination of **283** and **286** under Suzuki cross coupling conditions leads to the heterobiaryl **284**.

IV. Oxygen-Based DMGs

A. INTRODUCTION

DoM reactions on oxygen-based aromatics {DMG = OMe [79OR(26)1], OMOM (75SC65; 83T2031), OTHP [48JA4187; 78JOC3717; 79OR(26)1], OCONR$_2$ (83JOC1935), OSEM (88HCA957)} constitute well-established and powerful synthetic tools for the regioselective construction of ortho-substituted phenols, which circumvent problems of regioselectivity and harsh conditions normally used in classical electrophilic substitution methods to achieve similar purposes. In view of the ready availability of 2- and 4-pyridones and 3-pyridinols by classical routes and the nonregiolec-tive nature of their reactions in electrophilic substitution, their DoM reactions via suitable DMG-activated derivatives appeared to be a promising challenge. Although alkoxy groups are moderately activating [66JOC1221; 69JOM(20)251; 79JOC2480] and stronger than halogen DMGs in the aromatic series [79OR(26)1], this pattern is reversed for pyridine and quinoline systems. On the other hand, the poorer activation of alkoxy groups may be less significant than their LUMO-enhancing effect, which may make heteroaromatic ethers less prone to nucleophilic attack.

In 1982, Queguiner and Co-workers demonstrated that metalation of 3-alkoxypridines could be achieved with *n*-BuLi/TMEDA complex (82S235). More recently, Comins and LaMunyon showed that mesityllithium can be used to metalate 2-, 3-, and 4-methoxypyridines (88TL773). These workers also provided evidence by deuterium incorporation experiments that LDA metalation is inefficient on these poorly activated substrates [2-OMe: 18,7% (C-3); 4-OMe: 11.4% (C-3); 3-OMe: 6.3% (C-2), 3.8% (C-4)]. Thus, in general, alkyllithiums are used to effect alkoxypyridine deprotonation. The "catalyzed metalation" equilibrium technique with LDA/MeLi also leads to useful results (88JOC1367).

B. ALKOXYPYRIDINES

1. *2-Alkoxypyridines*

Under *n*-BuLi/THF conditions at a temperature range of 0–20°C, 2-methoxypyridine (**287**) undergoes competitive 3-lithiation **288** and nucleo-

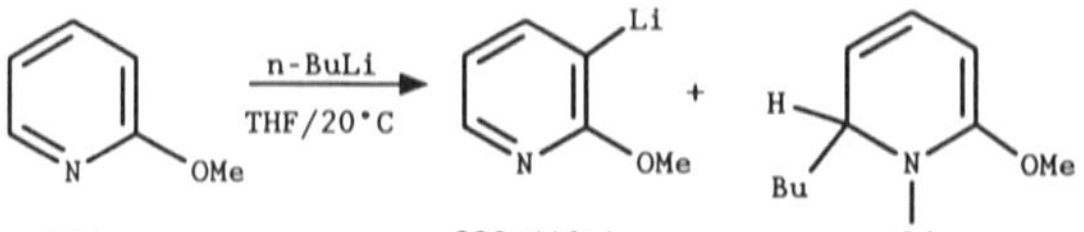

287   288 (40%)   Bu
                                                 |
                                                 Li
                                         289 (15%)

SCHEME 85

philic addition **289** (Scheme 85) (86JOC2184). The behavior of **287** parallels
that observed for 2-chloro and 2-fluoro pyridines (Section II) in that all
three systems undergo greater amounts of addition compared to their
respective 3- and 4-derivatives. The result can be rationalized by the lower
LUMO level of the 2-isomers (88JOC1367). The nucleophilic addition has
been used for synthetic purposes (86JOC2184).

Chemoselective metalation of 2-methoxypyridine has been achieved
using the equilibrium shift technique (concurrent addition of a threefold
excess of LDA and TMSCl in THF at room temperature) to give a quantitative yield of 2-methoxy-3-trimethylsilylpyridine (88JOC1367). Nevertheless, this method is not adaptable to other electrophiles. Attempts to
increase the concentration of the 3-lithio-2-methoxypyridine by addition of
alkyllithium (accumulation technique) before electrophile (DMF) quench
led to nonchemoselective results. The best results were observed using the
LDA catalyzed metalation technique, which involves conditions [(THF/
0°C/1.8 equiv. MeLi/(0.5 equiv. *i*-Pr)$_2$NH] that allow faster metalation to
give increased concentration of **288** compared to addition by an irreversible reaction between diisopropylamine and MeLi (Scheme 86). Using this
technique, deuteration (DC1, 70%), formylation (DMF, 55%), as well as
reaction with aldehyde, ketone, alkyl bromide and iodide, ethylene oxide,
1,3,5-trioxan, and TMSCl electrophiles have been achieved (88JOC1367).

When bromine, iodine, or allyl bromide electrophiles were used, 3-
methyl-2-methoxypyridine (**290**) was isolated, a result that can be explained by a fast exchange between the RX reagent and MeLi followed by
reaction of the resulting MeX with the 3-lithio species **288** (Scheme 87).

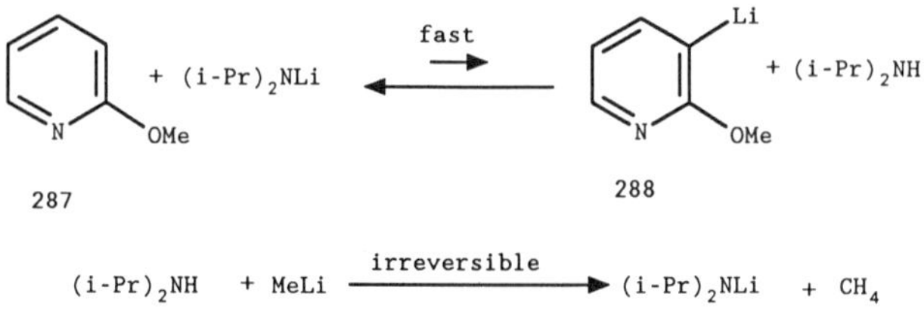

287   288

(i-Pr)$_2$NH + MeLi $\xrightarrow{\text{irreversible}}$ (i-Pr)$_2$NLi + CH$_4$

SCHEME 86

RX + MeLi ⇌ RLi + MeX

R = Br, I, Me, -CH₂-CH=CH₂.

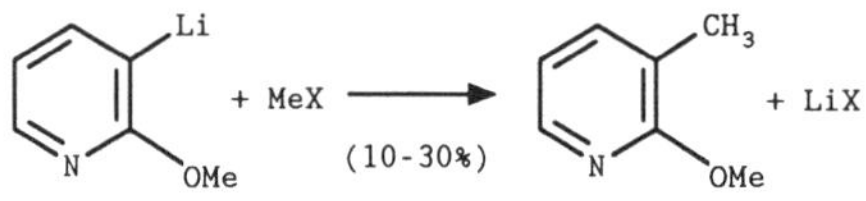

SCHEME 87

In a potentially useful development, Comins demonstrated that mesityllithium, prepared from bromomesitylene and *t*-BuLi (2 equiv.) or lithium metal, effects deprotonation of 2-methoxypyridine to give, after quenching with a variety of electrophiles (*n*-BuI, DMF, PhCHO, MeSSMe, PhSeSePh), 3-substituted products in good yields (88TL773). Nucleophilic attack is precluded by the highly crowded nature of the mesityllithium reagent. This strategy has been successfully extended to the C-2 lithiation of 3-methoxypyridines (90JOC69).

2. 3-Alkoxypyridines

Although *n*-BuLi metalation of 3-alkoxypyridine in cyclohexane leads to competitive DoM (**292**) and addition (**293**) (Scheme 88), addition is not observed, even at 60°C, using the same base in Et₂O or THF solution, and the lithiation is then regioselective and efficient at C-2 (82S235). With TMSCl as electrophile, other minor compounds could be observed with 3-ethoxypridine (**294**) (Scheme 89). Thus *n*-BuLi/TMEDA treatment of **294** at −40°C followed by TMSCl quench affords the 2-TMS derivative **295** together with minor amounts of 4-TMS **296** and 2,4-bis-TMS **297** products (86MI2). Increasing the reaction time and temperature (to 60°C) slowly shifts lithiation to favor the 4-position (84TH1), and use of MeLi as the metalating agent at 20°C followed by MeI quench affords the 4-methyl derivative **298** in good yield.

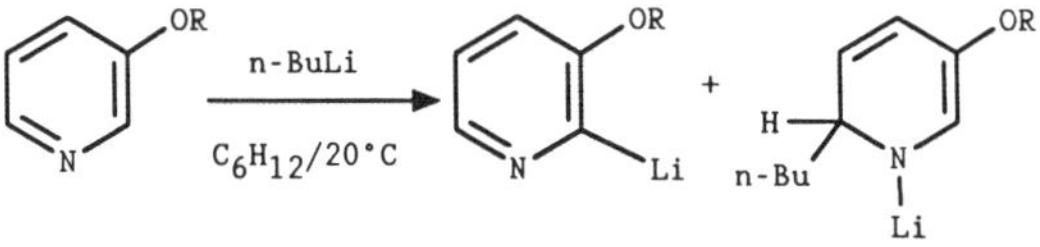

SCHEME 88

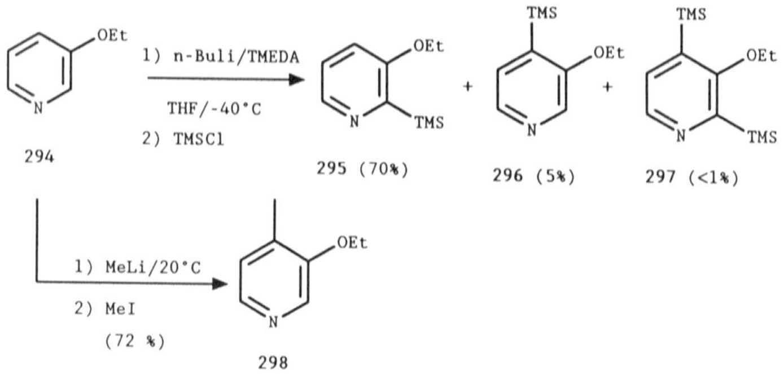

SCHEME 89

The formation of **295** prompts a mechanistic comparison with results observed in the metalation of 3-OMOM pyridine and N,N-dialkyl 3-carbamoyloxypyridine, in which only 4-metalation results were observed (see later). Whereas 2-deprotonation of 3-ethoxypyridine may be driven by the greater coordination ability of RLi base to the ring nitrogen **299,** such interaction may be less significant for the corresponding 3-OMOM and carbamoyloxy systems because of the greater coordination effects of two heteroatoms, **300** and **301,** respectively (Scheme 90).

Clearly, coordination effects are significant in alkoxypyridine metalation, and the significance of coordination and electron attracting effects are inverted compared to the halopyridines. Using the optimum n-BuLi/TMEDA/THF/$-40°$C conditions, a variety of 3-alkoxypyridine derivatives were tested for the synthesis of 2-substituted pyridines. The results of treatment of **291** with a variety of electrophiles to give products **302–305** are summarized in Scheme 91 (82S235). The best case 3-ethoxypyridine, was converted into derivatives **303—305.**

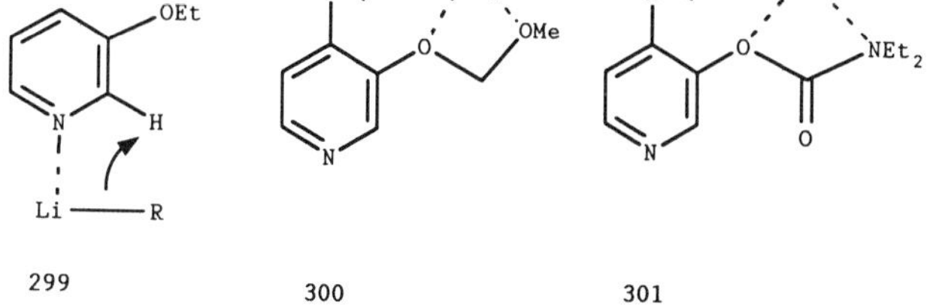

SCHEME 90

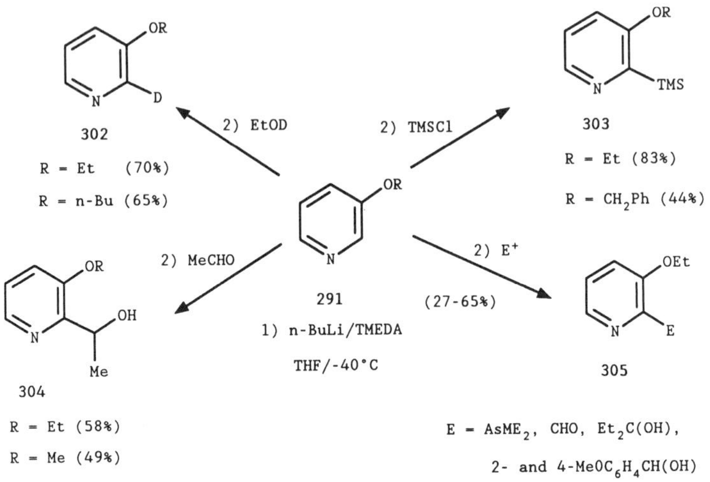

SCHEME 91

Metalation of 3-methoxypyridine using mesityllithium as a base to give 2-substituted products can also be used (88TL773). LDA metalation of 3-ethoxypyridine (294) followed by TMSCl quench at 0°C gives the 2- and 4-TMS derivatives 295 and 296, respectively (Scheme 92) (84TH1).

Simultaneous lithiation-trimethylsilylation (LDA/TMSCl/−42°C) of 3-methoxypyridine gave a mixture of 2-TMS (42%), 4-TMS (33%), and 2,4-bis-TMS (0.4%) derivatives (88TL773).

3. 4-Alkoxypyridines

Clean metalation of 4-methoxypyridine using mesityllithium has been demonstrated by quenching experiments with numerous electrophiles (88TL773). Less useful is the metalation result using LDA (THF/0°C) which, upon TMSCl quench, leads to a mixture of the 3-TMS (61%) and 3,5-bis-TMS (16%) derivatives (88TL773).

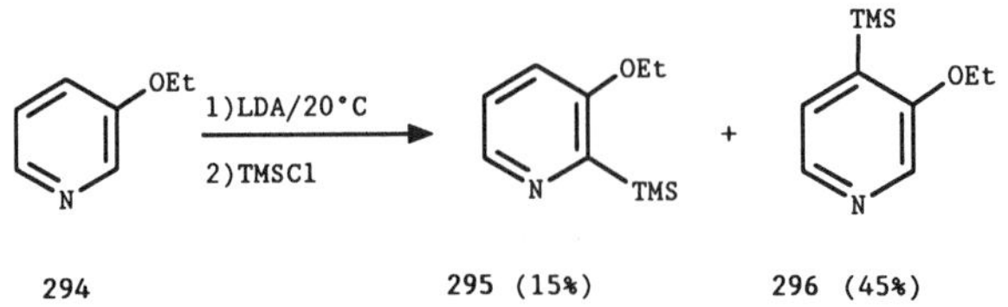

SCHEME 92

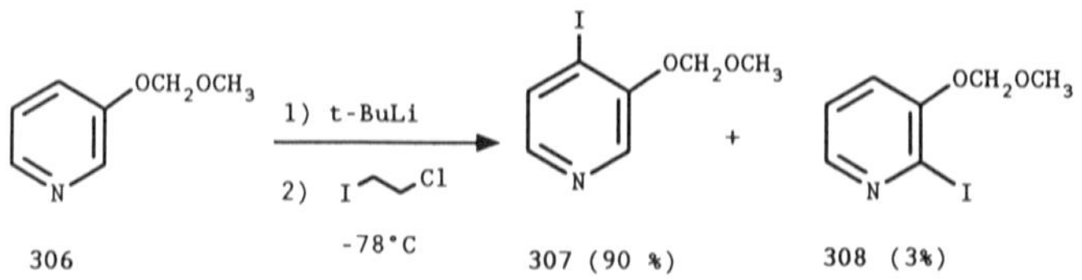

SCHEME 93

4. 3-Methoxymethoxypyridine

In a solitary study, 3-OMOM pyridine (306) has been shown to undergo clean *t*-BuLi mediated metalation as indicated by iodination (and deuteration) to give mainly the 4-iodopyridine 307 in addition to the 2-iodo derivative and a minor (1%) amount of *t*-Bu-adduct (Scheme 93) (82JOC2101; 83T2031).

5. 4-(2-Methoxyethoxy)pyridine

sec-BuLi metalation of 4-(2-methoxyethoxy)pyridine (309) followed by TMSCl quench affords products 310 and 311 in low combined yield (Scheme 94) (85CPB1016). While 310 is undoubtedly formed by direct addition, 311 can result by either addition and C-3 DoM reaction or by 2,5-addition of *sec*-BuLi and TMSCl. On the other hand, LDA metalation followed by silylation leads chemospecifically to the 3-TMS derivative 312 in modest yield. Under similar conditions, the 3-(2-methoxyethoxy)pyridine gave only a complex mixture of products.

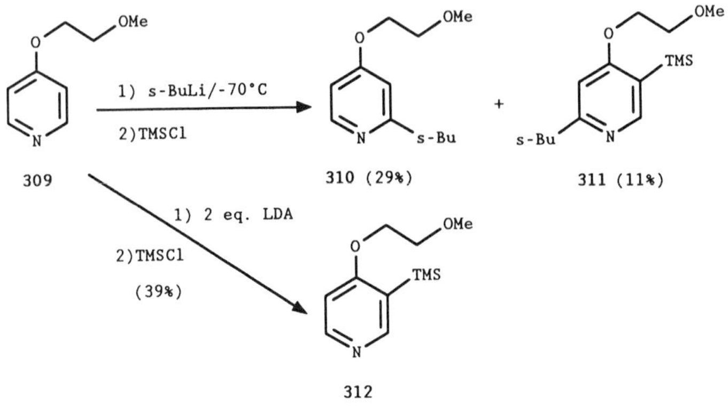

SCHEME 94

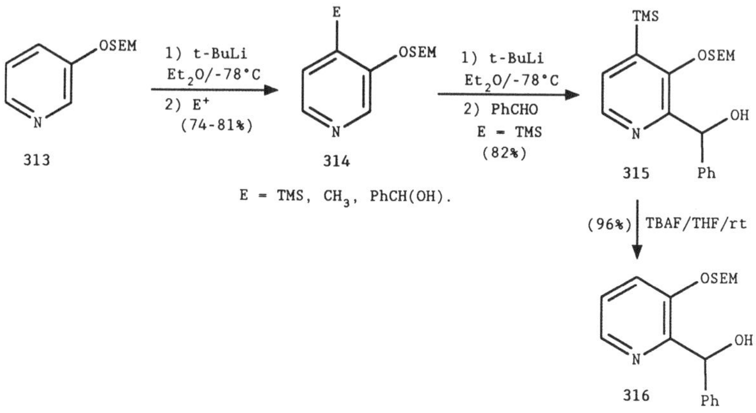

SCHEME 95

6. *3-OSEM Pyridines*

A further route to 2-substituted 3-oxygenated pyridines **315** begins with 3-OSEM pyridine (**313**) [90TL4267]. Metalation followed by electrophile quench leads to derivatives **314**. Using 4-TMS protection, a tactic developed in aromatic amide DoM chemistry (90CRV879), further metalation of 4-TMS derivative **314** and electrophile quench afford compounds **316**, which can be selectively C-4 desilylated to give product **315** (Scheme 95).

C. PYRIDYL *O*-CARBAMATES

The powerful *N,N*-diethylcarbamate DMG serves admirably for the synthesis of substituted oxygenated pyridines. Thus, metalation of all isomeric pyridyl *O*-carbamates **317** with *sec*-BuLi/TMEDA followed by quench with numerous electrophiles affords diversely substituted products **318** in good to excellent yields (Scheme 96) (85JOC5436). Base-induced hydrolysis provides access to pyridones and hydroxypyridines.

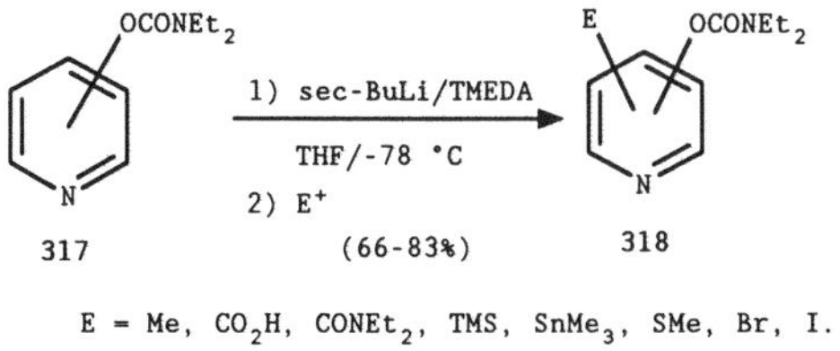

E = Me, CO₂H, CONEt₂, TMS, SnMe₃, SMe, Br, I.

SCHEME 96

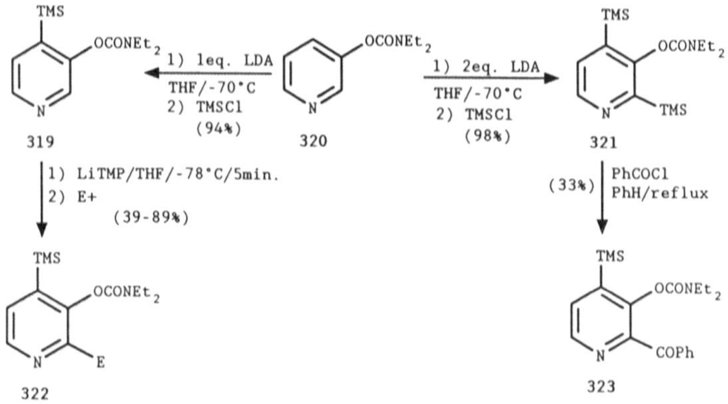

$$E = RCH(OH), Ph_2C(OH), COPh, SPh, SePh, Cl, SiEt_3.$$

SCHEME 97

The 3-pyridyl O-carbamate affords, under the sec-BuLi/TMEDA conditions, only 4-substituted products. A reinvestigation of LDA metalation (85JOC5436) has shown that high-yield conversion of 320 into the 4-TMS (319) and 2,4-bis-TMS (321) derivatives can be effected (Scheme 97) [90UP1]. Furthermore, LiTMP metalation of 319 followed by electrophile quench leads to derivatives 322, thus demonstrating the TMS protection route to 2-substituted 3-oxygenated pyridines. Another, potentially useful result is the 2-position selective ipso carbodesilylation of 321 with benzoyl chloride, yielding 323.

The achievement of anionic ortho-Fries type rearrangements (320→324 and 325→326) may open further DoM-based routes for the synthesis of oxygenated pyridines (Scheme 98) (85JOC5436).

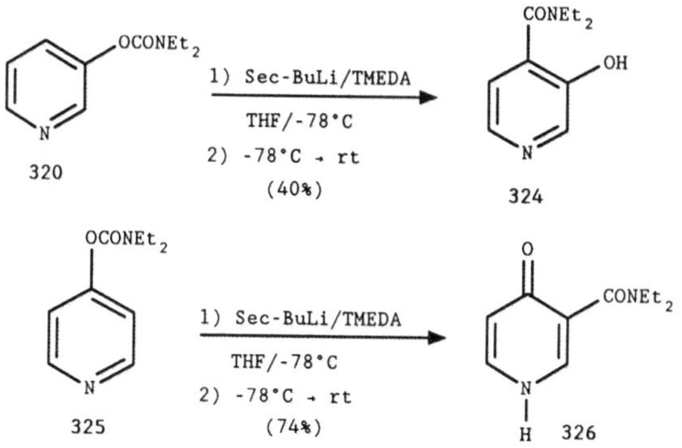

SCHEME 98

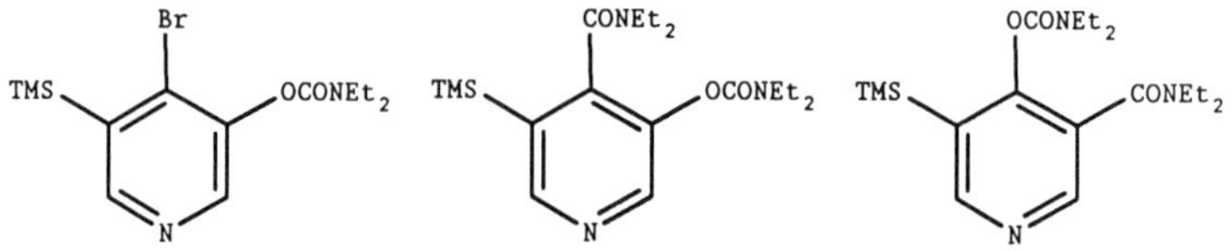

SCHEME 99

As broadly demonstrated in aromatic DoM chemistry (90CRV879), iterative metalation of pyridyl *O*-carbamates are synthetically useful processes. Thus, sequential metalation reactions of 3- and 4-pyridyl *O*-carbamates with electrophiles that provide incipient DMGs afford 3,4,5-trisubstituted pyridines **327, 328,** and **329,** respectively (Scheme 99) (85JOC5436).

Additional results of synthetic potential in pyridine chemistry concern the electrophile (MeCOCl, I$_2$) induced ipso destannylation of **330,** leading to 4-acetyl and 4-iodo products **331** (Scheme 100) (85JOC5436).

D. ALKOXYQUINOLINES

Quinoline undergoes nucleophilic attack with organolithium and organomagnesium reagents, owing to the low electron density and C-2 and C-4, a view reinforced by the low LUMO level of this heterocycle compared to that of pyridine.

Among his many pioneering studies in organometallic chemistry, Gilman showed in 1951 that *n*-BuLi metalation of 2-ethoxyquinoline (**332**) followed by carbonation gave only a small amount of the DoM product **333,** the major result being 2-substitution **334** (Scheme 101) (51JA32).

In a series of studies concerning the application of DoM reactions to the synthesis of furoquinoline alkaloids (see Section VIII), Narasimhan and co-workers effected the model metalations of 2-ethoxyquinoline (**332**) to give, after electrophile quench, low yields of 3-substituted products **335**

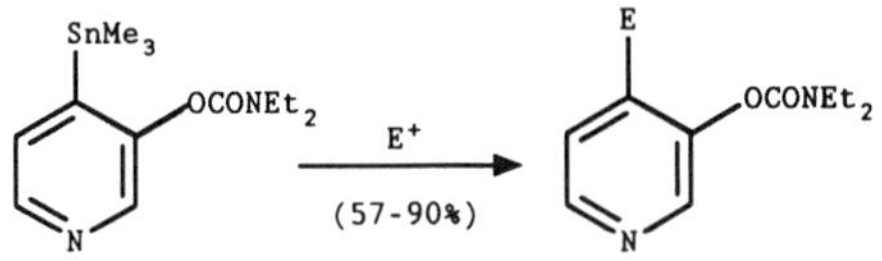

SCHEME 100

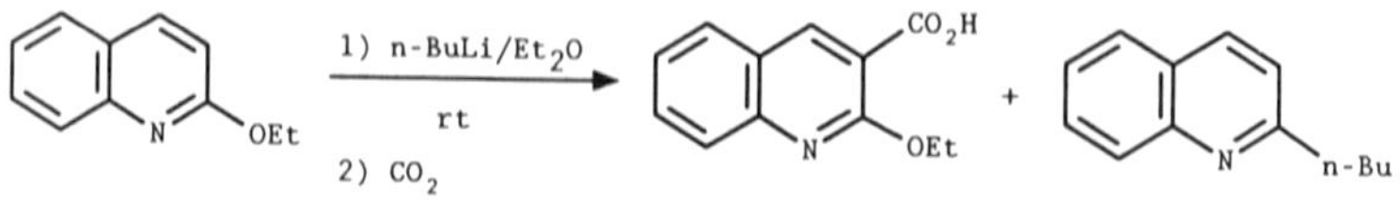

SCHEME 101

(Scheme 102) [67CI(L)831; 68CI(L)515; 71T1351]. Acid-catalyzed cyclization of the allyl and hydroxyethyl derivatives of **335** gave the dihydrofuro[2,3-*b*]quinolines **336**, completing the model investigation (71T1351).

Higher yields of 3-substituted products were observed in DoM reactions of 2,4-dimethoxy, 2,4,6-trimethoxy, and 2,4,8-trimethoxy quinolines. Furthermore, *n*-BuLi metalation of 2,4-dimethoxy, 2,4,6- 2,4,7- 2,4,8-trimethoxy, and 2,4,6,7-tetramethoxy quinolines **337** followed by *N*-methylformanilide [74T4153; 79IJC(B)115] or 3,3-dimethylallyl bromide [73JCS(P1)94] quench furnished good to excellent yields of the expected 3-substituted products **338** (Scheme 103).

Although metalation of 5- and 8-ethoxyquinolines led, under a variety of temperatures, mainly to addition products, the 2-*n*-butyl derivatives **339** and **342**, bearing DMGs in the benzene ring, respectively provided ring- (**340**) and ethoxy- (**343**) metalation directed results (Scheme 104) (83TH1). Application of the equilibrium shift procedure (combined LDA/TMSCl/ 0°C) on 8-ethoxy-5-chloroquinoline gave a low yield (6%) of product **341**.

E. QUINOLYL *O*-CARBAMATES

Comprehensive studies of DoM reactions of all isomeric quinolyl *O*-carbamates have been carried out by Queguiner [87JOM(336)1; 88CJC1135]. When the *N,N*-diethyl ˙O-(2-quinolyl)carbamate (**344**) was

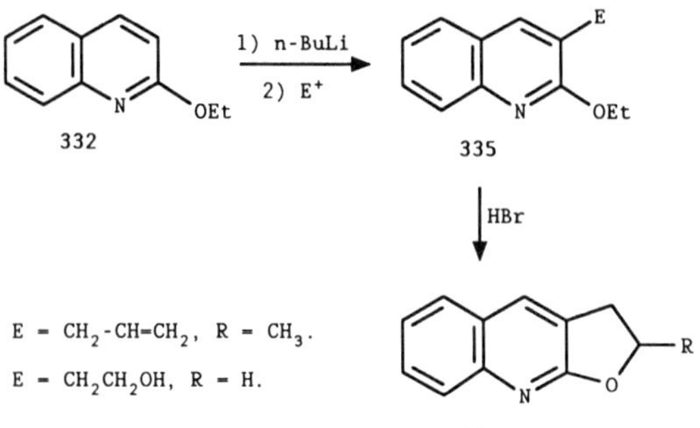

$E = CH_2-CH=CH_2$, $R = CH_3$.

$E = CH_2CH_2OH$, $R = H$.

SCHEME 102

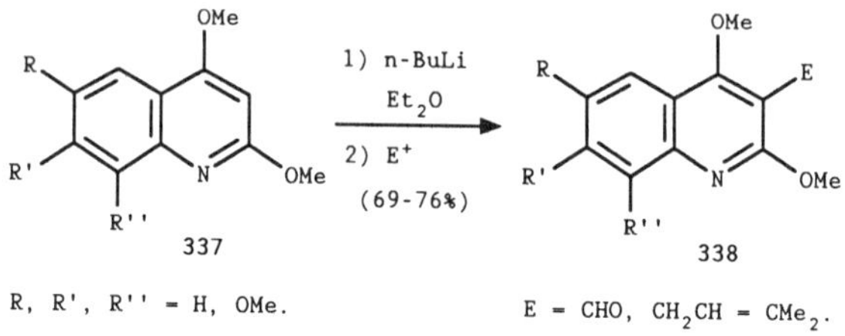

R, R', R'' = H, OMe.   E = CHO, $CH_2CH = CMe_2$.

SCHEME 103

subjected to LDA ($-78°C$) or *sec*-BuLi/TMEDA ($-105°C$) metalation followed by deuteriation, mixtures of 3-deuterated (**345**) and anionic ortho-Fries rearrangement products **346** were isolated in amounts that depended on conditions (Scheme 105) (88CJC1135). Under the LDA conditions, the analogous *N,N*-dimethyl carbamate **344** gave only the rearrangement product corresponding to **346**.

LDA metalation of the 2-carbamate **344** followed by aldehyde quench afforded mixtures of products **347** and **348** (Scheme 106). The formation of the carbostyryl derivative **348** can be rationalized by an intramolecular carbamoyl migration (**349**→**350**) (Scheme 107) driven by the leaving group ability of the 2-carbamoyl group.

The 3-carbamates **351** were lithiated at C-4; no Fries-like rearrangement was observed (Scheme 108) (88CJC1135).

The *N,N*-dimethylcarbamate **351,** when reacted with LDA and aliphatic or aromatic aldehydes, afforded surprisingly ortho-rearranged products **353** (Scheme 109) (87JHC1487).

Concerning the mechanism, the intermediate **354** could be isolated and easily led to **353** by loss of CO_2 (88CJC1135).

The *N,N*-dimethyl carbamate **351,** upon treatment with LDA followed by benzaldehyde, gave a mixture of dimethyl and diethylamino derivative

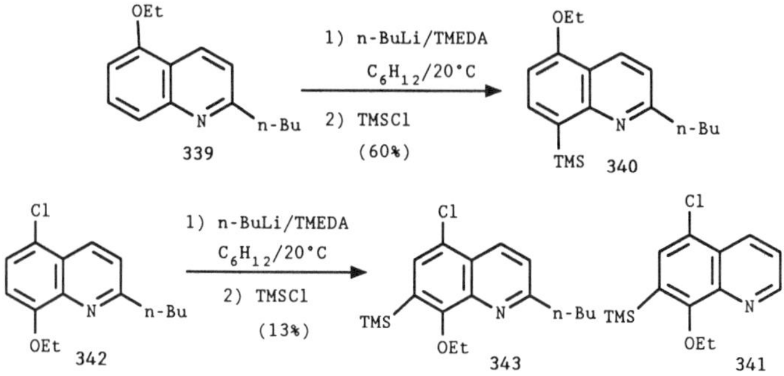

SCHEME 104

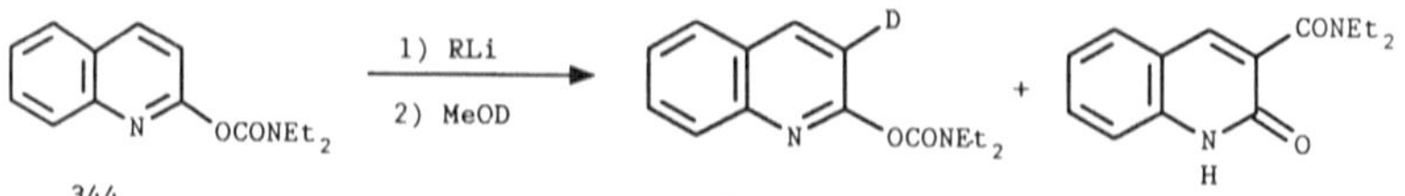

344 345 346

	345	346
RLi = LDA/-78°C	40%	60%
RLi = s-BuLi/-105°C	0%	80%

SCHEME 105

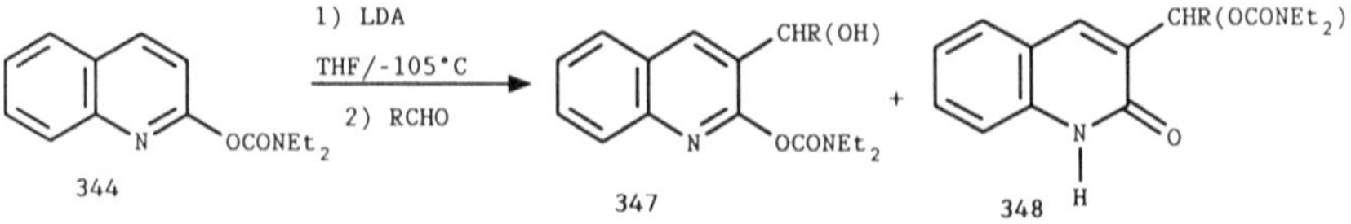

344 347 348

	347	348
R = Et	(30%)	(19%)
R = Ph	(28%)	(24%)

SCHEME 106

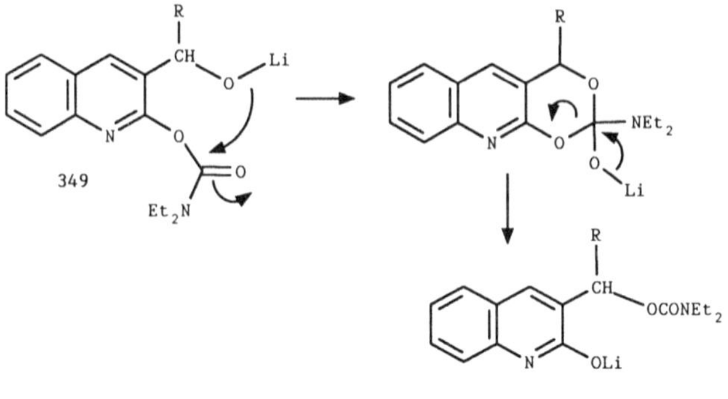

349 350

SCHEME 107

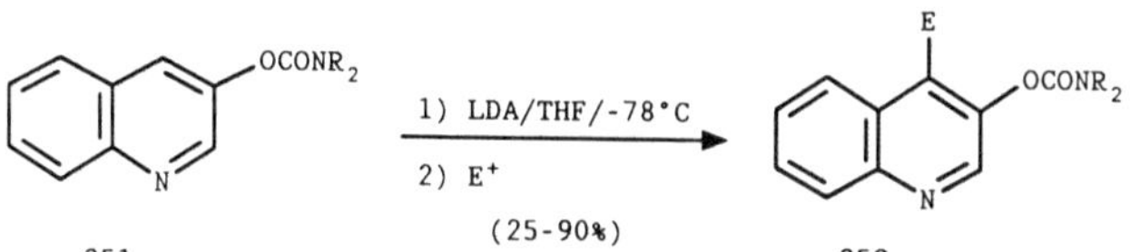

351 352

R = Me, Et.

E = D, TMS, MeCH(OH), EtCH(OH).

SCHEME 108

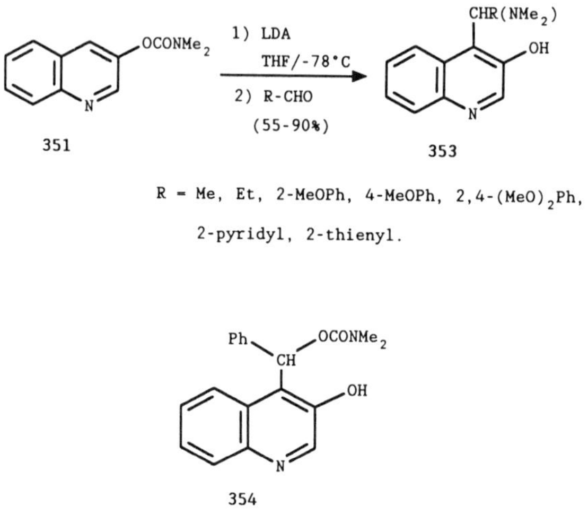

R = Me, Et, 2-MeOPh, 4-MeOPh, 2,4-(MeO)$_2$Ph,

2-pyridyl, 2-thienyl.

354

SCHEME 109

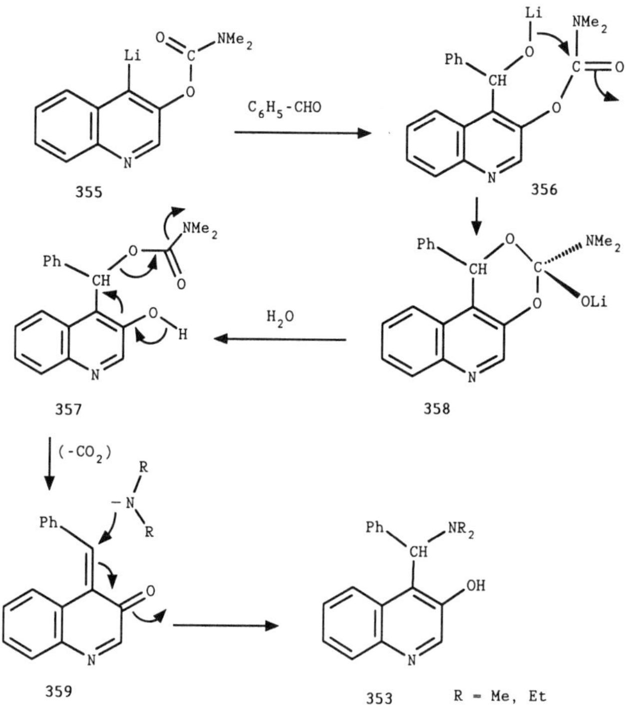

SCHEME 110

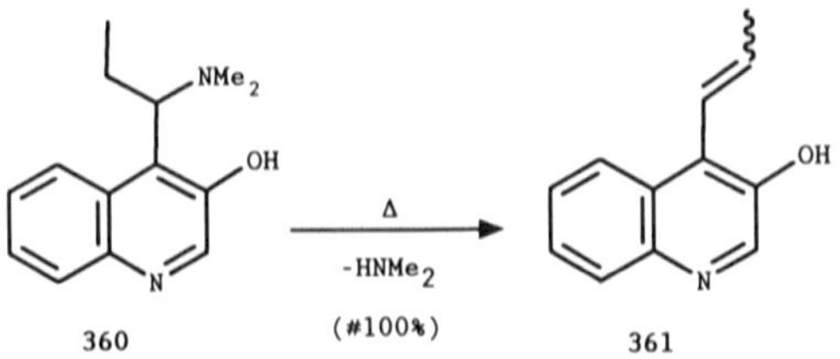

360 361

SCHEME 111

353 when diethylamine was added during the workup. This suggested the mechanism shown in Scheme 110 (88CJC1135).

The amine **360** obtained after lithiation of **351** and reaction of propionaldehyde affords, upon heating, the vinylquinoline **361** (88CJC1135) (Scheme 111).

The C-4 diethylcarbamate **362a** gave the derivatives **363.** With benzaldehyde, the quinolone **365** was isolated after basic treatment (refluxing 20% aq. NaOH) (Scheme 112) (88CJC1135). The dimethyl carbamate **362b** gave the Fries-like rearrangement to **364** (88CJC1135).

The *N,N*-dimethyl carbamates derived from 5-, 7-, and 8-hydroxyquinolines **366, 368,** and **370** were regioslectively converted into products **367, 369,** and **371,** respectively, using the LDA/TMSCl equilibrium shift procedure (Scheme 113) [87JOM(336)1].

Under these conditions, the 6-carbamate **372** showed poor regioselectivity, giving rise to a mixture of monosilylated C-5 (**373**) and C-7 (**374**) and bis-silylated (**375**) products in 2/2/1 relative amounts (Scheme 114).

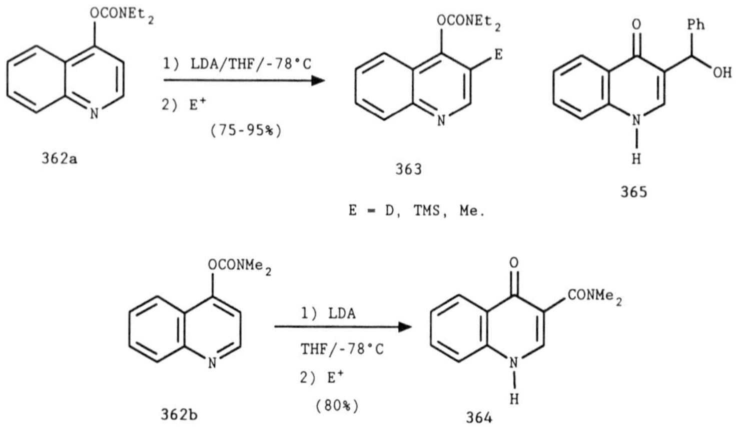

362a 363 365

E = D, TMS, Me.

362b 364

SCHEME 112

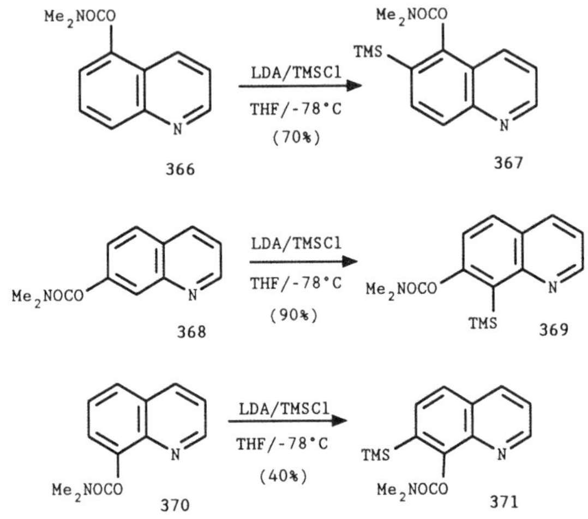

SCHEME 113

Attempts to use alkyllithium conditions to generate synthetically useful anions from the *N,N*-dimethyl 5-, 6-, and 7-*O*-carabamates **366, 368,** and **370** were unsuccessful, leading mainly to anionic ortho-Fries products. Under LDA conditions, the rearrangement (**376→377**) proceeded regiospecifically in modest yields at specific temperatures for each isomer to give products (temperature of rearrangement) **378** (−70°C), **379** (−70°C), **380** (−40°C), and **381** (20°C) (Scheme 115) [87JOM(336)1].

F. ALKOXYPYRIMIDINES

LiTMP metalation of 2,4-dimethoxypyrimidine (**382**) followed by quench with a variety of electrophiles leads to variable yields of the 5-substituted products **383** (Scheme 116) [87H(26)585].

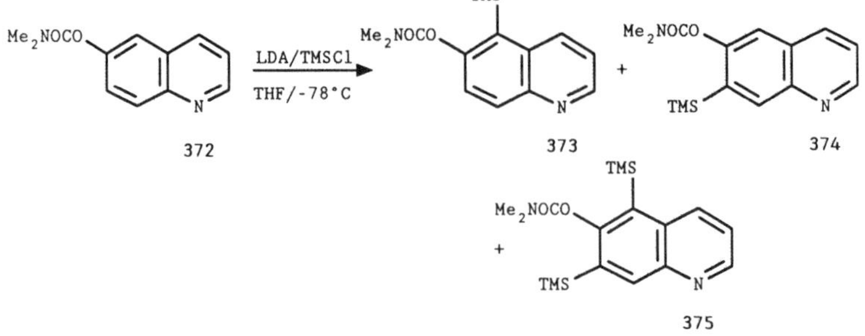

SCHEME 114

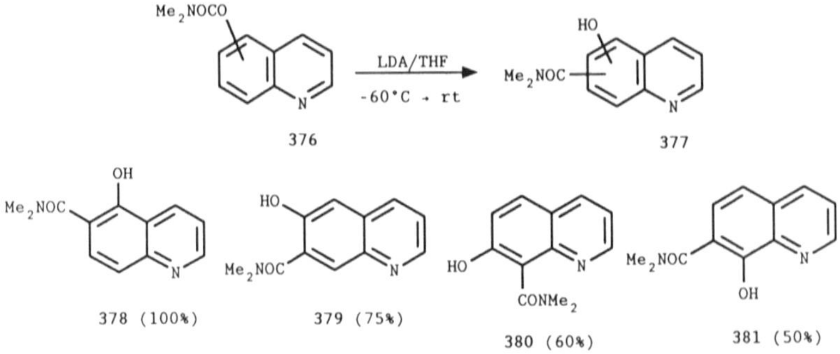

376   377

378 (100%)   379 (75%)   380 (60%)   381 (50%)

SCHEME 115

Application of these metalation conditions to a more diverse series of pyrimidines **384**, including a 4-*N*-pivaloyl derivative, with intervention of a TMSCl quench leads to 5-TMS products **385** in low yields (Scheme 117). Related metalations (*n*-BuLi) of 6-chloro-2,4-dimethoxypyrimidine (C-5 TMS and C-5 SnMe3 products) [88JOM(342)1] has also been done. 5-Methoxy, as well as 2,4,4,6-dimethoxy, and 2,4,6-trimethoxy pyrimides were lithiated under similar conditions (LiTMP/THF/−78°C), leading respectively to the 4- and 5-substituted derivatives in medium to high yields (90JOC3410).

Extensive work concerning DoM reactions of pyrimidinones in the context of nucleosides (85CL1401; 86T4187; 87CPB72, 87TL87; 88MI1) is considered to be outside the scope of this review. As an illustration of this emerging area of metalation chemistry, the 2'-deoxyuridine **386** has been investigated and affords C-5 **387** (major) and C-6 **388** substituted products (Scheme 118) (89TL2057).

G. ALKOXYPYRAZINES

In this as yet poorly investigated area, LiTMP metalation of 2-methoxy and 2,6-dimethoxypyrazines **389** and appropriate electrophile quench has been shown to give products **390** (Scheme 119) (90JOC3410, 90TH1; 91JOM).

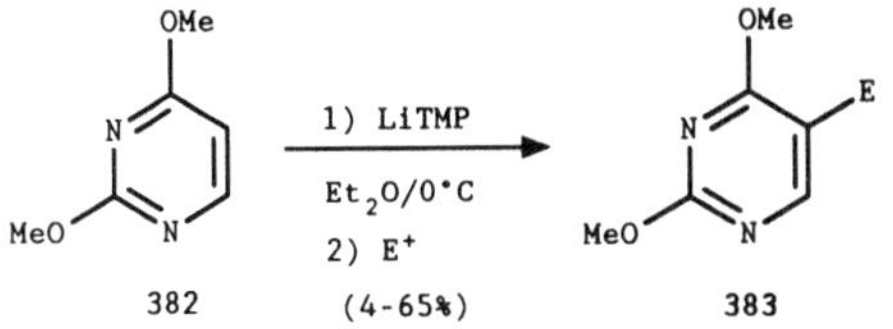

382   (4-65%)   383

E – TMS, CHO, COMe, CO$_2$H, CO$_2$Et, RCH(OH).

SCHEME 116

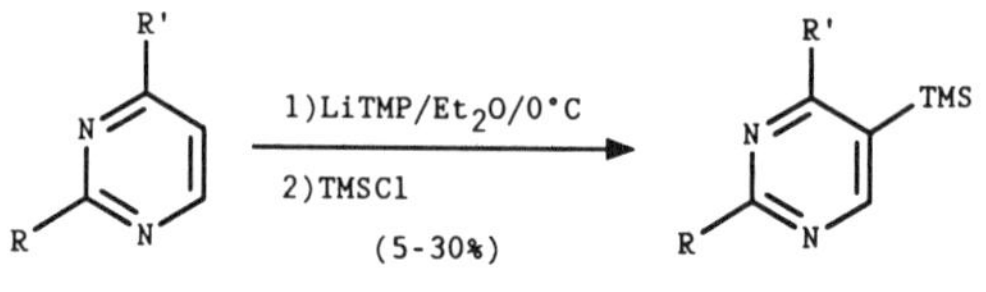

$$384 \xrightarrow[\substack{2)\text{TMSCl} \\ (5-30\%)}]{1)\text{LiTMP}/\text{Et}_2\text{O}/0^\circ\text{C}} 385$$

(R,R') = (H,OMe), (Cl,OMe), (OMEM,OMEM), (H,NHCOt-Bu)·

SCHEME 117

H. SYNTHETIC APPLICATIONS

1. *Alkoxypyridines*

2-Methoxypyridine-3-carboxaldehyde (148), prepared from 287 (88 JOC1367), serves as a key synthon for the construction of 8-azacoumarin (147) and the 1,9-diazaxanthone 392, via condensation, oxidation, and substitution product 391 (Scheme 120) [90JCS(P1)2409].

Metalation of 3-ethoxypyridine (294) followed by condensation with *o*-anisaldehyde affords 393 which, upon oxidation and cyclization, gives the 1-azaxanthone 394 (Scheme 121) (84TH1).

Synergism between OMe and 2-oxazolinyl DMGs allows clean C-2 metalation of 2-methoxy-4-oxazolinyl pyridine (395) with MeLi to give, after quench with aryl aldehydes, derivatives 396 (Scheme 122) (87S142). These were further converted into the pyridones 397, which served as intermediates for the preparation of biologically interesting condensed azaaromatics (see Section VIII).

2. *Alkoxyquinolines*

Metalation of 2-ethoxyquinoline (332) followed by sequential condensation with benzonitrile and acid-catalyzed cyclization furnished the py-

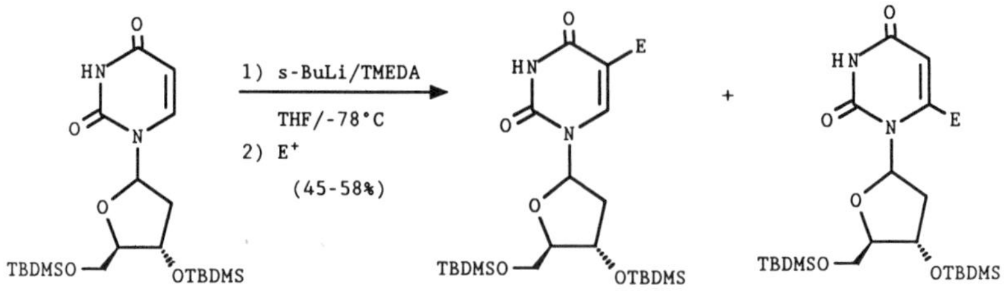

$$386 \xrightarrow[\substack{\text{THF}/-78^\circ\text{C} \\ 2)\ \text{E}^+ \\ (45-58\%)}]{1)\ \text{s-BuLi}/\text{TMEDA}} 387 \quad + \quad 388$$

E = D, Me, SH, COPh, SCH$_2$Ph.

SCHEME 118

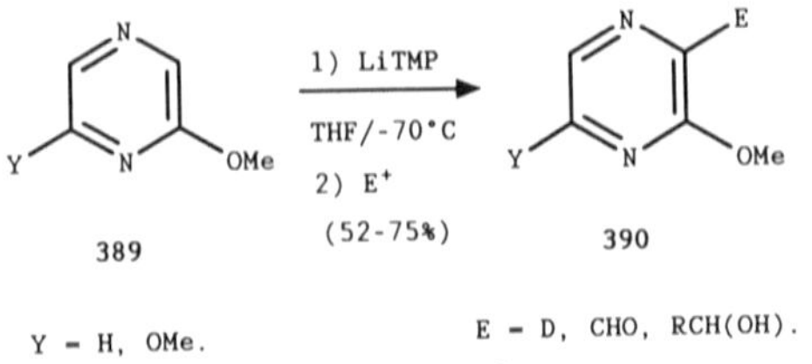

Y = H, OMe.

E = D, CHO, RCH(OH).

SCHEME 119

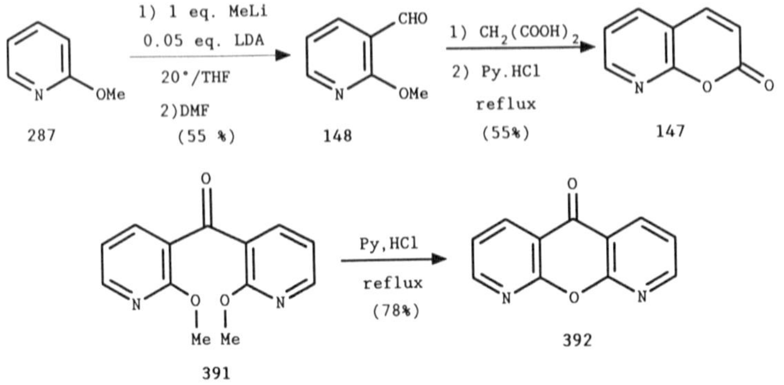

SCHEME 120

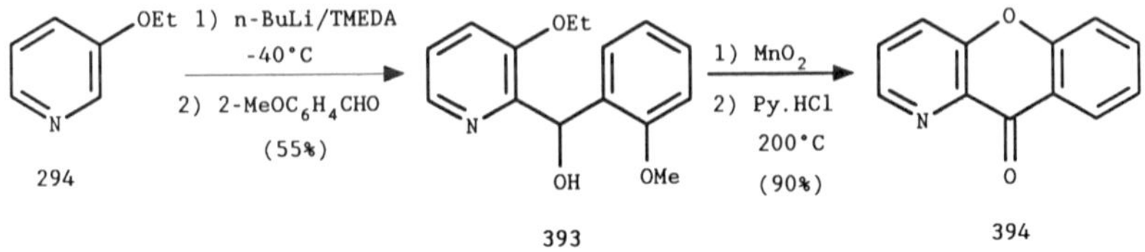

SCHEME 121

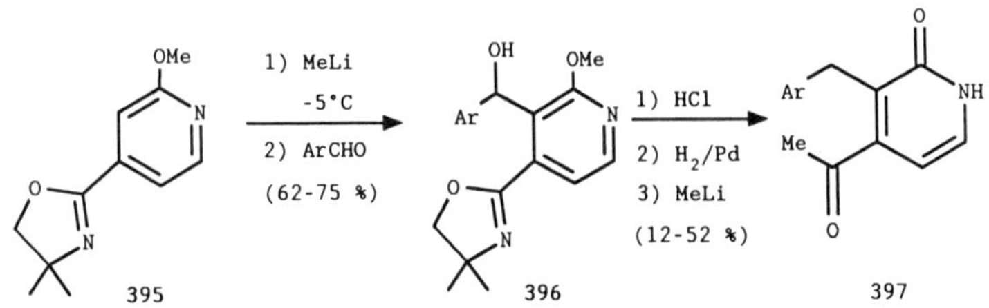

SCHEME 122

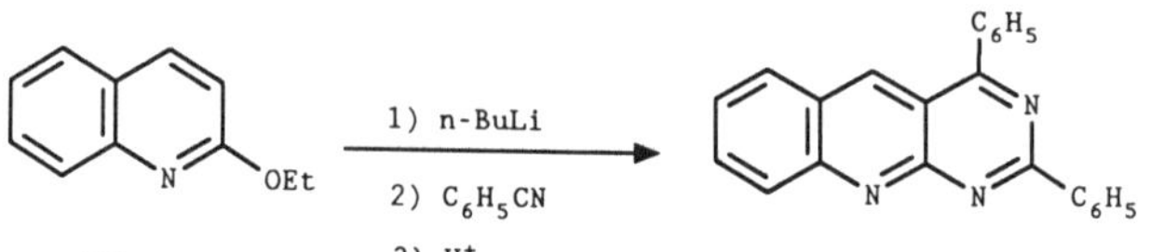

SCHEME 123

rimido[4,5-*b*]quinoline **398** (Scheme 123) [77CI(L)310]. A similar sequence of reactions on 2,4-dimethoxyquinoline (**399**) provided the pyrimido [5,4-*c*]quinoline isomer **400** (Scheme 124) [77CI(L)310].

3. *Quinolyl* O-*Carbamates*

LDA metalation of the 2-, 3-, and 4-quinolyl carbamates **401** followed by quenching with propionaldehyde gives respective carbinols **402** which, upon acid-catalyzed cyclization, furnish the dihydrofuroquinolines **336, 403,** and **404** (Scheme 125) (88JHC1053). In the case of product **336,** this route compares favorably with that based on metalation of 2-ethoxyquinoline (Scheme 102).

V. Carbon-Based DMGs

Carbon-based DMGs offer the potential for the regiospecific preparation of ortho-related heteroatom-carbon or, more significantly, carbon–carbon substituents. These tasks are challenging or impractical to achieve by classical methodologies, including *de novo* ring synthesis and electrophilic substitution reactions. Illustrative of the problem, which can be solved by DoM chemistry, is the construction of 2,3- and 3,4-dicarbon substituted pyridines.

In 1979, Ferles and Silhankva showed that ethyl nicotinate was converted into the self condensation product 4-(2-carboethoxypyridyl)-3-

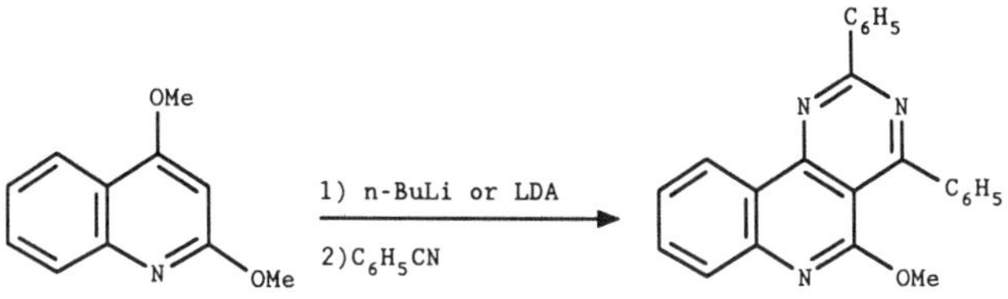

SCHEME 124

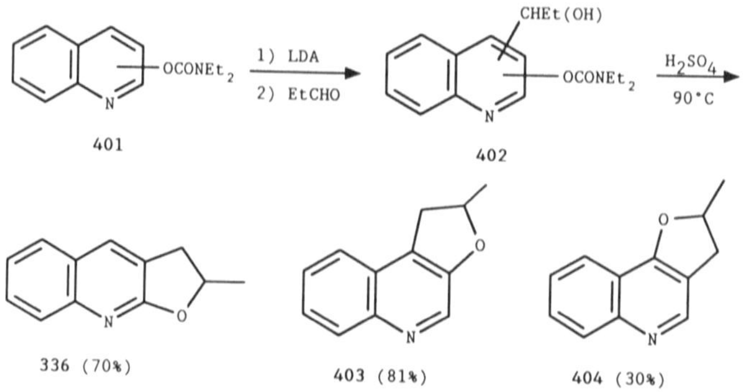

SCHEME 125

pyridyl ketone upon treatment with LDA (79CCC3137). The rapid and useful development of carbon-based DMG aromatic metalation chemistry (82ACR306) stimulated activity in the heterocyclic area. Consequently, methodologies based on 2-oxazolino (78TL227), secondary amide (81S127; 83TL4735), and tertiary amide (80JA1457, 80TL4739) DMGs in the pyridine series were developed.

A. PYRIDYL OXAZOLINES

Metalation studies on two of the three possible pyridyl oxazolines have been reported. Thus, treatment of 3-pyridyl oxazoline 405 with LiTMP resulted in the formation of the 4-lithiated species 406 as evidenced by the formation of carbinols 407 in good yields upon condensation with aldehydes and ketones (Scheme 126) [78H(11)133; 82JOC2633]. However, reactions of 406 with allylic and alkyl halides were less useful because of an onset of sequential bis metalation–alkylation pathways which gave 408 and 409, respectively. Carbinols 407 cyclized to the azaphthalides 410 under acidic conditions. Metalation of 405 with MeLi, n-BuLi, and PhLi reagents resulted solely in the formation of 1,4-dihydropyridine adducts which, however, have synthetic value [78CC615, 78H(11)133; 80JCS(P1)2070; 81TL5123; 82CJC1821, 82H(18)13, 82JOC2633; 84JCS (P1)2227].

Under n-BuLi or sec-BuLi conditions, 4-pyridyl oxazoline 411 behaves differently than the 3-isomer to give mixtures of ortho metalation (412) and addition (413) products, the latter type having been trapped upon workup by MeI treatment before its oxidation (Scheme 127) (78TL227; 82JOC2633). However, if metalation is carried out with MeLi at −78°C and

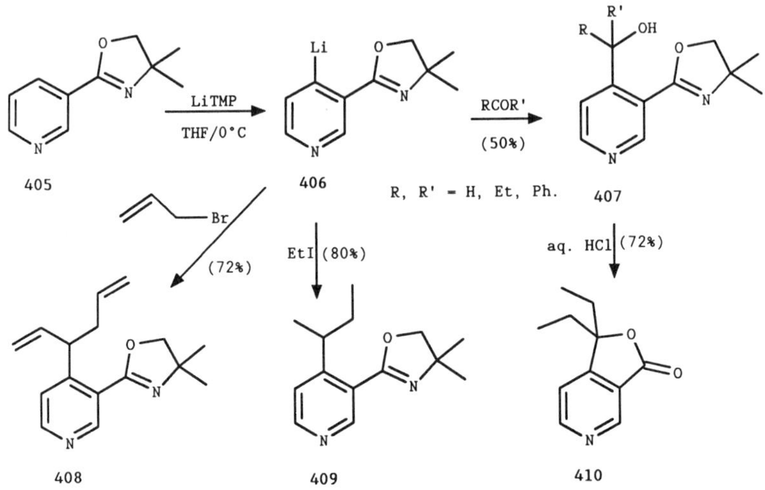

SCHEME 126

the solution warmed to 0°C before addition of electrophile, clean reactions were achieved, allowing the introduction of a variety of alkyl halide and carbonyl electrophiles to give products **414** in good yields. This result was attributed to the greater coordination of the smaller MeLi reagent to the DMG, allowing C-4 deprotonation by a proximity effect.

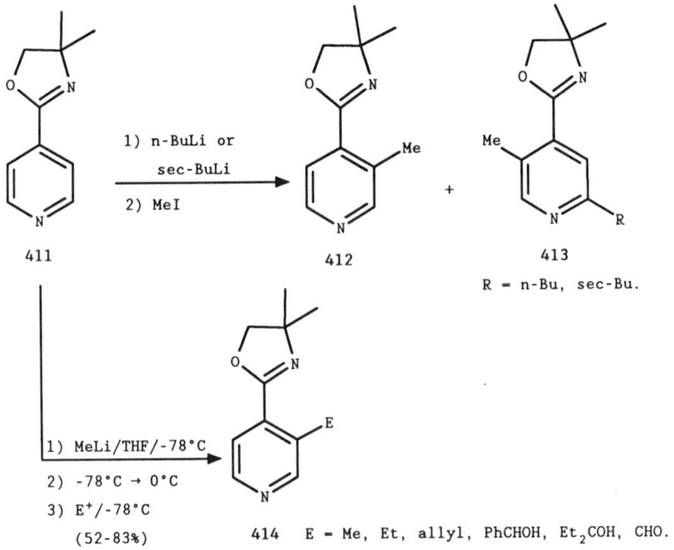

SCHEME 127

B. Pyridine Caraboxamides

1. Secondary Carboxamides

a. *DoM Reactions.* Metalation–deuteriation experiments on a series of picolinic (**415**) and isonicotinic (**418**) secondary amides show that the extent of dianion formation, **416** and **419** respectively as indicated by products **417** and **420**, varies as a function of the *N*-substitutent, the best results being achieved with the *N*-phenyl group (Scheme 128) [86JCR (S)20].

Reaction of the dilithiated species **416** and **419** with alkyl halides, aliphatic and aromatic aldehydes, and TMSCl leads to products **421** and **422** in modest yields (Scheme 129) [83TL4735; 86JCR(S)20]. In cases of the *N*-benzyl amides corresponding to **415** and **418**, small amounts of benzylic functionalization is observed (see also later). The methyl and carbinol derivatives corresponding to **421** and **422** were quantitively hydrolyzed into useful synthons **423** and **424**, and **425** and **426**, respectively.

Similarly, *n*-BuLi/TMEDA metalation of the 4-methoxypicolinic (**427a**) and 2-methoxyisonicotinic (**427b**) anilides followed by quenching with *p*-anisldehyde gave products **428a** and **428b**, respectively, which were transformed into the pyridones **429** and **430**, respectively (Scheme 130) (89T7469).

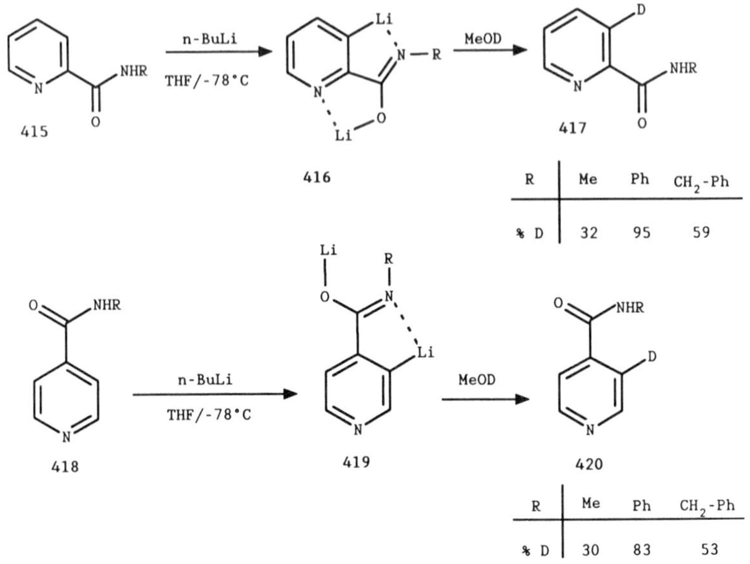

R	Me	Ph	CH$_2$-Ph
% D	32	95	59

R	Me	Ph	CH$_2$-Ph
% D	30	83	53

SCHEME 128

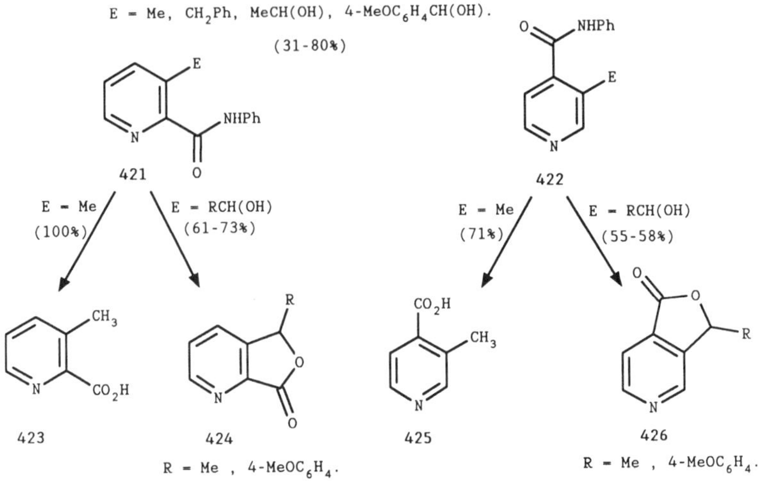

SCHEME 129

b. *Competitive Reactions*   If solutions of the dilithiated *N*-benzyl iso-nicotinic and picolinic amides (**431** and **434**) are warmed to 20°C, formation of the thermodynamically more stable benzylic lithiated species (**432** and **435**) is observed, as evidence by the formation of products **433** and **436**, respectively, upon quench with alkyl halides and carbonyl electrophiles (5–85%) (Scheme 131) [86JCR(S)20].

Treatment of *N*-methyl nicotinamide with MeLi, *n*-BuLi, and PhLi has been shown to give 1,4-dihydropyridine adducts (83TL4735).

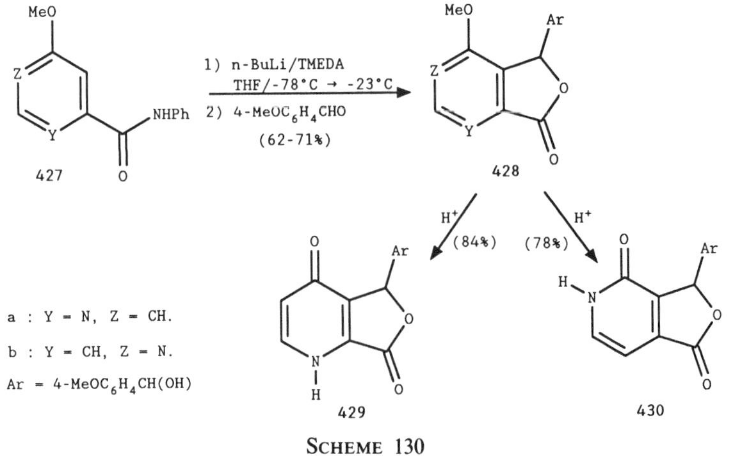

SCHEME 130

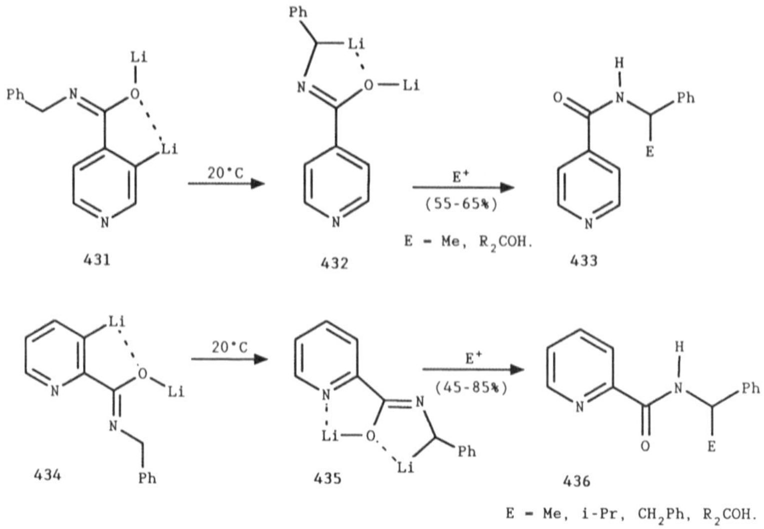

SCHEME 131

2. Tertiary Carboxamides

a. *DoM Reactions.* Early work on all three isomeric N,N-diisopropyl pyridine carboxamides showed that metalation under LDA/Et$_2$O/$-78°$C conditions followed by quench with DMF, PhCONMe$_2$, cyclopentanone, cyclohexanone, and benzophenone, but not alkyl halide and aliphatic aldehyde, electrophiles afforded the functionalized products **437–439** in 25–75% yields (Scheme 132) [80TL4739; 86JCR(S)18]. Acid-driven cyclization provided a series of azaphthalides (**440–442**) with the exception of phenyl-4-pyridylmethanol, which was converted into the keto acid **443**.

In another early experiment, the participation of N,N-diethyl isonicotinamide **444** in a tandem metalation sequence to give azathioquinone **446** was demonstrated (Scheme 133) (80JA1457). Thus, a sequence of metalation (*sec*-BuLi/TMEDA), condensation with 3-thiophene carboxaldehyde, metalation under the same conditions, and warming to room temperature furnished the quinone **446** in 20% yield, presumably via 2-thiophene DoM reaction of the intermediate **445**.

Systematic studies by Iwao on diethyl and diisopropyl amides **447** in reactions with benzamides to give products **448** are significant also in that optimum conditions for metalation (LiTMP/DME/$-78°$C/5–15 min.) on the recommended diisopropyl amide were established (Scheme 134) (83TL2649). The application of **448** in the synthesis of natural products is delineated in Section VIII.

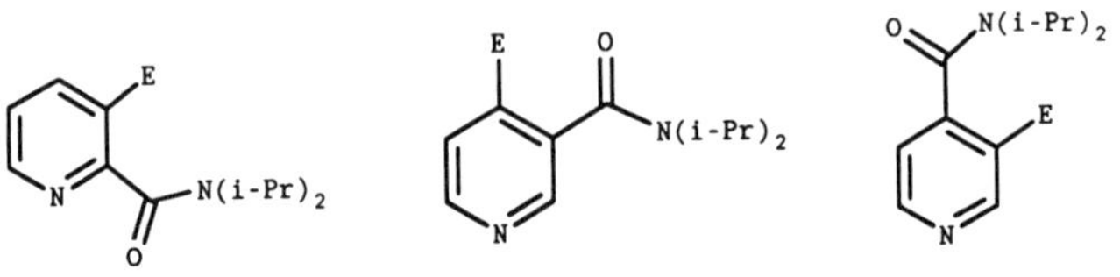

437   438   439

E = CHO, PhCHOH, Ph₂COH, PhCO, TMS.
(25-75%)

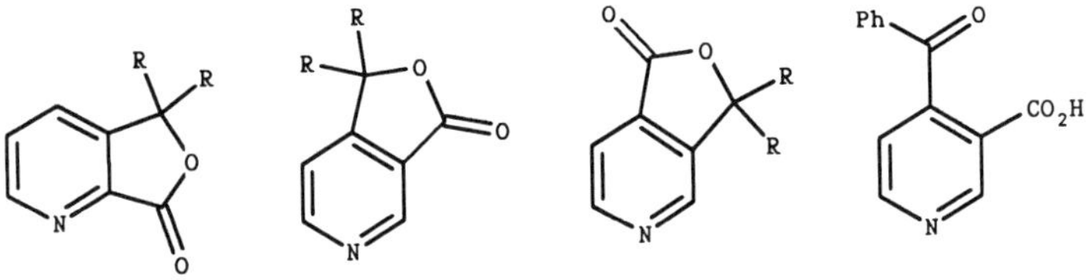

440
                441   442   443

R = Ph, -(CH₂)₅-, -(CH₂)₆-. (20-75%)   (24%)

SCHEME 132

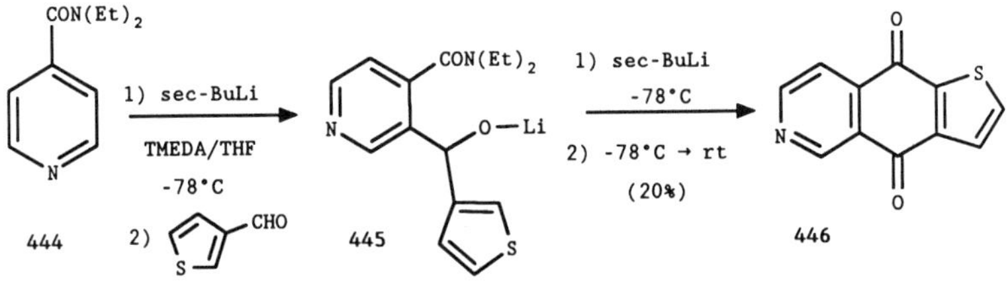

444   445   446

SCHEME 133

447   448

a : R = Et.
b : R = i-Pr.

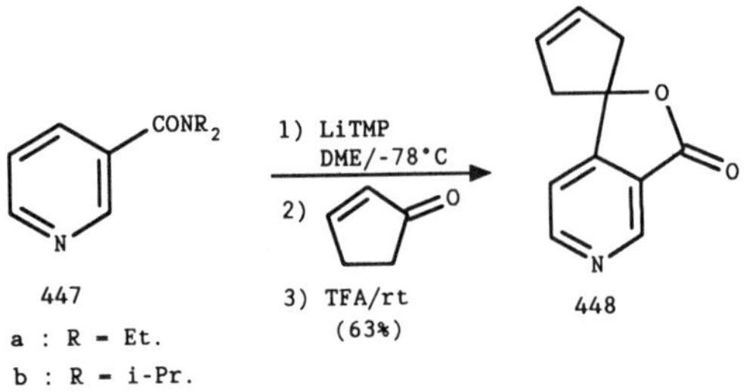

SCHEME 134

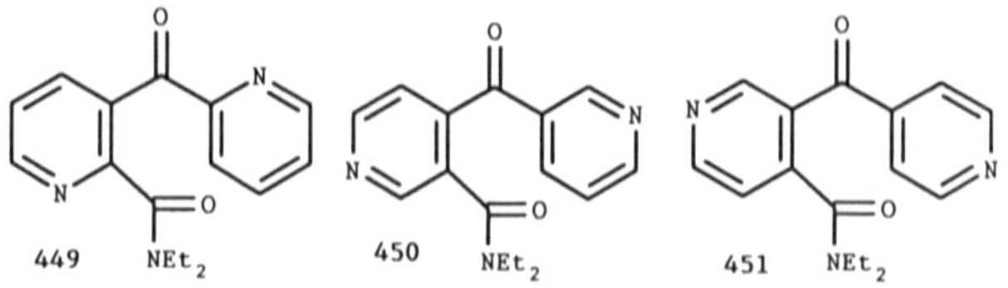

SCHEME 135

b. *Competitive Reactions.* The maintenance of low temperature conditions is of crucial significance for achieving pyridine amide DoM reactions. Thus, if all three isomeric N,N-dimethyl pyridine carboxamides are treated with lithium diethylamide in Et_2O at room temperature, they undergo transamidation to afford the corresponding N,N-diethyl amides (83TL4735). In turn, these can be converted into alkyl or aryl pyridyl ketones upon treatment with MeLi, n-BuLi, and PhLi in Et_2O solution.

Predominant self-condensation of all isomeric N,N-diethyl pyridine carboxamides to give products **449–451** is observed under $LDA/Et_2O/-78°C$ metalation conditions (Scheme 135) [80TL4739; 86JCR(S)18]. Similar results were obtained using LiTMP (83TL2649). This provides evidence for incomplete metalation under these conditions and dictates the use of the more highly hindered diisopropylcarboxamides to achieve useful DoM reactions.

Advantage has been taken of the propensity for equilibrium lithiation demonstrated for diethyl nicotinamide (**447a**) which has been induced under a double self-condensation to form the 2,7-diazaanthraquinone (**452**) (Scheme 136) (88S388). The quinone **452** was transformed by a reduction–aromatization sequence into the pyrido[3,4-*g*]isoquinoline (**453**) in high overall yield.

C. MASKED PYRIDINE ALDEHYDES AND KETONES

A limited number of studies concerning DoM reactions of masked aldehydes and ketones have been described [79OR(26)1]; this is particularly true of pi-deficient heterocycles.

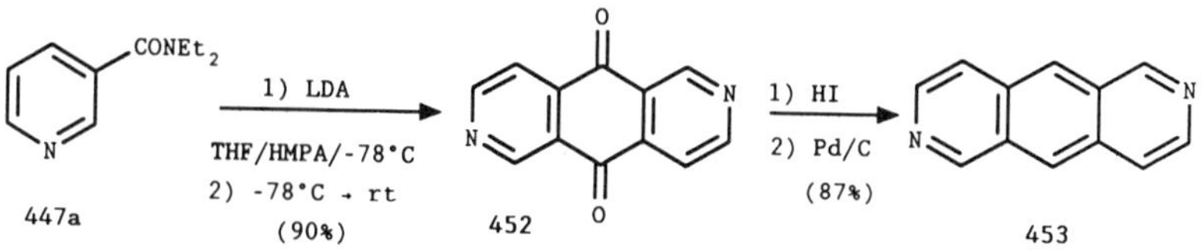

SCHEME 136

1. *Pyridyl α-Aminoalkoxides*

Comins and co-workers have systematically studied DoM processes that involve simultaneous ortho-activation and protection of aromatic aldehydes by the *in situ* generation of α-aminoalkoxides using lithium *N,N,N'*-trimethylethylenediamide (LiTMDA) and lithium *N*-methylpiperazide (LiNMP) (82TL3979; 83TL5465; 84JOC1078; 89JOC3730). Application of LiTMDA and LiNMP on pyridine-2-carboxaldehyde allows selective DoM reaction (85PC2).

Lithiation of methoxypyridinecarboxaldehydes using the α-aminoalkoxide strategy is highly regio-dependent on the amine component of the α-aminoalkoxide and the metalation conditions (90JOC69). 6-Methoxypyridine-6-carboxaldehyde (**454**) (Scheme 137) can be lithiated either at the C-3 or C-5 position using LiTMDA and *n*-BuLi or LiNMP and *t*-BuLi. Treatment of **454** by LiTMDA followed by metalation with *n*-BuLi (2 equiv.) and MeI quench gives the 3-methyl derivative **459** via intermediate **455** and **456**. On the other hand, reaction with LiNMP followed by *t*-BuLi (1.5 equiv.) metalation affords the 5-methyl product **458** via the bis-lithio α)-aminoalkoxide **457**.

2. *Pyridyl Ketone Acetals*

Although metalation of 4-acetylpyridine acetals cannot be achieved (83UP1), LDA treatment of the 2-fluoro and 2-chloro derivatives **460**, which display a combined activating effect of two DMGs, followed by quench with several electrophiles gives 3-substituted derivatives **461** (Scheme 138) [86PC1; 87CJC2027; 89H(29)1815; 91JOC(s1)].

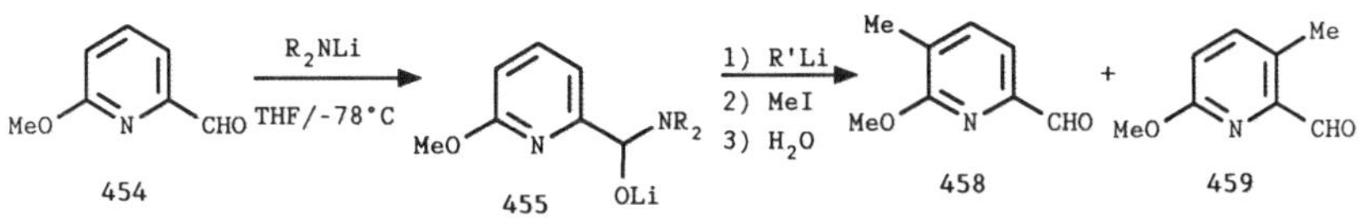

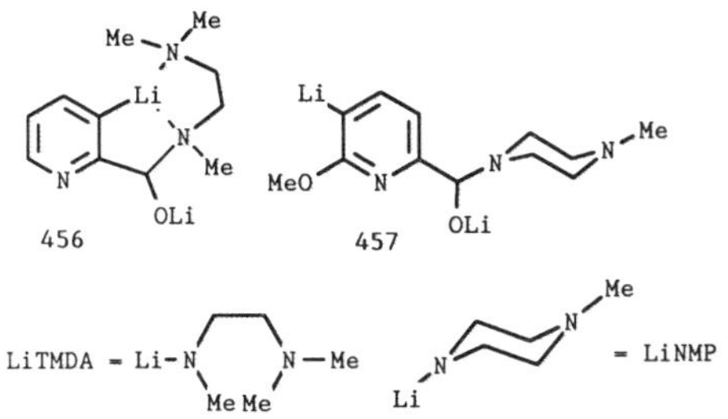

R_2NLi	LiTMDA	LiNMP
R'Li	2 n-BuLi -78°C/5h -42°C/5h	1.5 t-BuLi -78°C/1.25h -42°C/0.25h
458	3%	63%
459	72%	2%

SCHEME 137

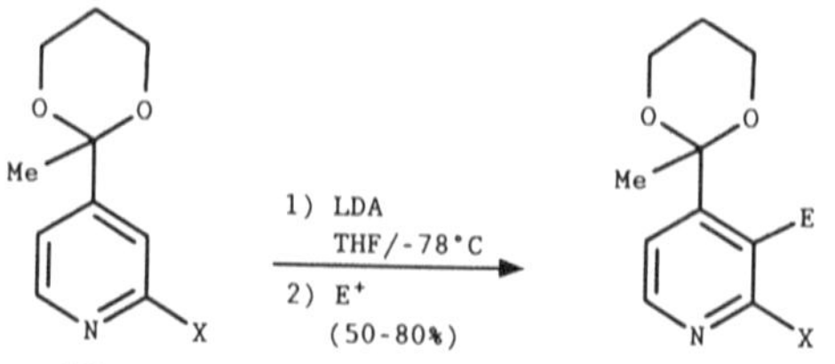

460 461

X = F, Cl. E = D, I, CHO, CO_2H, ArCH(OH).

SCHEME 138

In an interesting report, the *n*-BuLi metalation of the bispyridyl ketone acetal **462** at −70°C followed by deuteration or protonation has been shown to lead to products **463** and **464,** thus indicating the operation of competitive DoM (**463**) and metal–halogen exchange (**464**) processes (Scheme 139) (77JOC3524). Although at higher temperatures (−55°C), greater deuterium label incorporation was observed, competitive fragmentation–recyclization occurred presumably by the mechanism **465→466→467→468** (Scheme 140).

VI. Sulfur-Based DMGs

1. *Tertiary Sulfonamides*

DoM reactions of sulfur-based DMGs in aromatic systems has led to the development of a variety of useful DMGs {SR [83PS(12)779], SO_3Li (80JOC3728), SOR [83PS(16)167; 89JOC24], SO_2R (89JOC24), SO_2OR (86JOC2833), SO_2NHR (71JOC1843), and SO_2NR_2 (69CJC1543)}. The knowledge that some derivatives of the commercially available 3-pyridine

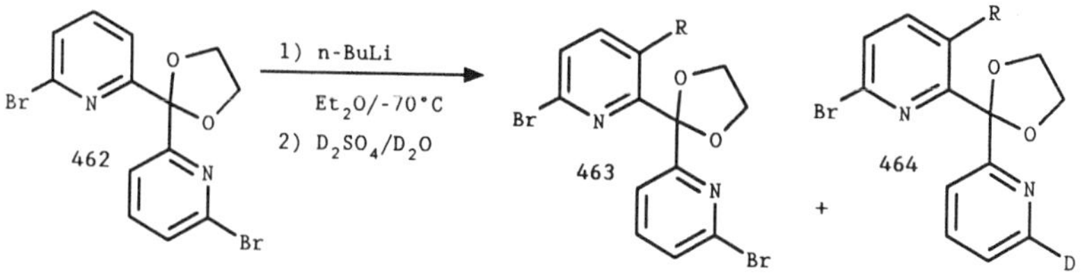

462 463 464

R	463 Yield, %	464
H	37	38
D	11	9

SCHEME 139

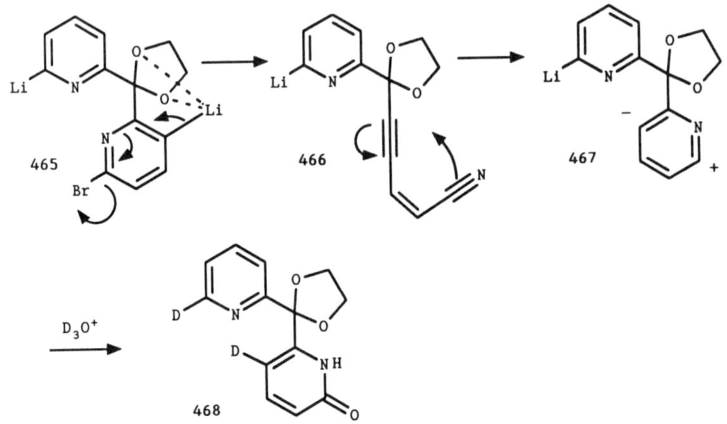

SCHEME 140

sulfonic acid, especially the sulfonamides, exhibit interesting antibacterial properties (65USP3202576) stimulate ongoing activity in DoM chemistry of sulfur-substituted heteroaromatics. In the pyridine series, DMG=SO_2-NR_2 (83S822; 87JOC1133) and SOPh (89TL7091) have been successfully metalated.

Early work by Abramovitch showed that treatment of 3-(N,N-diethylaminosulfonyl)pyridine with PhLi leads to C-2 addition (65CJC1752), a result which discouraged used of alkyllithiums for deprotonation of pyridyl sulfonamides. In a detailed investigation, Queguiner found that (a) the piperidine sulfonamide is the best DMG; (b) LDA is a useful metalation base in 2 equiv. amount (80% yield) rather than 1 equiv. (27% yield) on the basis of PhCHO quench (83S822).

In the case of 4-piperidinosulfonylpyridine (**469**), higher amounts of LDA are detrimental, leading to dimetalation (e.g., I_2 or PhCHO quenches led to mixtures of respective 3-substituted and 3,5-disubstituted products which favored the latter at longer metalation times). A 3,5-dilithio intermediate was ruled out by deuterium incorporation experiments. In fact, it appeared that the reaction of LDA and the electrophiles are slower than C-5 deprotonation of the monosubstituted product **470**. Scheme 141 summarizes results of LDA metalation experiments on the piperidyl sulfonamide **469** leading to products **470** and **471,** which support some of the previous facts (87JOC1133).

The 3-lithio-2-piperidinosulfonyl, 4-lithio-3-piperidinosulfonyl, and 3-lithio-4-piperidinosulfonyl pyridines are pale yellow powders that are quite stable at $-70°C$ in Et_2O and undergo reaction with a variety of electro-

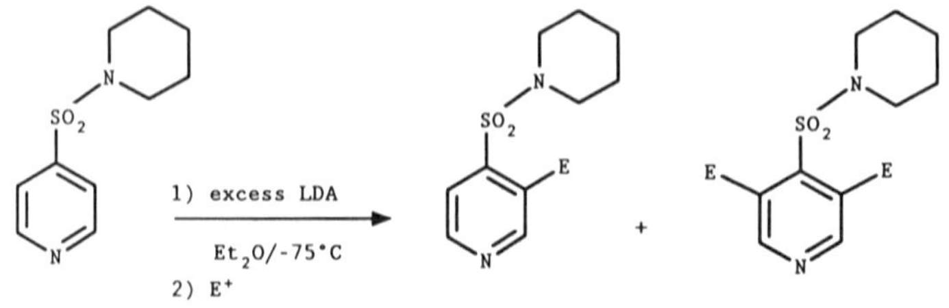

469 → 470 + 471

1) excess LDA
Et$_2$O/-75°C
2) E$^+$

E$^+$	E	LDA equiv.	470 Yield, %	471 Yield, %
MeOD/D$_2$O	D	2	100	0
MeOD/D$_2$O	D	3	100	0
I$_2$	I	2	86	7
PhCHO	PhCH(OH)	2 ^{a}	88	9
PhCHO	PhCH(OH)	3 ^{b}	66	32

a : the yield of (471) depends on the lithiation time :
   0.5 h (9%) and 3 h (20%).
b : a 3 h lithiation time was used.

SCHEME 141

philes, giving rise to respective substituted products **472–474** in good to excellent yields (Scheme 142) (83S822; 87JOC1133). Some of the carbinol derivatives of these products were oxidized using MnO$_2$ to the corresponding aldehydes and ketones (87JOC1133).

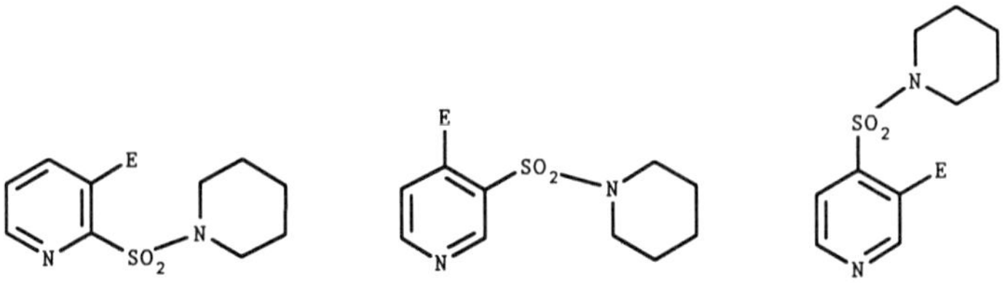

472 (55-95%)   473 (42-95%)   474 (75-95%)

E - D, I, TMS, PhS, CHO, CO$_2$H, Ph$_2$C(OH), Et$_2$C(OH), MeCH(OH), ArCH(OH).

SCHEME 142

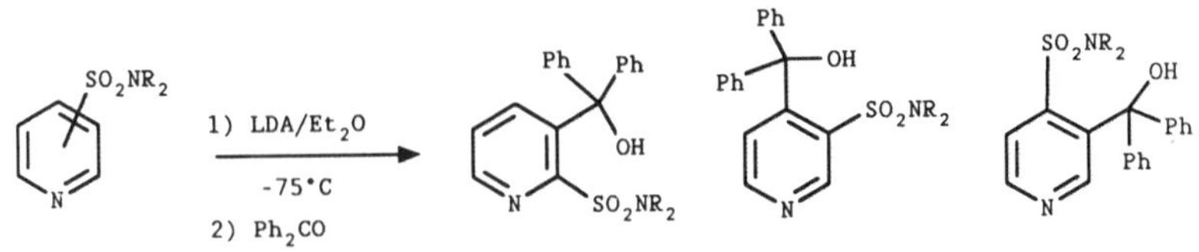

475 476 (87-95%) 477 (88-95%) 478 (85-95%)

NR_2 = 1-pyrrolidinyl, 1-morpholinyl.

SCHEME 143

All three isomeric pyrrolidinyl and morpholinyl sulfonamides **475,** upon sequential LDA metalation and condensation with benzophenone, afforded high yields of corresponding carbinols **476–478** (Scheme 143) (87JOC1133).

2. *Pyridyl Thioethers*

In an interesting study with potential for the synthesis of 3-substituted pyridines, Katritzky has shown that LDA metalation of the benzothiazol-2-yl thioether **479** gives the highly coordinated lithio species **480,** as demonstrated by quench with several electrophiles to give products **481** (Scheme 144) [87H(26)427].

3. *Pyridyl Sulfones and Sulfoxides*

As previously appreciated in the aromatic series, phenylsulfoxide (89TL7091) serves as useful DMG in the series. 2-, 3-, and 4-pyridylphe-

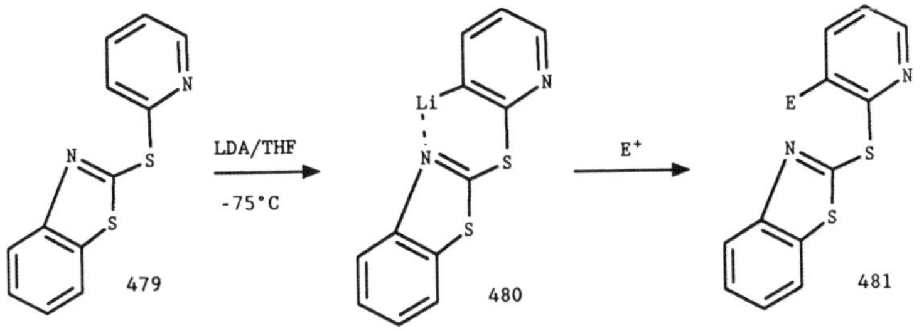

479 480 481

E = D, PhCH(OH), TMS.

SCHEME 144

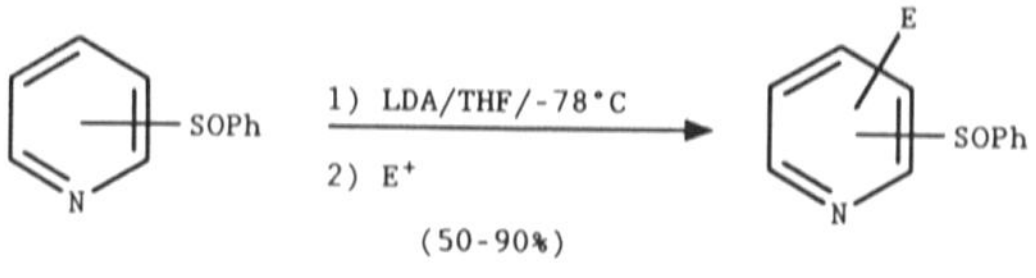

482 and 483a-c = 2-, 3- and 4-pyridyl.

E = TMS, PhS, Ph-CH(OH), Ph-CMe(OH).

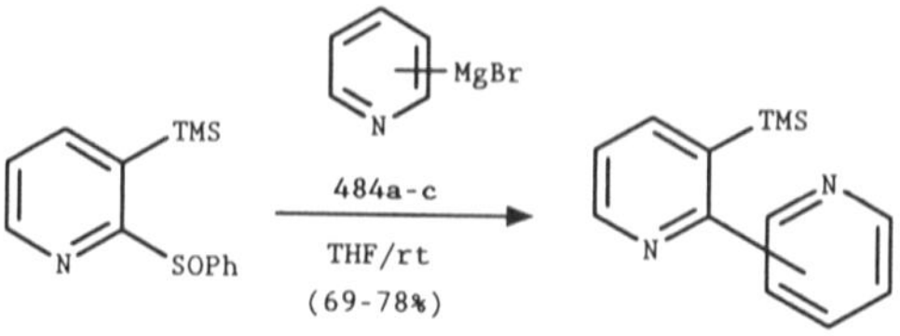

484 and 485a-c = 2-, 3- and 4-pyridyl.

SCHEME 145

nylsulfoxides (**482**) are regioselectively lithiated using LDA (THF/−78°C, giving rise to the 3-, 4-, and 3-substituted pyridylsulfoxides **483**. These sulfoxides were converted to the substituted bipyridines **485** in high yields via cross coupling reactions with 2-, 3-, and 4-pyridyl Grignard reagents (**484**) (Scheme 145).

t-Butyl-2- and 4-pyridylsulfoxides and sulfones have been regioselectively lithiated (LDA/THF/−78°C), leading to a great variety of ortho-functionalized derivatives in modest to high yields (89TH1).

VII. Heterocycles without Directing Metalation Groups

The classical studies of Ziegler (30CB1847) demonstrated that pi-deficient heteroaromatics suffer addition of alkyllithium reagents and that lithiated species cannot be obtained by treatment of halogen-substituted derivatives with lithium metal. In contrast, the original work of Gilman in 1940 (40JA2327) showed that such species could be prepared by metal–halogen exchange from the corresponding pyridyl bromides (51JOC1485). This method has been widely used in synthesis and has been comprehensively reviewed (74MI1).

A. Unsubstituted Heterocycles

The early H–D exchange studies of Abramovitch (67CC55) and Zoltewicz (69T4331) showed that major 4-pyridyl deprotonation occurred
under thermodynamic conditions (MeONa/MeOD, NaNH$_2$/NH$_3$) (*vide
infra* entry 3,4). Brandsma (84CC257, 84JOC3857) investigated the metalation of pyridine under the Schlosser–Lohmann superbase (*n*-BuLi/*t*-
BuOK) conditions at very low temperatures. The results (Table I), based
on D$_2$O and TMSCl quench experiments, show 90% of the 2-Li species in
Et$_2$O solution (entry 1) and, at the opposite extreme, 90% of the 4-Li
species in the more polar THF and HMPA solvent (entry 2). In THF, the
4/1/4 ratio of 2-/3-/4-Li species changed with reaction time to give the
4-isomer. These results strongly suggest that the 2-lithiopyridine is the
kinetic product, whereas the corresponding 4-lithio species is the thermodynamic one.

In contrast to the alkyllithiums, lithium amides show little tendency to
add to *N*-heteroaromatic systems in the absence of good leaving groups
2-and 4-related to the nitrogen. Nevertheless, lithium amides undergo
complex reactions with such substrates, as first demonstrated (74TL2373)
in reactions of pyridine, quinoline, isoquinoline, and 5-methylpyrimidine
with LDA/Et$_2$O/−78°C. The respectively observed dimeric products (**77**,
486, and **487**) (Scheme 146) were rationalized by the formation of the
lithiated intermediates, which undergo addition to another molecule of the
starting heteroaromatic. Addition of HMPA was shown to increase the
yields of the dimeric products. Attempts to trap the lithiated species with a
variety of electrophiles were unsuccessful.

In contrast, 5-methylpyrimidine (**488**) appeared to give the 6-lithiated
species, as evidenced by reaction of benzophenone to produce **489**
(Scheme 147) (74TL2373).

Analogous to the observations with pyridine, quinoline, and isoquinoline, 1,8-naphthyridine (**490**) was shown to produce a dimer **492** upon
treatment with 1-lithiodithiane (73CL1307). Electron spin resonance

TABLE I
REGIOSELECTIVITY IN PYRIDINE METALATION

Entry	Solvent	Reagent	Conditions	%Li-2	%Li-3	%Li-4
1	Et$_2$O	*n*-BuLi/*t*-BuOK	−105°C/90 min.	90	3	7
2	THF/HMPA	*n*-BuLi/*t*-*t*BuOK	−105°C/60 min.	4	9	87
3	MeOH	NaOMe	—	5	42	54
4	liq.NH$_3$	NaNH$_2$	—	0	23	77

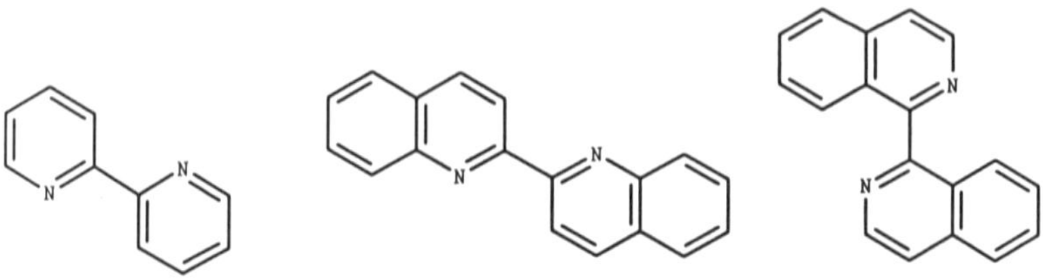

77 (50%)   486 (74%)   487 (55%)

SCHEME 146

(ESR) signals characteristic of the naphthyridine radical-anion **491** suggest a SET process (Scheme 148).

The preceding results and Ashby's observation (81JOC2429) that LDA acts as the single-electron donor towards aromatic hydrocarbons, stimulated Newkome (82JOC599) to study the reaction of pyridine with LDA (HMPA at 0°C) (Scheme 149). The products of this reaction, 2,4'- (**77**) and 4,4'- (**494**) bipyridines, obtained in the ratio 1 : 8, were rationalized as being formed from a pyridine radical-anion **493**. In support of this proposal, complex ESR signals corresponding to both pyridine radical-anion and the diisopropylamino radical were observed under these conditions.

In an other solvent such as THF, a strong signal characteristic of the diisopropylamino radical could be observed at −60°C (but no signal for the pyridine radical-anion). This suggests that HMPA stabilizes the pyridine radical-anion species (68JA6421). Product variation from that observed earlier occurred when the reaction was carried out in Et_2O in the presence of HMPA at −78°C (74TL2373). Under these conditions, only 2,2'-bipyridine (**77**) was isolated. Nevertheless, the overall mechanistic proposal has credence in that the dimerization of pyridine to 4,4'-bipyridine as major product under conventional alkali-metal reduction conditions has been shown to proceed via a radical-anion intermediate [62MP21; 79AHC(25)205]. On the basis of Ashby's results (81JOC2429), the diisopropylamino anion is expected to be a moderately strong reducing agent, although Marcus' theoretical treatment [84ACS(B)439] indicates

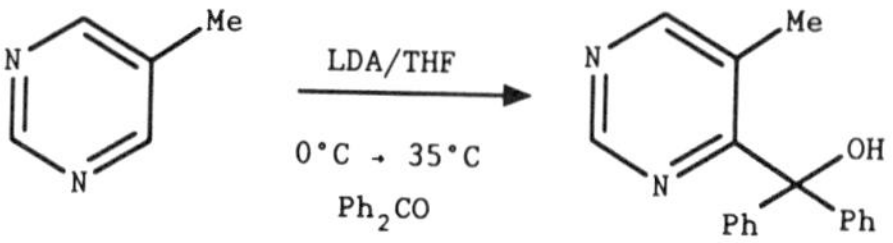

488   (30%)   489

SCHEME 147

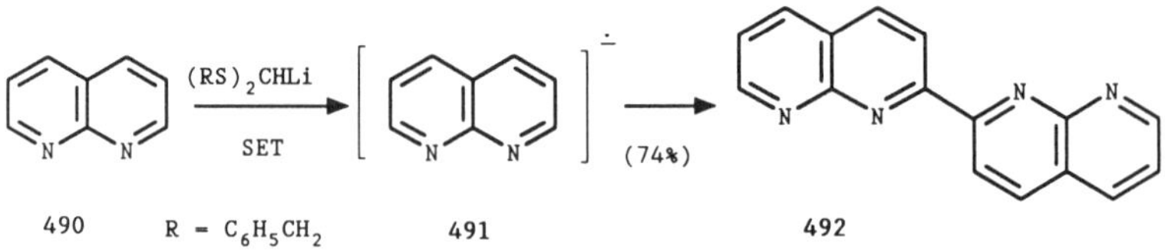

490   R = C$_6$H$_5$CH$_2$ 491   492

SCHEME 148

that the SET process for diisopropylamino anion with pyridine is not feasible.

In summary, although radical-anion formation in the dimerization of pyridine has been observed by ESR, the nature of the overall pathway (anionic or radical) remains to be elucidated.

As may be surmised from the earlier discussion, the metalation of the bare pyridine nucleus with alkyllithium or lithium amide bases lacks synthetic utility. However, the lithiation of 4-*t*-butylpyridine (**495**) has been achieved at room temperature with the *in situ* generated compatible base-electrophile combination of LiTMP (3 equiv.) and TMSCl (6 equiv.) (Scheme 150) (83JA6155, 83JOC4156). Under these conditions, the 2-TMS product **496** was obtained as the sole aromatic product while using a large excess of LiTMP (10 equiv.) and TMSCl (30 equiv.). The 2,6-disilylated pyridine **497** is formed in 30% yield. These results suggest the reaction proceeds by formation of equilibrium concentrations of anion, which are trapped by TMSCl.

B. HETEROCYCLE-*N*-OXIDES

In pyridine-*N*-oxide, 2-proton acidity is considerably enhanced by the inductive effect of the oxide and by the complexing capability of the lone pair on oxygen with lithium. Hence, 2-lithiation and sometimes 2,6-dilithiation with alkyllithium and lithium dialkylamide bases is feasible. In the case of ring substituted pyridine-*N*-oxides **498,** fair to good yields of

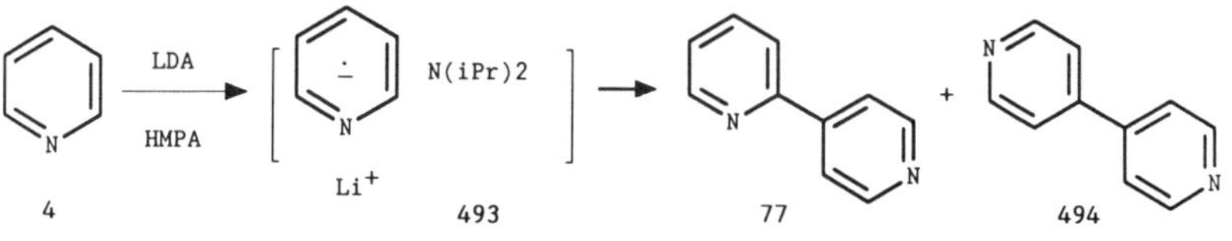

4   493   77   494

SCHEME 149

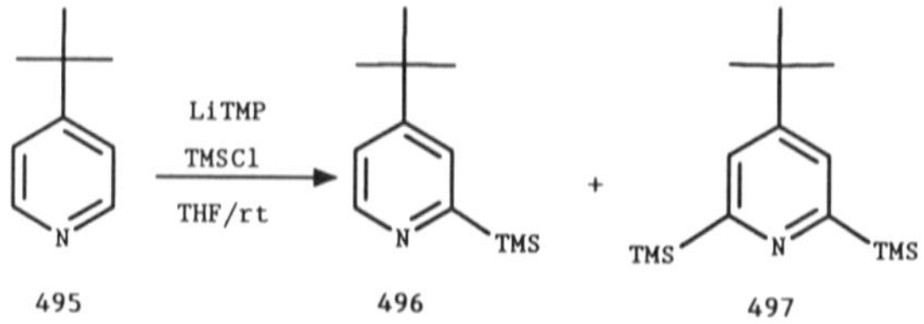

SCHEME 150

499 and **500** were obtained after quenching with electrophiles (72JOC1690, 72JOC3584; 74MI2) (Scheme 151). The compatability of LiTMP and hexafluoracetone allows the application of the *in situ* trap technique to form the carbinols **501** and **502** in useful yields (83JOC4156). Similar results were observed with TMSCl as the electrophile.

Lithiation of thiazolo[5,4-*b*]pyridine-*N*-oxides (**503**) by *n*-butyllithium at −65°C is selectively directed by the pyridine *N*-oxide moiety, whereas lithiation of the parent heterocycle by LDA at −78°C exclusively occurs at the C-4 position (89TL183). Interestingly, no metalation of the furan ring occurred (Scheme 152).

Application of this concept to the metalation with LDA of pyrazine-mono-*N*-oxides **505** and subsequent quenching with electrophiles led to the formation of substituted products **506** (Scheme 153) [83H(20)154].

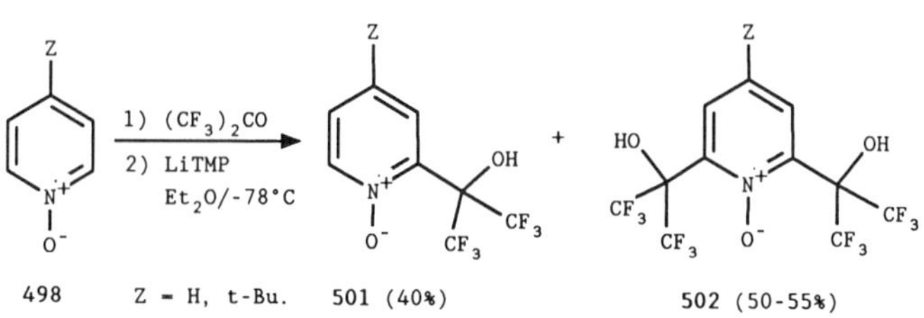

SCHEME 151

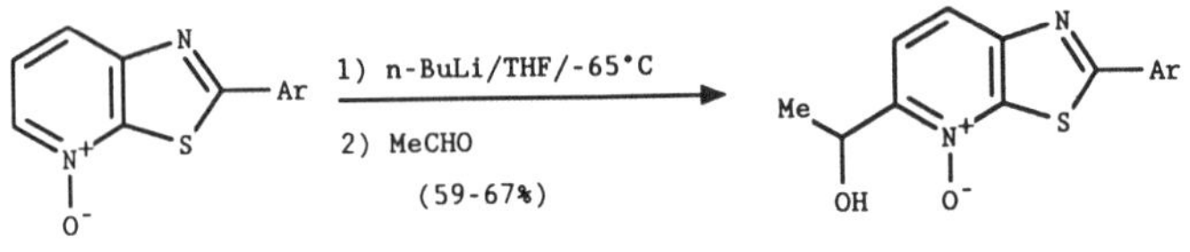

503   Ar = phenyl, 2-furyl.   504

SCHEME 152

C. N-ACTIVATED HETEROCYCLES

A behavior similar to that of pyridine and pyrazine N-oxides is exhibited by complexes of pyridines with hexafluoroacetone. Such complexes are expected to enhance the acidity of the pyridine hydrogens by oxygen–lithium chelating effects and therefore direct 2-deprotonation. In fact, treatment of t-butylpyridine (**495**) with LiTMP/THF-Et$_2$O/ $-107°C$ generated, via the known complex **507**, the stable 2-lithio species **508** which, upon reaction with electrophiles, furnished 2-substituted pyridines **509** (Scheme 154) (83JOC4156).

If complex **507** is warmed to $-78°C$ without addition of an electrophile, hexafluoro-2-(4-butylpyridin-2-yl)-2-propanol is isolated.

At $-107°C$, reaction of hexafluoroacetone-complexed pyridine with LiTMP followed by MeOD as the electrophile gave 90% C-2 deuteriation, whereas at $-78°C$, C-2 (48%) and C-4 (28%) deuterium incorporation was observed.

D. PYRROLO[3,4-c]PYRIDINE

n-BuLi metalation of the pyrrolo[3,4-c]pyridine **10** followed by D$_2$O quench leads to deuteriation at both C-2 and C-6 sites (78% and 50%, respectively) (85T1945).

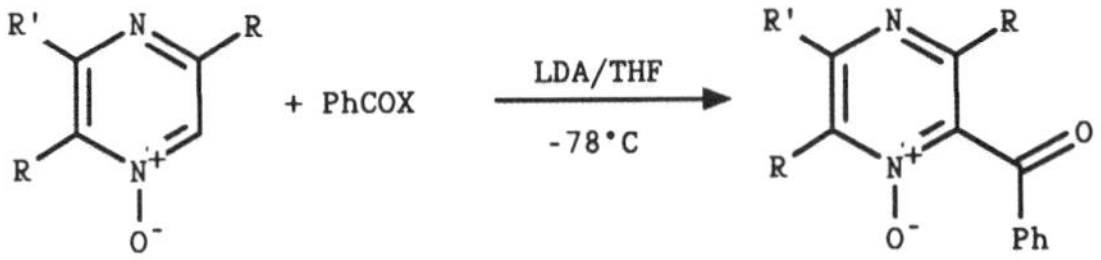

505   506

R = alkyl; R' = H, Cl; X = Cl, OMe.

SCHEME 153

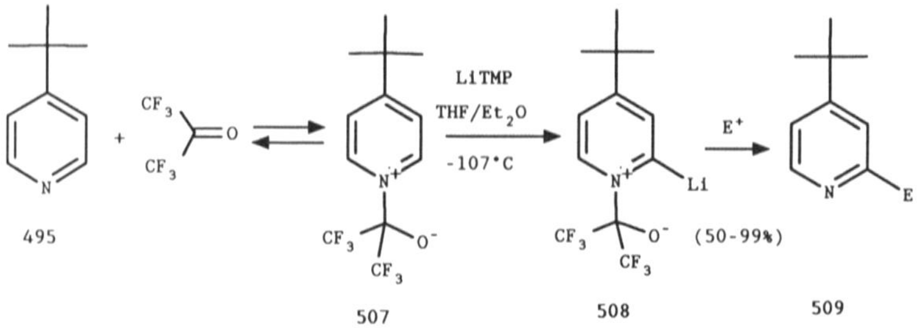

E - D, $(CF_3)_2C(OH)$, Br, I, PhCH(OH).

SCHEME 154

E. TRIAZOLO[1,5-a]PYRIDINES

Stimulated by their accessibility and conversion into useful pyridine and quinoline derivatives, the metalation of triazolopyridines and triazolo-quinolines has been systematically studied by Jones. Treatment of 1,2,3-triazolo[1,5-a]pyridine 511 with LDA/Et$_2$O/$-40°C$ results in 7-lithiation, as evidenced by the formation of alcohols 512 upon treatment with a variety of aldehydes and ketones (Scheme 155) (80TL4529). The resulting substituted systems were converted into the 6-substituted pyridine 2-carboxaldehydes 515 upon sequential treatment with bromine and silver nitrate. Furthermore, metalation followed by quench with ClCO$_2$Et gave the bis-(1,2,3-triazolo[1,5-a]-7-pyridinyl)methanone (513) and not the desired ester. Surprisingly, when DMF was used as electrophile, 1,2,3-triazolo[1,5-a]pyridin-7-ylmethanol (514) was obtained. No reaction was observed with methyl iodide, benzonitrile, or cyanogen bromide (Scheme 155) [82JCS(P1)967].

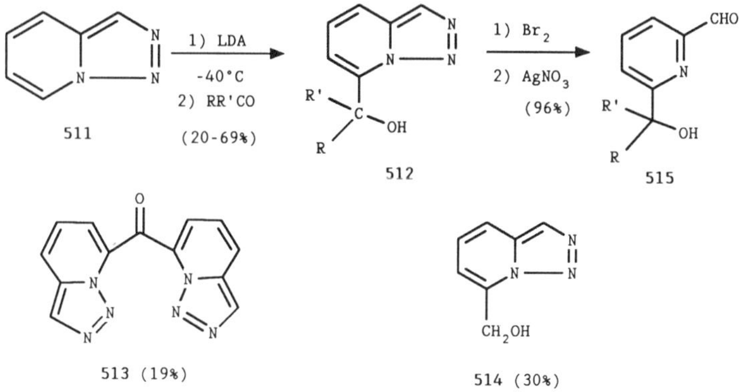

SCHEME 155

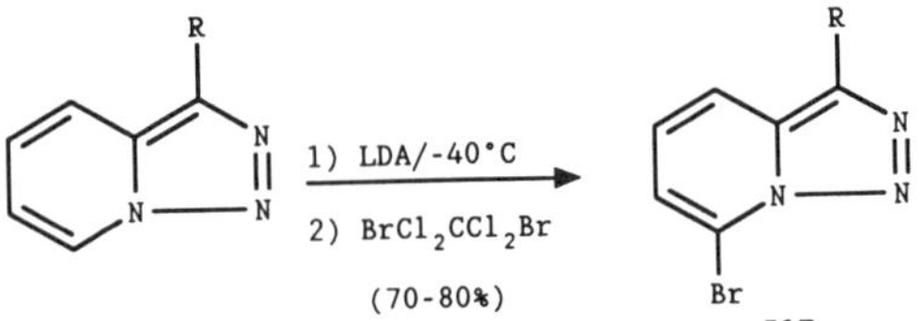

1) LDA/-40°C

2) BrCl₂CCl₂Br

(70-80%)

516   517

R = H, Me, CONEt₂

SCHEME 156

Analogous 7-lithiation was also demonstrated for 5-methoxy-1,2,3-triazolo[1,5-*a*]pyridine by the reaction with anisaldehyde to give the corresponding alcohol [83JCR(S)144]. Lithio derivatives of triazolo [1,5-*a*]pyridine, as well as the 3-methyl and 3-(*N*,*N*-diethylcarboxamido) derivatives **516,** gave the brominated products **517** when treated with bromine, but ring opening reactions of the triazolo ring were observed simultaneously. This could be avoided by using 1,2-dibromotetrachloroethane as brominating agent, which gave good yields of 7-bromo derivatives in all cases (Scheme 156) (86TL3543).

Further studies on 1,2,3-triazolo[1,5-*a*]pyridine (**511**) showed that competitive lithiations occur at the C-4 and C-7 positions with two different rates (fast lithiation at the C-7 position and slow at the C-4 position) [83JCR(S)144].

LDA metalation of 1,2,3-triazolo[1,5-*a*]quinoline (**518**) followed by carbonation leads to the 3-carboxylic acid, which can be converted into the corresponding amide by a standard procedure (Scheme 157) [84JCR(S)140]. The useful DMG properties of the CONEt₂ group is evidenced in the peri metalation leading, after quench with ketone electrophiles, to products **519.**

In 1,2,3-triazolo[5,1-*a*]isoquinoline (**520**), the N-3 atom appears to direct metalation to the peri C-4 position, as seen from the products **521** derived from electrophile quench experiments (Scheme 158) [85JCS(P1)1897]. With TMSCl, besides 60% of **521** (E=TMS), 23% of 1,5-disubstituted

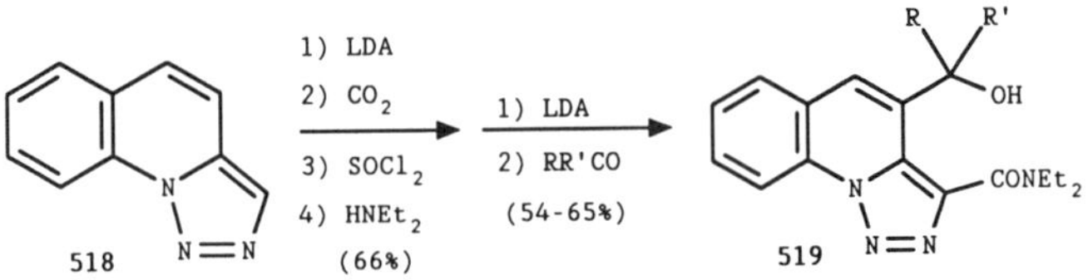

1) LDA

2) CO₂

3) SOCl₂

4) HNEt₂

(66%)

1) LDA

2) RR'CO

(54-65%)

518   519

SCHEME 157

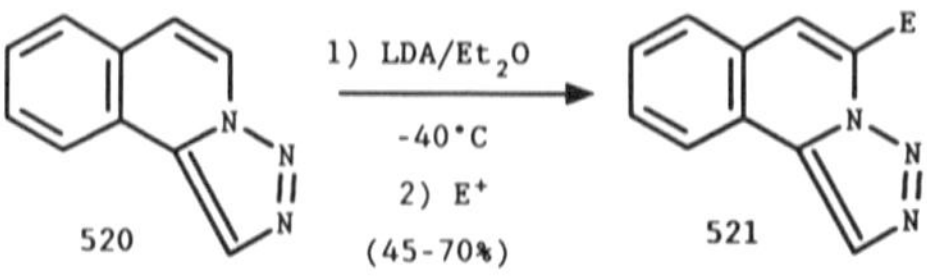

E - CONEt$_2$, TMS, MeCH(OH), 4-MeOC$_6$H$_4$CH(OH).

SCHEME 158

compound was obtained, which presumably arises via sequential met-alation–silylation processes. Contrary to the triazoloquinolines **519**, these derivatives were readily ring opened to the corresponding 3-substituted isoquinolin-1-ylmethanol.

In an ancillary study, the lithiation of triazolo[1,5-c]pyrimidine **522** under a variety of conditions proved to be unsuccessful (Scheme 159) [82JCS(P1)967].

VIII. Synthesis of Natural Products and Related Bioactive Molecules via Heterocyclic DoM Reactions

A. GENERAL

The application of DoM strategies to simple heteroaromatics for the construction of biomolecules and natural products is at an early stage of development. The normally tedious and lengthy preparation of hetero-cyclic synthons by *de novo* ring formation and (electrophilic) substitution methods suggests that increasing application of DoM tactics will be prof-itable. In view of the occurrence of pyridine, quinoline, isoquinoline, and diazines subunits in large and varied classes of natural products, the development of new DMGs and application of DoM chemistry borrowed from the well-tested aromatic series is expected to be fruitful for these systems. The vast and diverse area of pharmaceutical heterocyclic chem-istry stands to greatly benefit from the adaption of DoM methods. In this context, a cursory glance at this chapter shows that metalation chemistry

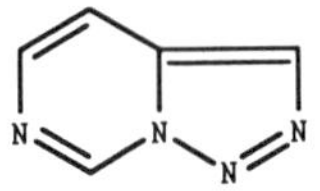

522

SCHEME 159

of the diazines is also in its infancy. Continuing advances in the discovery of new DMGs, development of new bases, base-electrophile combinations and conditions, and mechanistic insight project anticipated broader utility of the DoM reaction in pi-deficient heteroaromatic chemistry.

B. HALOGEN-BASED DMGS

As already described, with the exception of iodo, all halogens have proved useful DMGs in the pyridine series for the preparation of potential intermediates and final products related to biologically interesting molecules (Schemes 36, 41–47, 57). As specific applications, the reaction of chloro ketones **524** with ethylene diamine and 1,2-phenylenediamine give, respectively, pyrido[4,3-*e*]-1,4-diazepines **525** and pyrido[4,3-*b*]-1,5-benzodiazepines **523** in good yields (Scheme 160) (87AP704).

The preparation of thyroid hormone analogues **532** as potential thyromimetics involved fluoro- and bromopyridine metalation chemistry (Scheme 161) [88JCS(P1)3085]. Thus, monometalation of **526** followed by reaction with a benzaldehyde derivative afforded **527**, which was transformed by classical procedures into **528.** In the parallel manner, **529** was converted into **530** and thence into **531,** the latter stages involving a protodebromination via a metal–halogen exchange and thus constituting the use of Br as a latent DMG. Both **528** and **531** served as left-hand portions of the thyroid hormone analogues **532.**

A convergent approach to the 11-*nor*-ellipticines **536,** which constitute promising antitumor agents currently in clinical trials, begins with directed

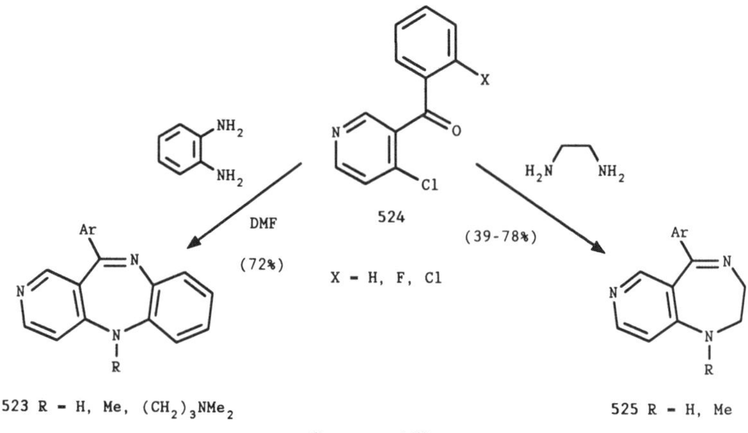

SCHEME 160

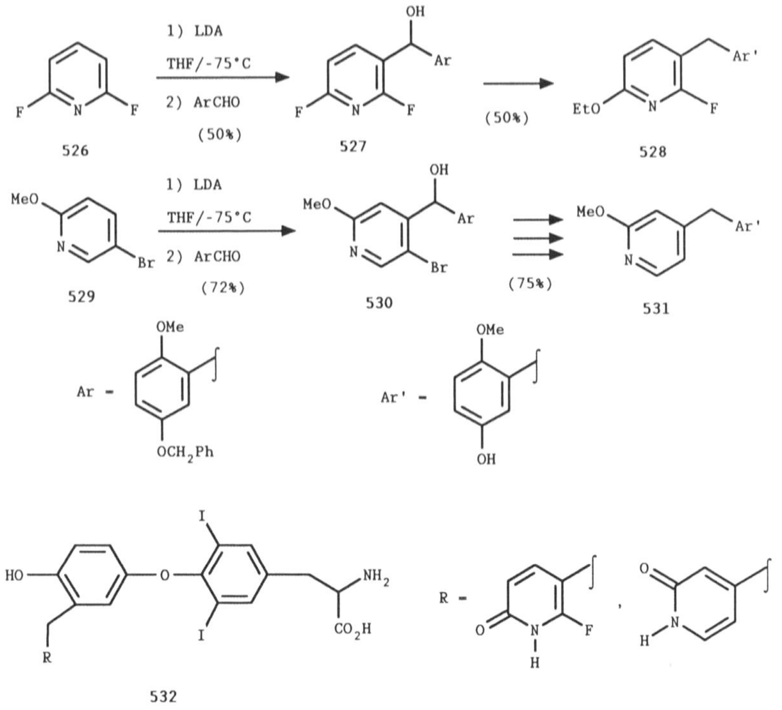

SCHEME 161

metalation of a 2-chloropyridine acetal **533** (Scheme 162) [87CJC2027; 88H(27)1671]. Thus, LDA metalation followed by quenching with indole or azaindole aldehydes provides the isolable alcohols **534.** Reduction and acid-catalyzed cyclization leads to **535,** which is converted into the amines **536** in the 28–35% yields required for animal tumor model studies (83JMC181).

Furthermore, synthon **533** is a common component of the polyhetero-cyclics **537–540,** prepared to further probe anticancer structure-activity relationships (Scheme 163) (87CJC2027; 88JOC5301).

A general route to substituted ellipticines has been studied, based on the synthesis of 2,3,4-trisubstituted pyridine (Scheme 164) [91JOC(s1)]. (4-Bromo-2-fluoropyridin-3-yl)ethanol (**621**) was obtained in good yield via a 2-step sequence (metalation and bromation, followed by metalation, bromine migration, and acetaldehyde quench), starting from 2-fluoropyridine (**11**). Condensation of the corresponding chloro derivative **622** with 1-indolylmagnesium iodide afforded coupled product **623.** Cross coupling reaction between the 4-bromo pyridine **623** and 1-ethoxyvinyltri-

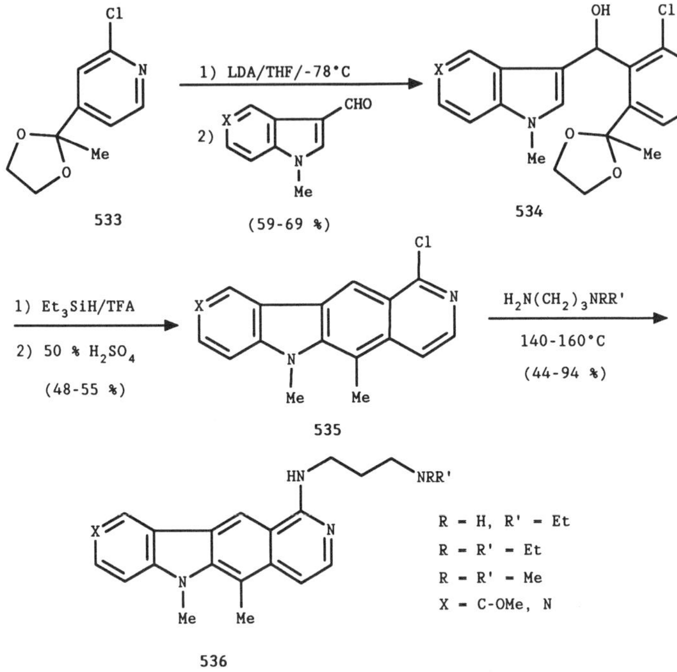
1) LDA/THF/-78°C
2)
CHO
(59-69 %)
533
534
1) Et_3SiH/TFA
2) 50 % H_2SO_4
(48-55 %)
535
$H_2N(CH_2)_3NRR'$
140-160°C
(44-94 %)
536
R = H, R' = Et
R = R' = Et
R = R' = Me
X = C-OMe, N

SCHEME 162

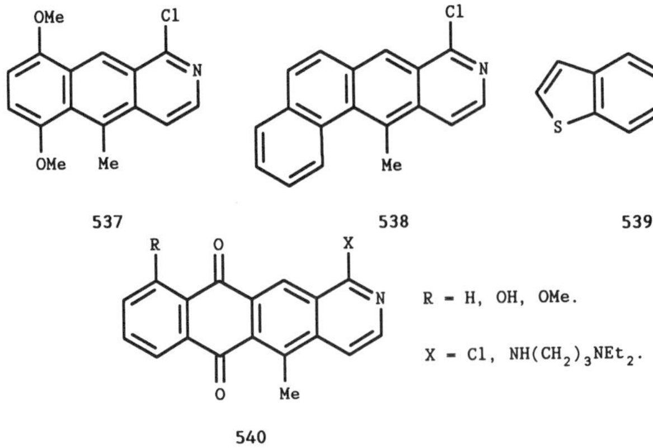
537
538
539
540
R = H, OH, OMe.
X = Cl, $NH(CH_2)_3NEt_2$.

SCHEME 163

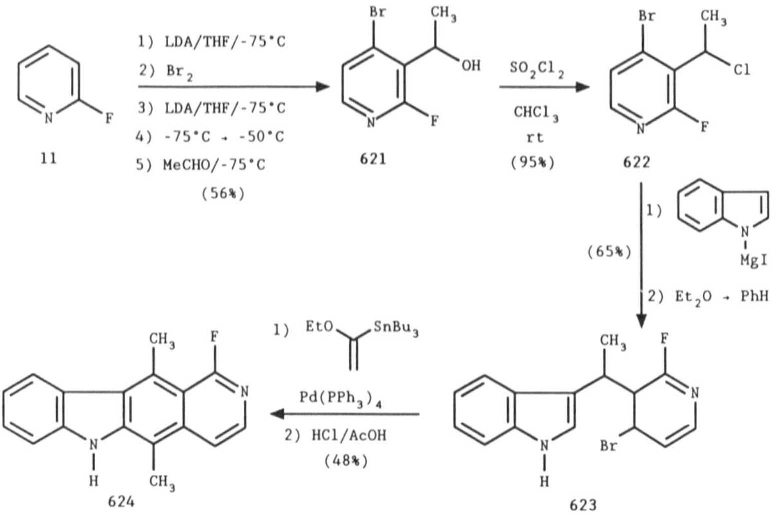

SCHEME 164

butyltin under palladium(0) catalysis, followed by acidic treatment allowed a convenient access to the unknown 1-fluoroellipticine **624**.

Certain structural indications of thromboxane A2 biosynthesis inhibition and hence potential therapeutic utility in arterial thrombosis prompted the synthesis of the pyridine prostanoid **544** (Scheme 165) (83TL3291). Brief metalation of **42** followed by DMF quench afforded aldehyde **541**, which upon Horner–Emmons chain extension, reduction, and protection gave **542**. Having served as a DMG, the bromo function was subjected to metal–halogen exchange, transmetalation (CuCN), and condensation with an iodo allene to furnish the 3,4-disubstituted pyridine **543**. The latter was transformed into two derivatives **544** (with and without double bond), which were shown to be effective inhibitors of thromboxane A2.

Several pyrimidine aminoketones **108** (Scheme 32) have been prepared for potential use as intermediates in the synthesis of pharmaceutically interesting pyrimido-1,4-diazepines (86S886; 87AP704).

C. Nitrogen-Based DMGs

Rather sparse utility of *N*-pivaloyl and *N-t*-Boc DMGs has been documented to date. A more direct preparation of 1,2-related aminoketo pyridines, of value as intermediates for pyrido-1,4-diazepine synthesis, to

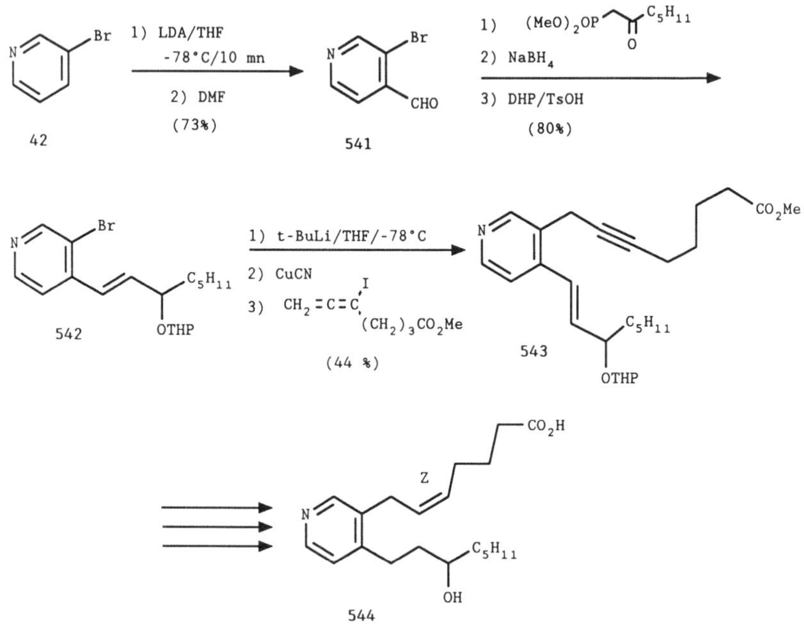

SCHEME 165

those using halopyridine precursors (Scheme 36) involves the directed metalation of *N*-pivaloyl or *N-t*-Boc pyridines **545** (Schemes 77 and 166) (87CJC1158). Condensation of the lithiated species of **545** with *N,N*-diethyl benzamide affords the benzoyl derivatives **546** in good yield. Alter-

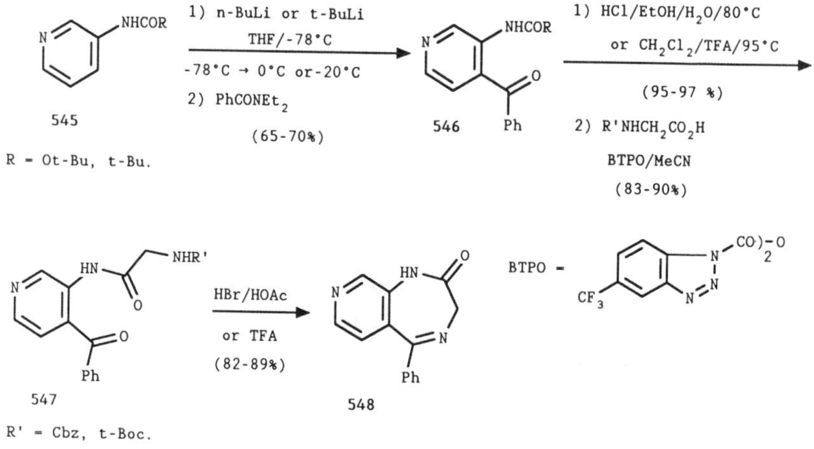

SCHEME 166

native routes to **546** via condensation with benzaldehyde and oxidation have also been achieved (82S499; 83JOC3401; 87CJC1158, 89JHC105). Deprotection followed by amidation afforded compounds **547**, which were readily transformed into the pyrido-1,4-diazepines **548**.

With the aim of developing new analogues of the cyclic AMP phosphodi-esterase inhibitor lixazinone (**554**), the synthesis of agents **553** and **555–557** was undertaken (Scheme 167) (88JMC2136). Thus, all possible 1,2-related *N-t*-Boc aldehydes **551** were prepared by directed metalation on isomers **549**, with the exception of that which required the use of a metal–halogen exchange reaction on the bromo precursor **550** (attempts to metalate 4-TMS-3-*N-t*-Boc pyridine proved inefficient). As exemplified for one particular isomer, conversion of **551** into **552** by reductive amination

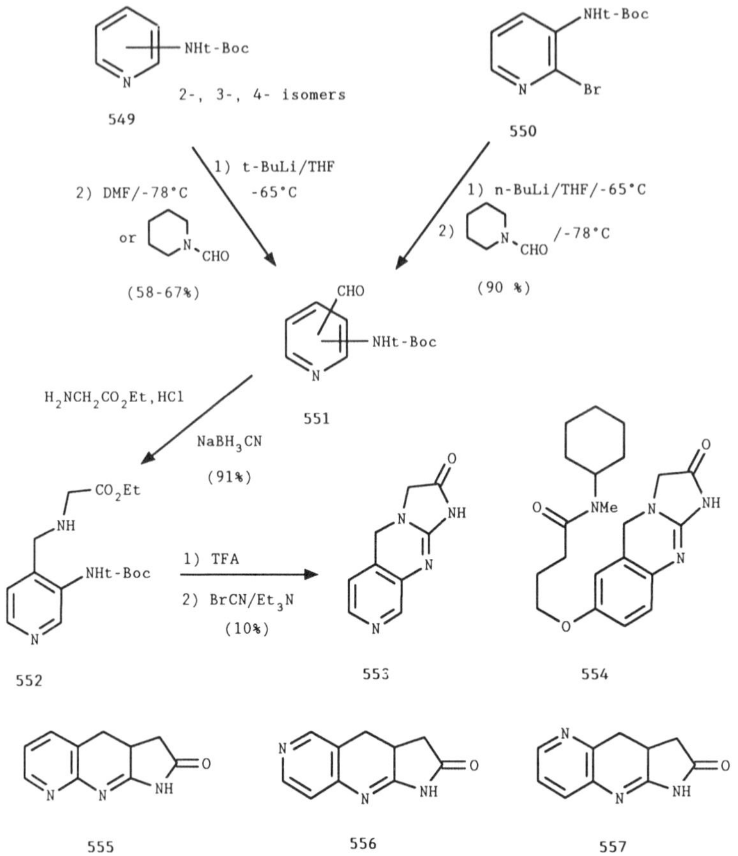

SCHEME 167

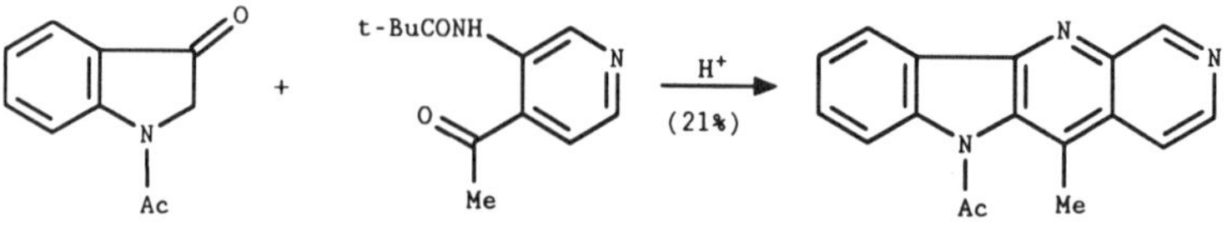

558 259 559

SCHEME 168

with glycine ethyl ester was followed by deprotection and one-carbon incorporation with cyanogen bromide to afford imidazo [2,1-*b*]quinazoline **553** in low yield. The other analogues **555–557** were obtained in better yields. All analogues showed poorer inhibitory activity compared to **554.**

Condensation of the readily available *N*-pivaloyl ketone **259** (Scheme 77) with *N*-acetyl indolin-3-one (**558**) under acid conditions provides good yields of 11-azapyrido[3,4-*b*]carbazole **559,** which represents an aza analogue of ellipticine (Scheme 168) (89JHC105).

The synthesis of **562,** the key intermediate for the preparation of an aza analogue of carteolol (**563**), a clinically useful *β*-receptor blocker, begins by the directed metalation of the readily available (Scheme 60) methoxy-*N*-pivaloyl pyridine **201** (Scheme 169) (82CPB1257). This regiospecific process, presumably favored by both DMGs, leads, upon DMF treatment, to the aldehyde **560.** Wittig chain extension provides **561,** which is transformed by unexceptional steps into the tetrahydro 1,7-naphthyridine **562.**

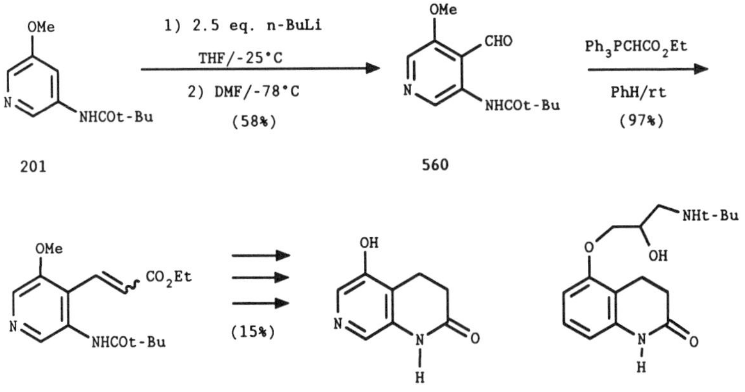

201 560

561 562 563 (Carteolol)

SCHEME 169

D. Oxygen-Based DMGs

The demonstration that 3-alkoxypyridines are metalated in the 2-position (Scheme 91) (82S235) allowed the preparation of a series of ribofuranosyl pyridines **565** as potential "deazapyrimidine" nucleosides for evaluation as thymidylate synthetase inhibitors (Scheme 170) (86MI2). Thus, metalation of the 3-alkoxypyridines **291** followed by condesation at lower temperatures with a protected D-ribose aldehyde afforded diastereoisomeric mixtures of compounds **564** which, upon mesylation and acid-catalyzed cyclization, delivered the ribofuranosyl pyridines **565** in high yields. Purification by affinity chromatography afforded the α- and β-anomers, which showed insignificant antileukemic activity.

Orelline (**570**), a metabolite of the toadstool, Cortinarius Orellanus, whose toxicity is apparently manifested in its di-N-oxide (orellanine), has been synthesized using various methodologies (86T1475; 87LA857), including the OSEM DMG (Scheme 171) (88HCA957). A homocoupling of the easily accessible **566** using Ni(0) catalysis afforded the bispyridine **567** which, upon dimetalation and oxygenation with 2-(phenylsulfonyl)-3-phenyloxaziridine, afforded a mixture of mono- (**568**) and di- (**569**) hydroxy products in equal amounts. Deprotection of the requisite **568** under acidic conditions afforded orelline (**570**).

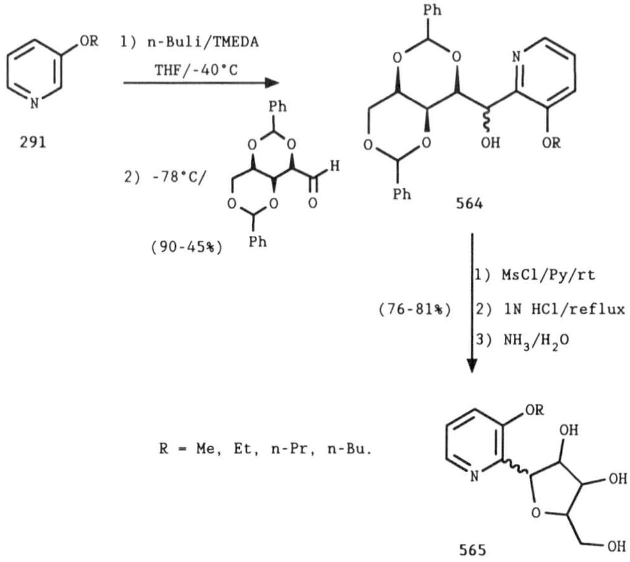

SCHEME 170

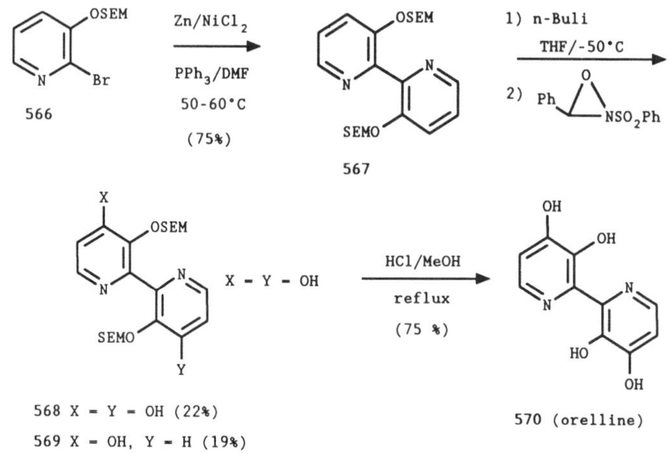

SCHEME 171

A number of furoquinoline alkaloids are available by taking advantage of in-between metalation of 2,4-dimethoxy quinoline derivatives, as established in model studies (Scheme 103). To illustrate, the trimethoxyquinoline **571,** upon metalation and ethylene oxide quench, afforded the carbinol **572** which, upon mild hydrolysis, furnished the alkaloid dihydro-γ-fagarine **(573)** together with the quinolone **574** (Scheme 172) (71T1351).

A different C-2 chain extension sequence on **575** led, via the interme-

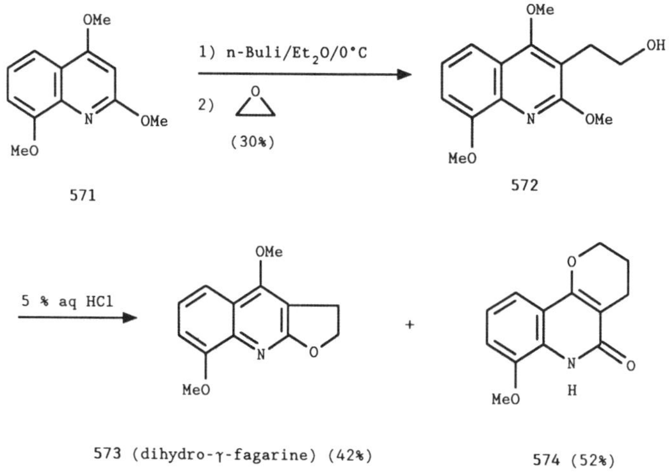

SCHEME 172

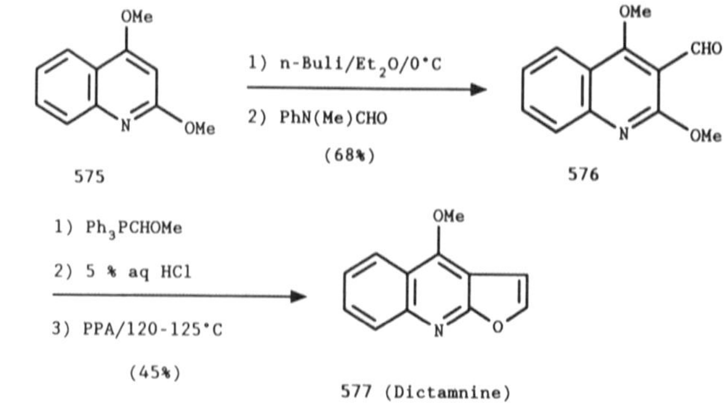

SCHEME 173

diate aldehyde **576** to dictamnine (**577**) in better (30%) overall yield (Scheme 173) (74T4153).

C-2 Chain extension via allylation (**578→579**) followed by oxidation to expose a latent aldehyde and cyclization constitutes another general route, in this case, leading to skimmianine (**580**) (Scheme 174) [73JCS(P1)94]. These routes, which overcome problems of poor 3-position reactivity by electrophilic substitution chemistry, were also applied to the furoquinoline

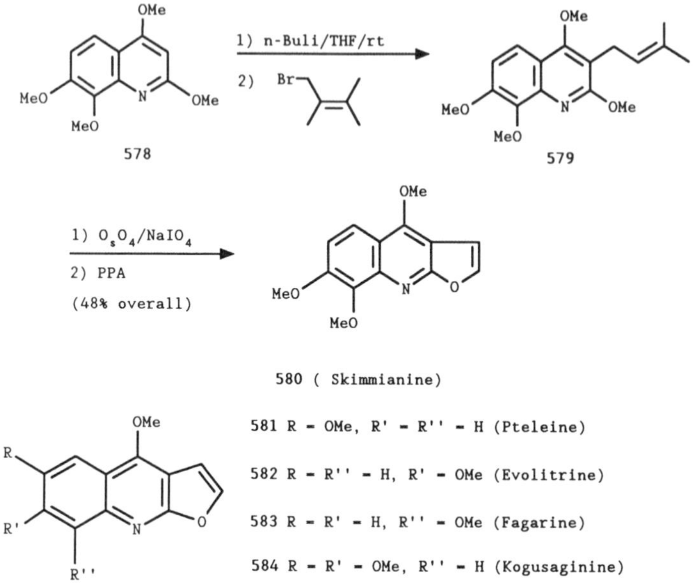

580 (Skimmianine)

581 R - OMe, R' - R'' - H (Pteleine)

582 R - R'' - H, R' - OMe (Evolitrine)

583 R - R' - H, R'' - OMe (Fagarine)

584 R - R' - OMe, R'' - H (Kogusaginine)

SCHEME 174

alkaloids, pteleine (**581**) (27% overall yield) (71T1351; 74T4153), evolitrine (**582**) (5.5% overall yield) (74T4153), fagarine (**583**) (15–40% overall yield) [73JCS(P1)94; 74T4153], dictamnine (**577**) (50% overall yield) [73JCS(P1)94], and kogusaginine (**584**) (10% overall yield) [79IJC(B)115].

Isomeric furoquinoline derivatives **336, 403, 404** (Scheme 125) related to these alkaloids have been obtained by taking advantage of the powerful carbamate DMG (88JHC1053).

E. CARBON-BASED DMGS

Compared to their use in the aromatic series, carbon-based DMGs have seen a disparate amount of application related to heteroaromatic natural products and biologically active molecules.

1. *Oxazoline DMG*

The provision of the pyridine lactone **585**, through the agency of the oxazoline DMG, was used in the construction of the antitumor alkaloid ellipticine (Scheme 175) [79JCR(M)4801]. Thus, metalation of **411** fol-

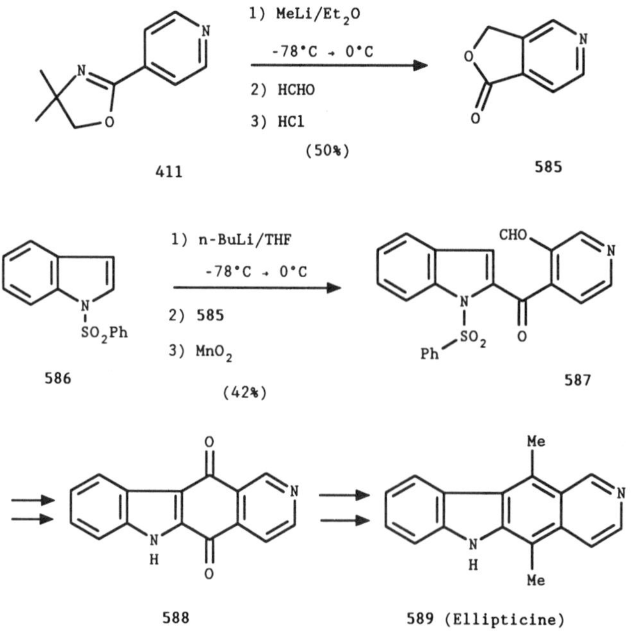

SCHEME 175

lowed by sequential reaction with formaldehyde, hydrolysis, and acid-catalyzed cyclization afforded **585** which, upon treatment with the 2-lithio species of indole **586** followed by oxidation, delivered the ketoaldehyde **587**. Conventional steps led first to the quinone **588** and then to ellipticine (**589**).

A series of aromatic and heteroaroatic fused quinolones **592–596** have been synthesized as part of a comprehensive structure–activity relationship study of analogues related to clinically siginificant antileukemic ellipticine alkaloids (Scheme 176) (87CJC2027, 87S142). The route to **592** is initiated by metalation of the readily available (Scheme 122) doubly activated **395** followed by quenching with 2,5-dimethoxybenzaldehyde to give the carbinol **590**. Treatment with acid resulted in cyclization and demethylation to afford the pyridone **591**. Zinc-mediated hydrogenolysis followed

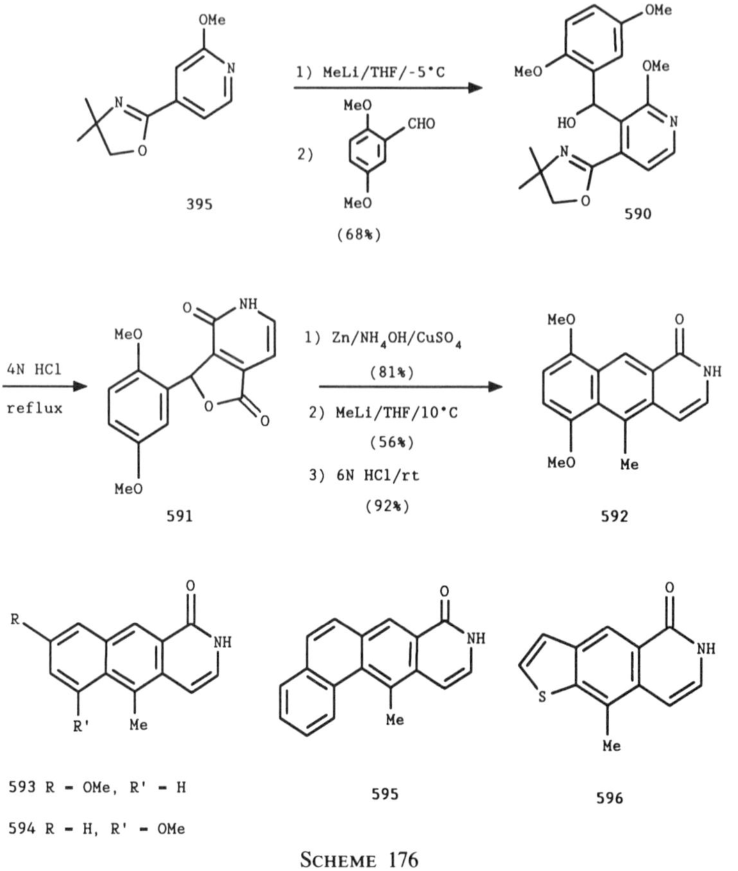

SCHEME 176

by a standard carboxylic acid into ketone conversion and cyclization furnished the target molecule **592** (87CJC2027). Analogues **592–596** and their corresponding condensed chloropyridine derivatives were similarly prepared starting with the key synthon **395** (87S142).

An alternative focus based on known antitumor activity of adriamycin-type systems stimulated the synthesis of the aza-anthraquinones **599** and **600** (Scheme 177) (84CC897). Thus, synergistic chloro-oxazoline directed metalation of **597** with methyllithium followed by treatment with 2,5-dimethoxybenzaldehyde and acid-promoted cyclization provided the lactone **598**. Radical bromination and base-induced hydrolysis gave an intermediate keto acid which, upon Friedel–Crafts cyclization with methanesulfonic acid, led to the aza-anthraquinone **599** in modest yield. The azanaphthacene dione **600** was prepared by an analogous series of reactions starting with **597**.

2. Secondary Amide DMG

The synthesis of Berninamycinic acid (**605**), a degradation product of the cyclic polypeptide antibiotic berninamycin, begins with an instructive differentiation of reactivity of the unsymmetrical pyridine diamide **601** (Scheme 178) [84TL(25)2127]. Metalation of **601** with 4.2 equiv. of *N*-BuLi followed by injection of 2.2 equiv. of the shown isothiocyanate afforded

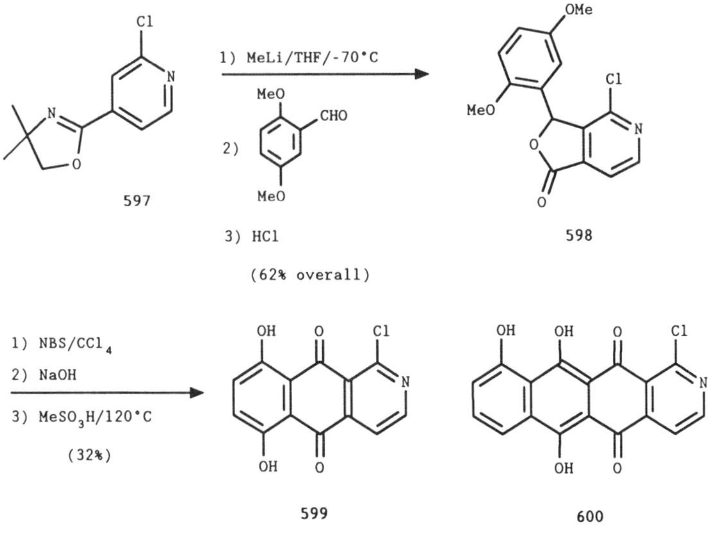

SCHEME 177

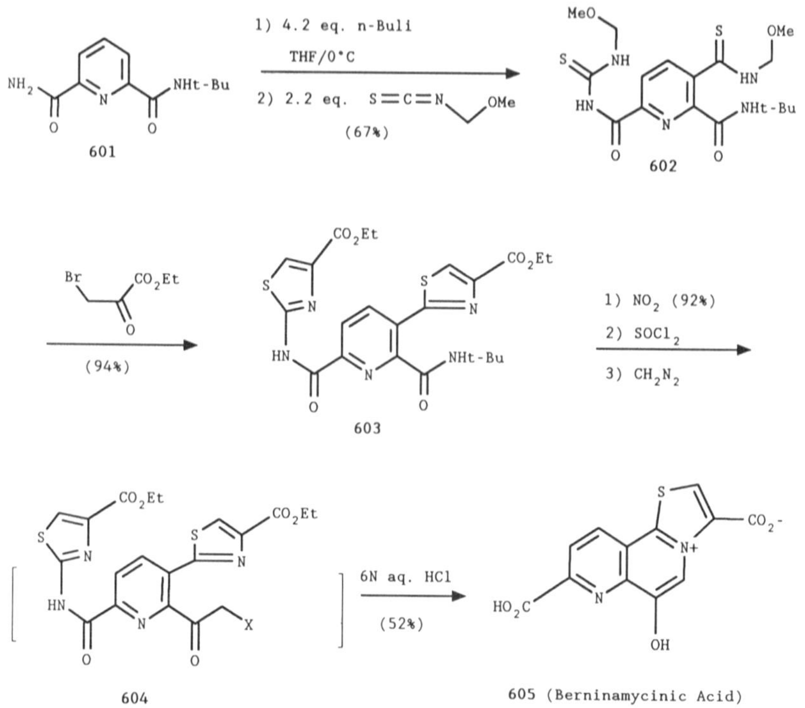

SCHEME 178

the highly functionalized pyridine **602** in a reaction which, by appropriate experiments, implicates a trianion intermediate. Condensation with ethyl bromopyruvate yields the dithiazole derivative **603**, the cleavage of the methoxymethylene groups occurring spontaneously during this reaction. The thiazole on the amide nitrogen of **603** provides differentiation of the two amides in the reaction with NO_2 to give an intermediate carboxylic acid which, upon homologation (**604**) and (without isolation) treatment with acid, furnished berninamycinic acid (**605**).

3. *Tertiary Amide DMG*

Sesbanine (**608**), a unique alkaloid isolated from Sesbania Drumondii seeds, which seems to exhibit potent antitumor activity, has been synthesized by prior carefull exploration for pyridine tertiary amide metalation (Scheme 179) (83TL2649). In the optimum case, LiTMP metalation of the diisolpropyl amide **447b** for 15 min. followed by condensation with 3-

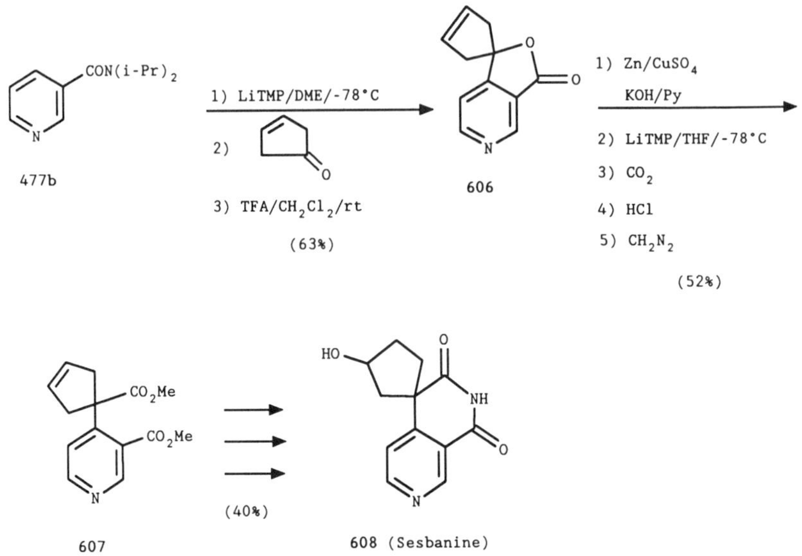

SCHEME 179

cyclopentenone and acid-catalyzed cyclization afforded the spirolactam **606.** Sequential hydrogenolysis followed by side chain metalation, carbonation, and esterification led to the diester **607,** which was efficiently converted into sesbanine (**608**).

A further utility of N,N-diisopropyl nicotinamide (**447b**) metalation is demonstrated in the synthesis of bostrycoidin (**611**), an aza-anthraquinone isolated from Fusarium bostrycoides, which exhibits antibiotic properties (Scheme 180) (87T5281). Considerable experimentation was invested (Scheme 134) in optimizing the condensation of the lithio species of **447b** with N,N-dimethyl 2,3,5-trimethoxybenzamide. The resulting keto amide was oxidized with MCPBA to give the N-oxide **609.** In a precedented procedure, compound **609** was sequentially treated with methyl 3-aminocrotonate and benzenesulfonyl chloride and aqueous HCl to give regioselectively the 2-methyl pryidine **610.** An alternative route to **610,** starting with a 6-methyl nicotinamide based on directed metalation chemistry, was, as expected, thwarted because of the observation of kinetic 2-methyl deprotonation (87T5281). Conversion of **610** into bostrycoidin (**611**) followed pathways similar to those used in the synthesis of other quinones (Scheme 176).

The synthesis of **614,** a conformationally rigid analogue of the tricyclic antidepressant imipramine, also begins with an N,N-diisopropyl nicotinamide ortho metalation (Scheme 181) [87H(26)3165]. Thus, LiTMP depro-

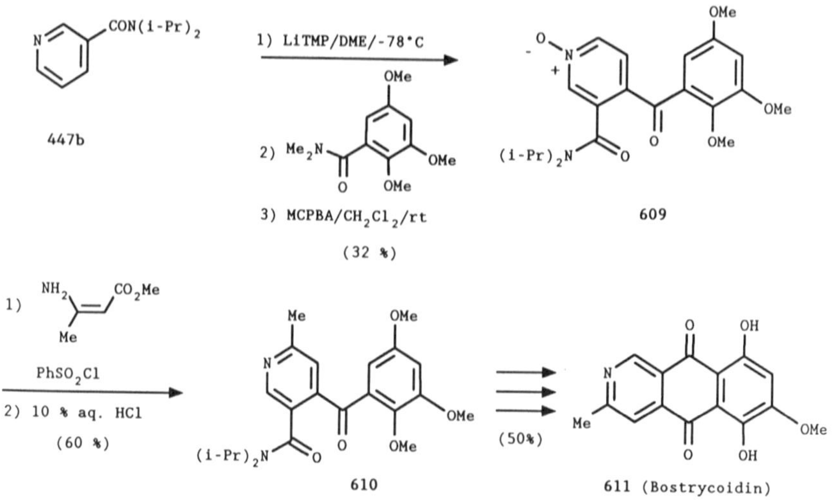

SCHEME 180

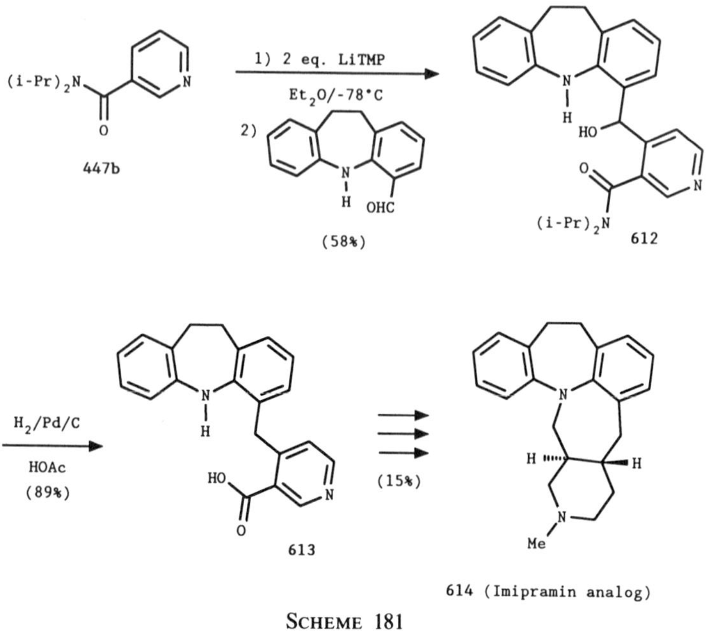

SCHEME 181

tonation of **447b** followed by condensation with a dibenzoazepine aldehyde, also obtained by directed metalation, furnished the alcohol amide **612**. Hydogenolysis under acidic conditions gave the acid **613**, a reaction which presumably proceeds via a lactone intermediate which, curiously, was rather irreproduceably converted into the same product. Unexceptional steps led to the imapramine analogue **614** as a 5:1 mixture of *cis*:*trans* isomers.

In a sequence that proceeds by tandem directed ortho metalation steps (Scheme 133) the *N*,*N*-diethyl isonicotinamide (**447a**) has been converted into the chemotherapeutic alkaloid ellipticine (**589**) (Scheme 182) (80JA1457). Thus, in a rapid, one-pot procedure, metalation of **447a** followed by condensation with *N*-protected indole-3-carboxaldehyde derivatives leads to the intermediates **615** which, upon second metalation and aerial oxidation affords the quinones **616** in modest to good yields. Established steps were used to convert **616, R = CH₂OMe** into ellipticine (**589**), concluding a route which complements that based on the oxazolino DMG (Scheme 175).

4. *Imidazolinone DMG*

MeLi DoM reactions on 2-(2-pyridyl)-4-imidazolinone (**618**), readily prepared from picolinic acid (**617**), lead to a number of 3-substituted products **619** in modest to good yields (Scheme 183) (85PC3;

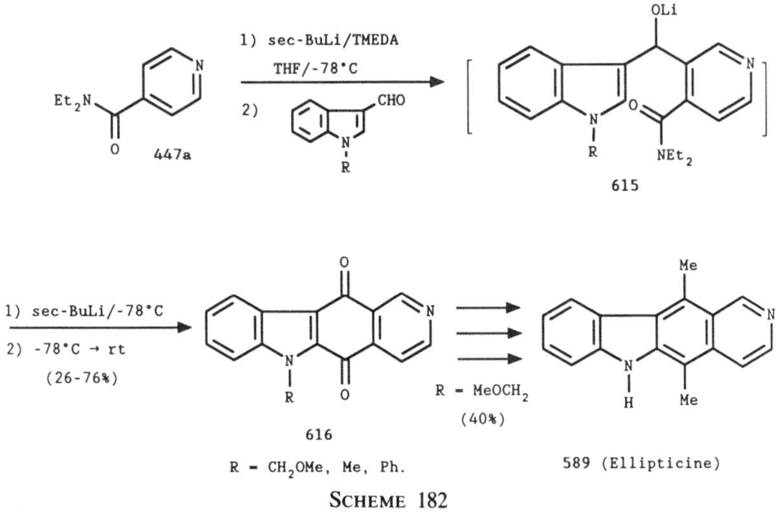

SCHEME 182

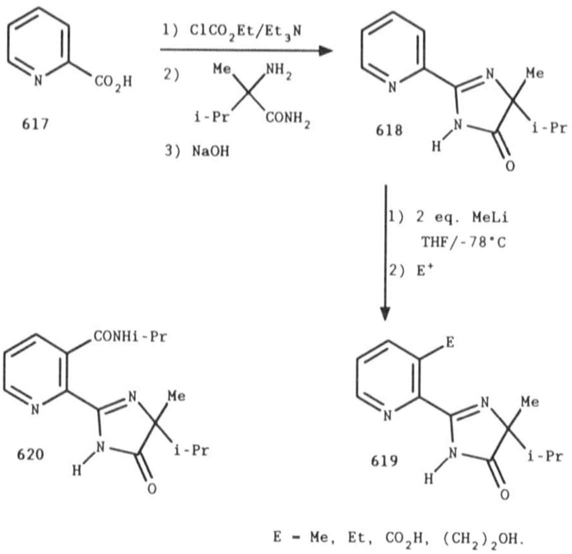

$E = Me, Et, CO_2H, (CH_2)_2OH.$

SCHEME 183

87EUP166907). The N-isopropyl nicotinamide **620** shows good herbicide activity.

ACKNOWLEDGMENTS

It is with admiration that we acknowledge all the co-workers for their contribution to the development of this review. Their effort, enthusiasm, and dedication are responsible for our studies and have continued to generate new avenues for investigation. We are also indebted to NATO for its financial support of this review. G. Queguiner and F. Marsais thank CNRS and V. Snieckus at the NSCRC. This chapter is dedicated to professor Paul Pastour.

References

12CB1128   W. Madelung, *Chem. Ber.* **46**, 1128 (1912).
30CB1847   K. Ziegler and H. Zieser, *Chem. Ber.* **63**, 1847 (1930).
39JA109   H. Gilman and R. L. Bebb, *J. Am. Chem. Soc.* **61**, 109 (1939).
40CB1197   G. Wittig and G. Fuhrman, *Chem. Ber.* **73**, 1197 (1940).
40JA2327   H. Gilman, W. Lagham, and F. W. Moore, *J. Am. Chem. Soc.* **62**, 2327 (1940).
48JA4187   W. E. Parham and E. L. Anderson, *J. Am. Chem. Soc.* **70**, 4187 (1948).
51JA32   H. Gilman and J. A. Beel, *J. Am. Chem. Soc.* **73**, 32 (1951).
51JOC1485   H. Gilman and S. M. Spatz, *J. Org. Chem.* **16**, 1485 (1951).
54OR(8)258   H. Gilman and J. W. Morton, *Org. React.* **8**, 258 (1954).

56JA(78)2217   H. Gilman and R. D. Gorsich, *J. Am. Chem. Soc.* **78**, 2217 (1956).
57JA(79)2625   H. Gilman and R. D. Gorsich, *J. Am. Chem. Soc.* **79**, 2625 (1957).
57JOC1715   H. Gilman and T. S. Soddy, *J. Org. Chem.* **22**, 1715 (1957).
60AG(E)91   R. Huisgen and J. Sauer, *Angew. Chem., Int. Ed. Engl.* **72**, 91 (1960).
62MP21   A. Carrington and J. Dos Santos-Veiga, *Mol. Phys.* **5**, 21 (1962).
64JOC853   W. H. Puterbaugh and C. R. Hauser, *J. Org. Chem.* **29**, 853 (1964).
64TL3207   R. J. Martens, H. J. Den Hertog, and M. van Ammers, *Tetrahedron Lett.*, 3207 (1964).
65ACS1741   S. Gronowitz and J. Röe, *Acta Chem. Scand.* **19**, 1741 (1965).
65CJC1752   R. A. Abramovitch, K. S. Ahmed, and C. S. Giam, *Can. J. Chem.* **41**, 1752 (1965).
65JA3530   B. Frydman, M. E. Despuy, and H. Rapoport, *J. Am. Chem. Soc.* **87**, 3530 (1965).
65JCS5045   R. D. Chambers, F. G. Drakesmith, and W. K. R. Musgrave, *J. Chem. Soc.*, 5045 (1965).
65JOC2531   R. R. Lorenz, B. F. Tullar, C. F. Koelsch, and S. Archer, *J. Org. Chem.* **30**, 2531 (1965).
65USP3202576   E. F. Rogers and R. L. Clark, U.S. Pat. 3,202,576 (1965) [*CA* **63**, 14641 (1965)].
66JOC1221   B. H. Graybill and D. A. Shirley, *J. Org. Chem.* **31**, 1221 (1966).
66JOC2047   A. R. Lepley, W. A. Khan, A. B. Giumanini, and A. Giumanini, *J. Org. Chem.* **31**, 2047 (1966).
66JOC3852   E. A. Fehnel and D. E. Cohn, *J. Org. Chem.* **31**, 3852 (1966).
67CC55   R. A. Abramovitch, G. M. Singer, and A. R. Vinutha, *J. C. S. Chem. Commun.*, 55 (1967).
67CI(L)831   N. S. Narasimhan and M. V. Paradkar, *Chem. Ind. (London)*, 831 (1967).
67JA1537   R. A. Abramovitch, M. Saha, E. M. Smith, and R. T. Coutts, *J. Am. Chem. Soc.* **89**, 1537 (1967).
67JOM(10)171   C. Eaborn, P. Golborn, and R. Taylor, *J. Organomet. Chem.* **10**, 171 (1967).
67MI1   R. W. Hoffman, "Dehydrobenzene and Cycloalkynes." Academic Press, New York, 1967.
67USP3314941   R. Littell and D. S. Allen, Jr., U.S. Pat. 3,314,941 (1967) {*CA* **67**, 64453 (1967)].
68CI(L)515   N. S. Narasimhan and R. H. Alurkar, *Chem. Ind. (London)*, 515 (1968).
68JA810   J. F. Bunnett and R. R. Victor, *J. Am. Chem. Soc.* **98**, 810 (1968).
68JA6421   J. Chanduri, S. Kume, J. Jagur-Grodzinski, and M. Szwarc, *J. Am. Chem. Soc.* **90**, 6421 (1968).
69CJC1543   H. Watanabe, R. A. Schwarz, C. R. Hauser, J. Lewis, and D. W. Slocum, *Can. J. Chem.* **47**, 1543 (1969).
69JA5501   J. A. Zoltewicz, G. Grahe, and C. L. Smith, *J. Am. Chem. Soc.* **91**, 5501 (1969).
69JCS(C)1700   R. D. Chambers, C. A. Heaton, and W. K. R. Musgrave, *J. Chem. Soc. C*, 1700 (1969).
69JOM(20)251   D. A. Shirley and C. F. Cheng, *J. Organomet. Chem.* **20**, 251 (1969).

69T4331   J. A. Zoltewicz and C. L. Smith, *Tetrahedron* **25**, 4331 (1969).
70JCS(D)478   C. A. Giam and J. L. Stout, *J. C. S. Dalton*, 478 (1970).
70JOC1288   R. E. Ludt, G. P. Crowther, and C. R. Hauser, *J. Org. Chem.* **35**, 1288 (1970).
71AG(E)20   T. Kauffmann and R. Wirthwein, *Angew. Chem., Int. Ed. Engl.* **10**, 20 (1971).
71JOC1843   J. G. Lombardino, *J. Org. Chem.* **36**, 1843 (1971).
71T1351   N. S. Narasimhan, M. V. Paradkar, and R. H. Alurkar, *Tetrahedron* **27**, 1351 (1971).
71TL1875   H. N. M. van der Lans, N. J. den Hertog, and H. van Veldhuizen, *Tetrahedron Lett.*, 1875 (1971).
72ACR139   J. F. Bunnett, *Acc. Chem. Res.* **5**, 139 (1972).
72CC505   E. Ager, G. E. Chivers, and H. Suschitzky, *J. C. S. Chem. Commun.*, 505 (1972).
72CR(C)(274)719   M. Mallet, G. Queguiner, and P. Pastour, *C. R. Hebd. Seances Acad. Sci., Ser. C* **274**, 719 (1972).
72CR(C)(275)1439 M. Mallet, F. Marsais, G. Queguiner, and P. Pastour, *C. R. Hebd. Seances Acad. Sci., Ser. C* **275**, 1439 (1972).
72CR(C)(275)1535 F. Marsais, M. Mallet, G. Queguiner, and P. Pastour, *C. R. Hebd. Seances Acad. Sci., Ser. C* **275**, 1535 (1972).
72JOC1690   R. A. Abramovitch, E. M. Smith, E. E. Knaus, and M. Saha, *J. Org. Chem.* **37**, 1690 (1972).
72JOC3584   R. A. Abramovitch, R. T. Coutts, and E. M. Smith, *J. Org. Chem.* **37**, 3584 (1972).
73CL1307   T. Taguchi, M. Nishi, K. Watanabe, and T. Mukaiyama, *Chem. Lett.*, 1307 (1973).
73JCS(P1)94   J. F. Collins, G. A. Gray, M. F. Grundon, D. M. Harrison, and C. G. Spyropoulos, *J. C. S. Perkin 1*, 94 (1973).
73RTC304   H. J. den Hertog and D. J. Burrman, *Recl. Trav. Chim. Pays-Bas* **92**, 304 (1973).
73TL1887   H. N. M. van der Lans and H. J. den Hertog, *Tetrahedron Lett.*, 1887 (1973).
74JA5601   K. Utimoto, N. Sakai, and H. Nozaki, *J. Am. Chem. Soc.* **96**, 5601 (1974).
74JOC3565   C. S. Giam, E. E. Knaus, and F. M. Pasutto, *J. Org. Chem.* **39**, 3565 (1974).
74JOM(69)161 N. J. Foulger and B. J. Wakefield, *J. Organomet. Chem.* **69**, 161 (1974).
74MI1   B. J. Wakefield, "The Chemistry of Organolithium Compounds." Pergamon, Oxford, 1974.
74MI2   R. A. Abramovitch and E. M. Smith, in "Pyridine and Its Derivatives" (R. A. Abramovitch, ed.), Part 2, p. 1. Wiley, New York, 1974.
74MI3   R. H. Yale, in "Pyridine and Its Derivatives" (R. A. Abramovitch, ed.), Part 2, p. 489. Wiley, New York, 1974.
74T4153   N. S. Narasimhan and R. S. Mali, *Tetrahedron* **30**, 4153 (1974).
74TL2373   A. J. Clarke, S. McNamara, and O. Meth-Cohn, *Tetrahedron Lett.*, 2373 (1974).
75AG(E)713   T. Kaufmann, B. Greving, J. König, A. Mitschker, and A. Woltermann, *Angew. Chem., Int. Ed. Engl.* **14**, 713 (1975).

75SC65	H. Christensen, *Synth. Commun.* **5,** 65 (1975).
76JOC1564	F. E. Ziegler and K. W. Fowler, *J. Org. Chem.* **41,** 1564 (1976).
76JSP(59)216	S. D. Sharma and S. Doraiswamy, *J. Mol. Spectrosc.* **59,** 216 (1976).
77CI(L)310	A. C. Ranade, R. S. Mali, R. M. Gidwani, and H. R. Deshpande, *Chem. Ind. (London),* 310 (1977).
77JOC3524	G. R. Newkome, J. D. Sauer, and S. K. Staires, *J. Org. Chem.* **42,** 3524 (1977).
78CC615	C. S. Giam and A. E. Hauck, *J. C. S. Chem. Commun.,* 615 (1978).
78H(11)133	A. I. Meyers and R. A. Gabel, *Heterocycles* **11,** 133 (1978).
78JOC3227	R. F. Francis, C. D. Crews, and B. S. Scott, *J. Org. Chem.* **43,** 3227 (1978).
78JOC3717	G. Buchi and P. S. Chu, *J. Org. Chem.* **43,** 3717 (1978).
78TL227	A. I. Meyers and R. A. Gabel, *Tetrahedron Lett.,* 227 (1978).
78ZC382	M. Weissenfels and B. Ulrici, *Z. Chem.* **18,** 382 (1978).
79AG(E)1	T. Kauffmann, *Angew. Chem., Int. Ed. Engl.* **18,** 1 (1979).
79AHC(25)205	P. Hanson, *Adv. Heterocycl. Chem.* **25,** 205 (1979).
79CCC3137	M. Ferles and A. Silhankva, *Collect. Czech. Chem. Commun.* **44,** 3137 (1979).
79IJC(B)115	N. S. Narasimhan, R. S. Mali, and A. M. Gokhale, *Indian J. Chem., Sect. B* **18B,** 115 (1979).
79CR(M)4801	D. A. Taylor, M. M. Baradarani, S. J. Martinez, and J. A. Joule, *J. Chem. Res., Miniprint,* 4801 (1979).
79JOC1133	W. Fuhrer and H. W. Gschwend, *J. Org. Chem.* **44,** 1133 (1979).
79JOC1519	P. S. Anderson, M. E. Christy, C. D. Colton, W. Halczenko, G. S. Ponticello, and K. L. Shepard, *J. Org. Chem.* **44,** 1519 (1979).
79JOC2081	T. J. Kress, *J. Org. Chem.* **44,** 2081 (1979).
79JOC2480	G. A. Kraus and J. O. Pezzanite, *J. Org. Chem.* **44,** 2480 (1979).
79JOC4612	N. J. Loenard and J. D. Bryant, *J. Org. Chem.* **44,** 4612 (1979).
79JOM(171)273	F. Marsais, E. Bouley, and G. Queguiner, *J. Organomet. Chem.* **171,** 273 (1979).
79OR(26)1	H. W. Gschwend and H. R. Rodriguez, *Org. React.* **26,** 1 (1979).
79PC1	F. Marsais, E. Bouley, and G. Queguiner, personal communication, ECHC, Louvain, Belgium (1979).
80H(14)1649	V. Snieckus, *Heterocycles* **14,** 1649 (1980).
80JA1457	M. Watanabe and V. Snieckus, *J. Am. Chem. Soc.* **102,** 1457 (1980).
80JCS(P1)2070	A. E. Hauck and C. S. Giam, *J. C. S. Perkin 1,* 2070 (1980).
80JOC1546	R. R. Bard and J. F. Bunnett, *J. Org. Chem.* **45,** 1546 (1980).
80JOC3728	G. D. Figuly and J. C. Martin, *J. Org. Chem.* **45,** 3728 (1980).
80JOC4798	J. Muchowski and M. C. Venuti, *J. Org. Chem.* **45,** 4798 (1980).
80TL1943	R. Beugelmans, B. Boudet, and L. Quintero, *Tetrahedron Lett.* **21,** 1943 (1980).
80TL4137	G. W. Gribble and M. G. Saulnier, *Tetrahedron Lett.* **21,** 4137 (1980).
80TL4529	G. Jones and D. R. Sliskovic, *Tetrahedron Lett.* **21,** 4529 (1980).

80TL4739   J. Epsztajn, Z. Berski, J. Z. Brzezinski, and A. Jozwiak, *Tetrahedron Lett.* **21**, 4739 (1980).
81JOC2429   E. C. Ashby, A. B. Goel, and R. N. De Priest, *J. Org. Chem.* **46**, 2429 (1981).
81JOC3564   Y. Tamura, M. Fujita, L. C. Chen, M. Inoue, and Y. Kita, *J. Org. Chem.* **46**, 3564 (1981).
81JOC4494   F. Marsais, P. Granger, and G. Queguiner, *J. Org. Chem.* **46**, 4494 (1981).
81JOM(215)139   T. Güngör, F. Marsais, and G. Queguiner, *J. Organomet. Chem.* **215**, 139 (1981).
81JOM(216)139   F. Marsais, P. Breant, A. Ginguene, and G. Queguiner, *J. Organomet. Chem.* **216**, 139 (1981).
81S127   A. R. Katritzky, J. Rahimi-Rastgoo, and N. R. Ponkshe, *Synthesis*, 127 (1981).
81SC513   N. Miyaura, T. Yanagi, and A. Suzuki, *Synth. Commun.* **11**, 513 (1981).
81TL5123   A. I. Meyers, N. R. Natale, D. G. Wettlaufer, S. Rafii, and J. Clardy, *Tetrahedron Lett.* **22**, 5123 (1981).
82ACR306   P. Beak and V. Snieckus, *Acc. Chem. Res.* **15**, 306 (1982).
82ACR395   T. Hayashi and M. Kumada, *Acc. Chem. Res.* **15**, 395 (1982).
82CJC1821   M. Gamal El Din, E. E. Knaus, and C. S. Giam, *Can. J. Chem.* **60**, 1821 (1982).
82CPB1257   Y. Tamura, L. C. Chen, M. Fujita, and Y. Kita, *Chem. Pharm. Bull.* **30**, 1257 (1982).
82H(18)13   A. I. Meyers and N. R. Natale, *Heterocycles* **18**, 13 (1982).
82JCR(S)278   F. Marsais, B. Laperdrix, T. Güngör, M. Mallet, and G. Queguiner, *J. Chem. Res. Synop.*, 278 (1982).
82JCS(P1)967   G. Jones and D. R. Sliskovic, *J. C. S. Perkin 1*, 967 (1982).
82JOC599   G. R. Newkome and D. C. Hager, *J. Org. Chem.* **47**, 599 (1982).
82JOC2101   M. R. Winkle and R. C. Ronald, *J. Org. Chem.* **47**, 2101 (1982).
82JOC2633   A. I. Meyers and R. A. Gabel, *J. Org. Chem.* **47**, 2633 (1982).
82S235   F. Marsais, G. Le Nard, and G. Queguiner, *Synthesis*, 235 (1982).
82S499   T. Güngör, F. Marsais, and G. Queguiner, *Synthesis*, 499 (1982).
82T1169   A. R. Katritzky, D. Winwood, and N. E. Grzeskowiak, *Tetrahedron* **38**, 1169 (1982).
82T3035   M. Mallet and G. Queguiner, *Tetrahedron* **38**, 3035 (1982).
82TL3979   D. L. Comins, J. D. Brown, and N. B. Mantlo, *Tetrahedron Lett.*, **23**, 3979 (1982).
83H(20)154   A. Ohta, Y. Akiba, and M. Inoue, *Heterocycles* **20**, 154 (1983).
83JA6155   T. D. Krizan and J. C. Martin, *J. Am. Chem. Soc.* **105**, 6155 (1983).
83JCR(S)144   B. Abarca, D. J. Hayles, G. Jones, and D. R. Sliskovic, *J. Chem. Res., Synop.*, 144, (1983).
83JMC181   C. Rivalle, F. Wendling, P. Tambourin, J. M. Lhoste, E. Bisagni, and J. C. Chermann, *J. Med. Chem.* **28**, 181 (1983).
83JOC1935   M. P. Sibi and V. Snieckus, *J. Org. Chem.* **48**, 1935 (1983).
83JOC3401   J. A. Turner, *J. Org. Chem.* **48**, 3401 (1983).
83JOC4156   S. L. Taylor, D. Y. Lee, and J. C. Martin, *J. Org. Chem.* **48**, 4156 (1983).

83PS(12)779	L. Horner, A. J. Lawson, and G. Simons, *Phosphorus Sulfur* **12,** 779 (1983).
83PS(16)167	N. Furukawa, F. Takahashi, T. Kawai, K. Kishimoto, S. Ogawa, and S. Oae, *Phosphorus Sulfur* **16,** 167 (1983).
83S822	P. Breant, F. Marsais, and G. Queguiner, *Synthesis,* 822 (1983).
83S957	N. S. Narasimhan and R. S. Mali, *Synthesis,* 957 (1983).
83S987	A. J. Guilford, M. A. Tometzki, and R. W. Turner, *Synthesis,* 987 (1983).
83T2009	F. Marsais and G. Queguiner, *Tetrahedron* **39,** 2009 (1983).
83T2031	R. C. Ronald and M. R. Winkle, *Tetrahedron* **39,** 2031 (1983).
83TH1	D. Defebvre, Thesis, University of Rouen, France (1983).
83TL2649	M. Iwao and T. Kuraishi, *Tetrahedron Lett.* **24,** 2649 (1983).
83TL3291	E. J. Corey, S. G. Pyne, and A. I. Schafer, *Tetrahedron Lett.* **24,** 3291 (1983).
83TL4735	J. Epsztajn, A. Bieniek, J. Z. Brzezinski, and A. Jozwiak, *Tetrahedron Lett.* **24,** 4735 (1983).
83TL5465	D. L. Comins and J. D. Brown, *Tetrahedron Lett.* **24,** 5465 (1983).
83UP1	F. Marsais, unpublished results (1983).
84ACS(B)439	L. Eberson, *Acta Chem. Scand., Ser. B* **B38,** 439 (1984).
84CC257	J. Verbeek, A. V. E. George, R. L. P. de Jong, and L. J. Brandsma, *J. C. S. Chem. Commun.,* 257 (1984).
84CC897	M. Croisy-Delsey and E. Bisagni, *J. C. S. Chem. Commun.,* 897 (1984).
84CC1304	C. W. G. Fishwick, R. C. Storr, and P. W. Manley, *J. C. S. Chem. Commun.,* 1304 (1984).
84JCR(S)140	B. Abarca, E. Gomez-Aldaravi, and G. Jones, *J. Chem. Res. Synop.,* 140 (1984).
84JCS(P1)2227	A. E. Hauck and C. S. Giam, *J. C. S. Perkin 1,* 2227 (1984).
84JOC1078	D. L. Comins and J. D. Brown, *J. Org. Chem.* **49,** 1078 (1984).
84JOC3857	J. Verbeek and L. Brandsma, *J. Org. Chem.* **49,** 3857 (1984).
84TH1	F. Marsais, Thesis, University of Rouen, France (1984).
84TL(25)495	E. J. Corey and A. W. Gross, *Tetrahedron Lett.* **25,** 495 (1984).
84T(25)2127	T. R. Kelly, A. Echavarren, N. S. Chandrakumar, and Y. Köksal, *Tetrahedron Lett.* **25,** 2127 (1984).
84TL(40)2107	P. Ribereau and G. Queguiner, *Tetrahedron Lett.* **40,** 2107 (1984).
85CL1401	H. Hayakawa, H. Tanaka, T. Maruyama, and T. Miyasaka, *Chem. Lett.,* 1401 (1985).
85CPB1016	A. Wada, S. Kagatomo, and S. Nagai, *Chem. Pharm. Bull. Tokyo* **33,** 1016 (1985).
85HC	S. Gronowitz, in "The Chemistry of Heterocyclic Compounds" (A. R. Weisberger and E. C. Taylor, ed.), Vol. 44. Wiley, New York, 1985.
85JCS(P1)1897	B. Abarca, R. Ballesteros, E. Gomez-Aldaravi, and G. Jones, *J. C. S. Perkin 1,* 1897 (1985).
85JOC2746	S. L. Crump, J. Netka, and B. Rickborn, *J. Org. Chem.* **50,** 2746 (1985).
85JOC5436	M. A. J. Miah and V. Snieckus, *J. Org. Chem.* **50,** 5436 (1985).
85PC1	D. M. Roland and W. Macchia, personal communication, 10th ICHC, Waterloo, Canada (1985).

85PC2	D. L. Comins, M. O. Killpack, and N. B. Mantlo, personal communication, 10th ICHC, Waterloo, Canada (1985).
85PC3	M. Los, R. K. Russell, P. C. Lauro, B. L. Lences, S. K. Hadden, T. E. Brady, and P. J. Wepplo, personal communication, 10th ICHC, Waterloo, Canada (1985).
85T837	M. Reuman and A. I. Meyers, *Tetrahedron* **41**, 837 (1985).
85T1945	C. Robaut, C. Rivalle, M. Rautureau, J. M. Lhoste, E. Bisagni, and J. C. Chermann, *Tetrahedron* **41**, 1945 (1985).
85T3433	M. Mallet and G. Queguiner, *Tetrahedron* **41**, 3433 (1985).
85TH1	P. Breant, Thesis, University of Rouen, France (1985).
85TL5997	M. J. Sharp and V. Snieckus, *Tetrahedron Lett.* **26**, 5997 (1985).
867AG(E)508	J. K. Stille, *Angew. Chem., Int. Ed. Engl.* **25**, 508 (1986).
86JCR(S)18	J. Epsztajn, J. Z. Brzezinski, and A. Jozwiak, *J. Chem. Res., Synop.*, 18 (1986).
86JCR(S)20	J. Epsztajn, A. Bieniek, and M. W. Plotka, *J. Chem. Res., Synop.*, 20 (1986).
86JOC2184	E. W. Thomas, *J. Org. Chem.* **51**, 2184 (1986).
86JOC2833	J. N. Bonfiglio, *J. Org. Chem.* **51**, 2833 (1986).
86MI1	L. Brandsma and H. Verkruijsse, "Preparative Polar Organometallic Chemistry 1." Springer-Verlag, Berlin, 1986.
86MI2	I. Vrijens, M. Belmans, E. L. Esmans, R. Dommisse,. J. A. Lepoivre, F. C. Alderweireldt, L. L. Wotring, and L. B. Townsend, *Nucleosides Nucleotides* **5**, 207 (1986).
86PC1	P. Pineau, F. Marsais, and G. Queguiner, personal communication, 12th ECHC, Reims, France (1986).
86PC2	F. Marsais, G. Queguiner, A. Martin, S. Chapelle, and P. Granger, personal communication, 12th ECHC, Reims, France (1986).
86S670	J. M. Jacquelin, F. Marsais, A. Godard, and G. Queguiner, *Synthesis*, 670 (1986).
86S886	R. Radinov, M. Haimova, and E. Simova, *Synthesis*, 886 (1986).
86T1475	M. Tiecco, M. Tingoli, L. Testaferri, D. Chianelli, and E. Wenkert, *Tetrahedron* **42**, 1475 (1986).
86T2253	M. Mallet and G. Queguiner, *Tetrahedron* **42**, 2253 (1986).
86T4027	A. R. Katritzky, W. Q. Fan, and K. Akutagawa, *Tetrahedron* **42**, 4027 (1986).
86T4187	H. Tanaka, H. Hayakawa, K. Obi, and T. Miyasaka, *Tetrahedron* **42**, 4187 (1986).
86TH1	P. Breant, Thesis, University of Rouen, France (1986).
86TL3543	B. Abarca, R. Ballesteros, G. Jones, and F. Mojarrad, *Tetrahedron Lett.* **27**, 3543 (1986).
87AP704	R. Radinov, M. Haimova, E. Simova, N. Tyutyulkova, and J. Gorantcheva, *Arch. Pharm. (Weinheim, Ger.)* **320**, 704 (1987).
87CJC1158	C. Y. Fiakpui and E. E. Knaus, *Can. J. Chem.* **65**, 1158 (1987).
87CJC2027	E. Bisagni, M. Rautureau, M. Croisy-Delcey, and C. Huel, *Can. J. Chem.* **65**, 2027 (1987).
87CPB72	H. Hayakawa, K. Haraguchi, H. Tanaka, and T. Miyasaka, *Chem. Pharm. Bull.* **35**, 72 (1987).

302   GUY QUEGUINER et al.   [Refs.

87EUP166907	M. Los, Eur. Pat. 166,907 (1987) [*CA* **106**, 50206 (1987)].
87H(26)427	A. R. Katritzky, J. M. Aurrecoechea, and L. M. Vazquez de Miguel, *Heterocycles* **26**, 427 (1987).
87H(26)585	A. Wada, J. Yamamoto, and S. Kanamoto, *Heterocycles* **26**, 585 (1987).
87H(26)3165	P. G. Dunbar and A. R. Martin, *Heterocycles* **26**, 3165 (1987).
87JHC1487	Y. Robin, A. Godard, and G. Queguiner, *J. Heterocycl. Chem.* **24**, 1487 (1987).
87JOC792	D. J. Pollart and B. Rickborn, *J. Org. Chem.* **52**, 792 (1987).
87JOC1133	F. Marsais, A. Cronnier, F. Trecourt, and G. Queguiner, *J. Org. Chem.* **52**, 1133 (1987).
87JOM(336)1	A. Godard, Y. Robin, and G. Queguiner, *J. Organomet. Chem.* **336**, 1 (1987).
87LA857	E. V. Dehmlow and H.-J. Schulz, *Liebigs Ann. Chem.*, 857 (1987).
87MI1	N. S. Narasimhan and R. S. Mali, *Top. Curr. Chem.* **138**, 63 (1987).
87MI2	K. Tamao and M. Kumada, *in* "The Chemistry of the Metal-Carbon Bond. The Use of Organometallic Compounds in Organic Synthesis" (F. R. Hartly, ed.) Wiley London, 1987.
87S142	E. Bisagni and M. Ratureau, *Synthesis,* 149 (1987).
87T5281	M. Watanabe, E. Shinoda, Y. Shimizu, and S. Furukawa, *Tetrahedron* **43**, 5281 (1987).
87TL87	H. Hayakawa, H. Tanaka, K. Obi, M. Itoh, and T. Miyasaka, *Tetrahedron Lett.* **28**, 87 (1987).
87UP1	P. Geffray, F. Marsais, and G. Queguiner, unpublished results, University of Rouen, France (1987).
88AHC(44)199	D. L. Comins and S. O'Connor, *Adv. Heterocycl. Chem.* **44**, 199 (1988).
88BSCF67	V. Snieckus, *Bull. Soc. Chim. Fr.,* 67 (1988).
88CI(L)302	K. Smith, C. M. Lindsay, and I. K. Morris, *Chem. Ind. (London),* 302 (1988).
88CJC1135	J. M. Jacquelin, Y. Robin, A. Godard, and G. Queguiner, *Can. J. Chem.* **66**, 1135 (1988).
88HCA957	H. A. Hasseberg and H. Gerlach, *Helv. Chim. Acta* **71**, 957 (1988).
88H(27)1671	E. Bisagni, M. Rautureau, and C. Hucl, *Heterocycles* **27**, 1671 (1988).
88H(27)2643	S. P. Khanapure and E. R. Biehl, *Heterocycles* **27**, 2643 (1988).
88JCS(P1)3085	P. D. Leeson and J. C. Emmett, *J. C. S. Perkin 1,* 3085 (1988).
88JHC81	F. Marsais, F. Trecourt, P. Breant, and G. Queguiner, *J. Heterocycle. Chem.* **25**, 81 (1988).
88JHC1053	A. Godard, J. M. Jacquelin, and G. Queguiner, *J. Heterocycl. Chem.* **25**, 1053 (1988).
88JMC2136	M. C. Venuti, R. A. Stephenson, R. Alvarez, J. J. Bruno, and A. Strosberg, *J. Med. Chem.* **31**, 2136 (1988).
88JOC1367	F. Trecourt, M. Mallet, F. Marsais, and G. Queguiner, *J. Org. Chem.* **53**, 1367 (1988).
88JOC2740	L. Estel, F. Marsais, and G. Queguiner, *J. Org. Chem.* **53**, 2740 (1988).

88JOC5301   M. Croisy-Delsey, M. Rautureau, C. Huel, and E. Bisagni, *J. Org. Chem.* **53,** 5301 (1988).
88JOM(342)1   C. Parkanyi, N. S. Cho, and G. S. Yoo, *J. Organomet. Chem.* **342,** 1 (1988).
88JOM(354)273   A. Godard, J. M. Jacquelin, and G. Queguiner, *J. Organomet. Chem.* **354,** 273 (1988).
88MI1   H. Hayakawa, H. Tanaka, K. Haraguchi, M. Kazuhiro, M. Nakajima, T. Sakamari, and T. Miyasaka, *Nucloesides Nucleotides* **7,** 121 (1988).
88S388   V. Bolitt, C. Mioskowski, S. P. Reddy, and J. R. Falck, *Synthesis,* 388 (1988).
88S881   A. Türck, L. Mojovic, and G. Queguiner, *Synthesis,* 881 (1988).
88TH1   L. Estel, Thesis, University of Rouen, France, (1988).
88TL773   D. L. Comins and D. H. LaMunyon, *Tetrahedron Lett.* **29,** 773 (1988).
88TL5725   J. N. Reed, J. Rotchford, and D. Strickland, *Tetrahedron Lett.* **29,** 5725 (1988).
89H(29)1815   E. Bisagni, M. Rautureau, and C. Huel, *Heterocycles* **29,** 1815 (1989).
89JHC105   L. Estel, F. Linard, F. Marsais, A. Godard, and G. Queguiner, *J. Heterocycl. Chem.* **26,** 105 (1989).
89JHC1589   F. Marsais, A. Godard, and G. Queguiner, *J. Heterocycl. Chem.* **26,** 1589 (1989).
89JOC24   M. Iwao, T. Iihama, M. M. Mahalanabis, H. Perier, and V. Snieckus, *J. Org. Chem.* **54,** 24 (1989).
89JOC3730   D. L. Comins and J. D. Brown, *J. Org. Chem.* **54,** 3730 (1989).
89T7469   J. Epsztajn, A. Bieniek, M. W. Plotka, and K. Suwald, *Tetrahedron* **45,** 7469 (1989).
89TH1   H. Perrier, M.Sc. Thesis, University of Waterloo, Canada (1989).
89TL183   A. Couture, E. Huguerre, and P. Grandclaudon, *Tetrahedron Lett.* **30,** 183 (1989).
89TL2057   R. W. Amstrong, S. Gupta, and F. Whelihan, *Tetrahedron Lett.* **30,** 2057 (1989).
89TL7091   N. Furukawa, T. Shibutani, and H. Fujihara, *Tetrahedron Lett.* **30,** 7091 (1989).
90CRV879   V. Snieckus, *Chem. Rev.* **90,** 879 (1990).
90JCS(P1)2409   F. Trecourt, F. Marsais, T. Güngör, and G. Queguiner, *J. C. S. Perkin 1,* 2409 (1990).
90JCS(P1)2611   F. Marsais, J. C. Rovera, A. Turck, A. Godard, and G. Queguiner, *J. C. S. Perkin 1,* 2611 (1990).
90JHC563   A. S. Katner and R. F. Brown, *J. Heterocycl. Chem.* **27,** 563 (1990).
90JHC1377   A. Turck, N. Ple, L. Mojovic, and G. Queguiner, *J. Heterocycl. Chem.* **27,** 1377 (1990).
90JOC69   D. L. Comins and M. O. Killpack, *J. Org. Chem.* **55,** 69 (1990).
90JOC3410   R. J. Mattson and C. P. Sloan, *J. Org. Chem.* **55,** 3410 (1990).
90JOM(382)319   M. Mallet, G. Branger, F. Marsais, and G. Queguiner, *J. Organomet. Chem.* **382,** 319 (1990).
90TH1   D. Trohay, Thesis, INSA of Rouen, France, (1990).
90TL1665   J.-M. Fu and V. Snieckus, *Tetrahedron Lett.* **31,** 1665 (1990).

90TL4267   S. Sengupta and V. Snieckus, *Tetrahedron Lett.* **31,** 4267 (1990).
90UP1   M. Tsukazaki and V. Snieckus, unpublished results, University of
            Waterloo, Canada (1990).
91JHC283   N. Ple, A. Turck, E. Fiquet, and G. Queguiner, *J. Heterocycl.
            Chem.* **88,** 283 (1991).
91JOC(s1)   F. Marsais, P. Pineau, M. Mallet, A. Türck, F. Nivolliers, A.
            Godard, and G. Queguiner, *J. Org. Chem.* (accepted for pub-
            lication) (1991).
91JOM   A. Turck, D. Trohay, L. Mojovic, N. Ple, and G. Queguiner,
            J. Organomet. Chem. (accepted for publication) (1991).